TIC

Technology
Innovation
Centre

FOR REFERENCE ONLY

NOT TO BE TAKEN AWAY

Dictionary of Metallurgy

Dictionary of Metallurgy

Colin D. Brown

C. Eng., M.I.M., Sen. M. Weld. I
Independent Consultant Metallurgist,
Loughborough, UK

JOHN WILEY AND SONS
Chichester • New York • Weinheim • Brisbane • Singapore • Toronto

Other Wiley Editorial Offices

John Wiley & Sons, Inc., 605 Third Avenue,
New York, NY 10158-0012, USA

Wiley-VCH Verlag GmbH, Pappelallee 3,
D-69469 Weinheim, Germany

Jacaranda Wiley Ltd, 33 Park Road, Milton,
Queensland 4064, Australia

John Wiley & Sons (Asia) Pte Ltd, Clementi Loop #02-01,
Jin Xing Distripark, Singapore 129809

John Wiley & Sons (Canada) Ltd, 22 Worcester Road,
Rexdale, Ontario M9W 1L1, Canada

Library of Congress Cataloging-in-Publication Data
Brown, Colin D.
    Dictionary of metallurgy/Colin D. Brown.
        p.   cm.
    ISBN 0-471-96155-8
    1. Metallurgy–Dictionaries.   I. Title.
TN609.B76   1998
669'. 03–dc21   97-27383
                                                        CIP

British Library Cataloguing in Publication Data

A catalogue record for this book is available from the British Library

ISBN 0 471 96155 8

Typeset in 9/10pt Times by Keytec Typesetting Ltd, Bridport, Dorset.
Printed in Great Britain by Bookcraft (Bath) Ltd, Somerset
This book is printed on acid-free paper responsibly manufactured from sustainable foresta-
tion, for which at least two trees are planted for each one used for paper production

Contents

Preface

This extended dictionary has been developed for the benefit of a wide range of readers. For practitioners and students of applied metallurgy it will provide a quick reminder of basics as well as a listing of terms and their usage. For students and practitioners of other technologies, it will give an introduction to physical and engineering metallurgy and its interaction with their speciality. For professionals outside the engineering fields the book will provide a readily assimilated overview of metallurgy and its impact on many areas of modern life. It will, therefore, be of value to readers from diverse backgrounds, non-technical as well as technical, including engineers and mechanics, welding technologists and welders, plumbers, journalists, health and safety advisers, lawyers, insurance brokers, loss adjusters and surgeons developing prostheses.

With this wide readership in mind the extended form of presentation has been adopted. It is based on the usual alphabetical entry having each term defined individually but with, in many cases, a further explanation and comment on usage. Related terms defined within entries are highlighted in **bold** to aid quick identification, for example the alphabetical entry on **Fatigue** includes a definition of **Endurance** and the entry on Endurance is then cross-referenced back to Fatigue. Where possible, each entry is self-sufficient but frequently it is necessary to use terms defined at their alphabetical position. Such cross-referenced terms are indicated in *italics* at their first occurrence in each entry. Finally, a major feature is that within the alphabetical entries attention is drawn to other entries or tables providing supplementary information. Some of these, such as those on Steel, Fatigue and Tensile test, have been recognized as key topics and are dealt with at some length but always in a manner accessible to the non-specialist. Individually, these key topics explain the relationship between groups of kindred terms and collectively they provide a broad overview of engineering metallurgy and physical metallurgy. Where useful, diagrams and sketches are provided. By these means the basic dictionary format has been extended to provide a summary of the basics of the science and practice of applied metallurgy which should be adequate for most non-specialists.

Inevitably, some of the terms in the dictionary have multiple meanings in a technical as well as a lay context and some phenomena can be described in different terms. Perhaps surprisingly, this is not usually a result of geographical or national variations across the English-speaking world except in such trivia as spelling or the units of measurement. Rather, the differences reflect the context of the reference, such as the industry or scientific establishment involved or even the generation of the user. This dictionary aims to recognize all reasonable usage, indicate potential ambiguities and, in a few cases, suggest that certain usage be avoided. The areas covered are, primarily, physical metallurgy and engineering metallurgy, including appropriate areas of welding technology, since these impinge directly on the world at large, affecting the choice and performance of all metallic components. They are, therefore, matters of interest and concern for the anticipated wide readership. Other areas of more specialized interest, such as extraction metallurgy, are not covered apart from brief entries considered to be of interest to the intended reader.

With regard to alloy names the choice between inclusion and exclusion is inevitably arbitrary. At one extreme it would be ridiculous to offer a metallurgical dictionary that excluded terms such as Brass and Solder or even Admiralty brass and Sterling silver. At the other extreme it would be impossible, even assuming it was desirable, to include all the immense number of names introduced, and often discarded, at the whim of manufacturers. The aim therefore has been to include generic names such as those mentioned above together with the small number of trade or proprietary names such as Dural and Nimonic that have become virtually common currency. Finally, the dictionary includes a number of tables in the end pages. Most of these summarize various mechanical and physical properties of metals, alloys and other materials. Again it would be impossible for any single volume to provide a fully comprehensive listing of all material properties so the aim has been to give a broad overview of the properties that can be achieved by a cross-section of alloys in various heat treatments in each of the main alloy series. Further tables provide conversion factors, list the Greek mathematical symbols and present the Periodic Table.

Å Ångstrom unit (10^{-10} metre).

A, A₁, Ac, etc. Designations of critical temperatures in the transformation of *Steel*.

Ablation Removal of material. In medical contexts the term usually refers to the surgical removal of body parts, in geological contexts to erosion. In metallurgical contexts it usually refers to the process by which a protective, but intentionally consumable, coating is progressively removed by an aggressive environment. The usually quoted example is the **Ablative polymer** forming the heat shield on the nose of space vehicles. The material, a low-density, low thermal conductivity silicone fibre, is consumed on re-entry but protects the underlying vessel surface from the high temperature generated by frictional heating.

Abnormal steel A *Steel* with variable *Hardenability* leading to local soft spots when components are hardened. In the *Annealed* or *Normalized* conditions these areas comprise abnormally coarse and irregular pearlite. The effect is associated with segregation of alloy elements or variation in aluminium (deoxidizer) content.

Abrasion, abrasive wear Removal of material from a metal surface by hard particles or a hard surface sliding across it. The mechanism is predominantly a mechanical cutting action as opposed to the repeated welding/tearing action involved in *Adhesive wear*.

Absolute temperature The temperature related to absolute zero. On the Celsius (Centigrade, C) scale, 0° Kelvin (K) is minus 273.15° Celsius and one Kelvin degree interval equals one Celsius degree interval. On the Fahrenheit (F) scale, 0° Rankine (R) is minus 459.67° Fahrenheit and one Rankine degree interval equals one Fahrenheit degree interval.

Absorber (1) A material capable of physically assimilating others. In a nuclear context an absorber is a material capable of accepting and retaining a large number of *Neutrons* without *Fission*.

Absorber (2) A Material or component intended to absorb *Kinetic energy* as in the shock absorber or a vehicle.

Absorption The process whereby material, usually vapour or liquid, is taken within the body of some other material. Unless otherwise stated it is usually implicit that the interaction is physical, there is no chemical reaction or alloying.

AC Alternating current.

Acceptor (in semiconductor) See Semiconductor.

ACPD Alternating current, potential drop, a technique for measuring crack propagation, see Potential drop.

Accumulator (1) A reversible *Electrolytic* cell in which electricity is stored as chemical energy.

Accumulator (2) A vessel serving as a reservoir for a fluid power source. A gas, often air, would be stored compressed, a liquid, often water, might be externally pressurized or not.

Accumulator metal Alloys based on lead with small quantities of tin and other elements used for electrical storage batteries.

Acetylene A fuel gas, chemical formula C_2H_2, burnt with oxygen for heating, welding and cutting. Acetylene cannot be compressed to more than about 2 atmospheres so it is commonly stored in cylinders which are packed with a suitable porous material such as carbon or kapok and then filled with acetone. The acetone is capable of absorbing about twenty five times its own volume of acetylene for every atmosphere increase in pressure.

Acicular Needle-shaped.

Acid An aqueous solution which produces hydrogen ions, H^+, is termed acid. If it produces hydroxyl ions, OH^-, it is termed alkaline. See also pH. In a metallurgical context, the term indicates processes in which the environmental conditions such as the furnace refractory lining and the slag are acidic as opposed to basic. Siliceous refractories are acid.

Acid cleaning The use of acid to dissolve undesirable materials, such as *Scale*, from a metal surface. Dissolution of the underlying metal is intended to be minimal. Also termed *Pickling* although this term also includes the use of alkaline solvents.

Acid dip The use of acid solutions to clean components. The term tends to imply an intent to removing slight tarnishing rather than severe scale, i.e. a less aggressive attack than implied by *Acid cleaning*.

Acid embrittlement Any *Embrittlement* resulting from contact with acid, for example during *Pickling*. It is a form of *Hydrogen embrittlement* caused when hydrogen, released by the chemical reaction with the acid, enters the metal in quantities large enough to cause damage such as cracking or a tendency to non-*Ductile* failure during forming operations.

Acid etching The use of acid to dissolve, within controlled limits, the surface of a metal. It is used in *Metallography* to reveal details of the microstructure of a

metal, in printing to provide a relief effect or in industry to prepare a metal surface for treatments such as painting or electroplating.

Acid lining A refractory furnace lining, usually *Silica*, which produces an acid reaction with the *Charge*.

Acid steel *Steel* made in a furnace with an acid refractory lining and acid slag reaction. Low sulphur and phosphorous ores are required for this process.

Acoustic fatigue Cracking by *Fatigue* induced by atmospheric vibration associated with noise, air turbulence or *Vortex shedding*. Also called Sonic fatigue.

Acitinium A radioactive metallic element with no major commercial application. See Table 15 for physical properties.

Activated Any material treated in some way to enhance its response to its environment.

Activated alumina Partly dehydrated aluminium hydroxide. It is used as a desiccant and catalyst.

Activated carbon A carbon-rich material produced by heating carbonaceous material in the absence of oxygen. It is used to absorb contaminants etc.

Activated sintering See Sintering.

Activation energy The additional energy necessary to initiate a thermally activated chemical reaction or physical process. Mathematically, it is the slope of the natural logarithm of the reaction rate plotted against the reciprocal of the absolute temperature.

Active (material) Reacting, or capable of reacting, with its environment.

Active mass The molecular concentration.

Addition agent Any substance added to a system to modify its characteristics. Examples include Brighteners added to *Electroplating* baths to improve lustre, or materials added to *Pickling* baths to inhibit pitting.

Adhesion (1) Bonding of components by means of some glue-like substance in the interface.

Adhesion (2) The bonding or locking of

two metal surfaces usually under heavy interfacial pressure. If the bond forms primarily by diffusion effects that are a function of time and temperature adhesion is a form of *Cold welding*. If relative movement and friction effects are involved, terms such as *Galling* and *Seizure* are appropriate.

Adhesive wear Removal of material from a metal surface by a repeated welding and tearing action. When two surfaces move relative to each other in unlubricated contact the high points on the two surfaces can weld together as a result of local high temperatures and disruption of surface films. Continuing movement immediately breaks the weld but the resultant surface damage becomes a site for further welding and tearing cycles. As the damage increases it may be referred to, progressively, as *Scuffing*, *Galling* and, finally, *Seizure*. The damage may also be described as pitting, scoring or gouging but these terms are also used for damage by other mechanisms.

Admiralty brass Alloy of 70% copper, 29% zinc, 1% tin having good corrosion resistance particularly in sea water, hence its use for condenser tubing in steam-driven vessels. See Table 8.

Admiralty gunmetal Alloy of 88% copper, 10% tin, 2% zinc having good strength, corrosion resistance and casting characteristics, historically used for naval guns. See Table 8.

Adsorption The process whereby material, initially present as a vapour, liquid or solid, attaches to the surface of some other material forming a fairly loosely bonded molecular layer.

Aeration cell corrosion Same as Differential aeration corrosion, see entry on this topic.

Afterblow In the basic *Bessemer* steelmaking process the main blow produces a strong flame resulting from the carbon removal. Continuation of the blow after carbon removal, termed the afterblow, produces a lesser flame associated with phosphorus removal.

Age Any process taking place over a period of time. Examples include an increase in hardness, as in *Strain age hardening*, or an increase in the *Hysteresis* losses in magnetic steels. The term is also a common contraction of *Age harden*.

Age hardening An increase in *Hardness* occurring spontaneously over a period of time, particularly that observed in alloys that have been solution treated. See Precipitation hardening and Strain age hardening.

Agricultural steel A largely obsolete term implying plain carbon *Steels* as used for simple farming implements such as ploughs or spades possibly with a *Hard face* deposit.

Air The gas atmosphere in which we live. It comprises about 80% nitrogen, 20% oxygen, small quantities of carbon dioxide, rare gases such as argon, and pollutants such as sulphur dioxide.

Air arc cutting Any cutting processes in which the material to be cut is melted by an electric arc struck between an electrode and the workpiece and the molten material is ejected by a high-velocity air jet. Terms such as Air carbon arc cutting and Air metal arc cutting indicate the material of the electrode.

Air bearing A bearing for a shaft or sliding interface where air is induced to keep the faces separate and act as a lubricant.

Air circulation furnace An *Air furnace* that has some system, in addition to natural convection, to circulate air to ensure even heating of the contents.

Air furnace A furnace in which the atmosphere is untreated as opposed to fuel combustion products, some gas deliberately introduced or a vacuum.

Air gap (1) (in welding) The distance between the *Electrode* and the workpiece over which an electric arc is struck.

Air gap (2) (in an electromagnetic context) A gap crossed by the magnetic flux to complete the magnetic circuit. In a generator or motor it is the radial gap between rotor and stator and even when the machine is hydrogen cooled and the gap is filled with hydrogen during service it is still referred to as the air gap.

Air hardening (steel) A steel containing a sufficient quantity of alloying elements to cause it to fully harden when cooled in still air from the austenitic range. Unless specified to the contrary it is usually implicit that full hardening to the centre should be achieved in a round bar of 50 mm (2-inch) diameter. See Steel and Hardenability.

Alabaster The naturally occurring crystalline form of *Gypsum*.

Alclad A proprietary form of *Clad* aluminium sheet.

Alcomax A series of proprietary alloys with strong permanent *Magnet* characteristics. Iron based with about 20% cobalt, 15% nickel, 10% aluminium and small quantities of other elements such as titanium, Niobium and copper, they are hard, brittle and can only be formed by casting or powder metal processes. **Alnico** alloys are similar.

Alkali A soluble base which dissociates in water producing OH$^-$ ions. See also pH.

Alkali cleaner/pickle Alkaline solutions or salts used for cleaning and pickling.

Alkali metals Metals which react with water to form strongly alkaline hydroxides. Specifically Group 1A of the Periodic Table: lithium, sodium, potassium, rubidium, caesium and francium. See Table 16.

Alkaline earths/earth metals The metals which form strongly alkaline oxides, i.e. earths. Specifically Group IIA of the Periodic Table: beryllium, magnesium, calcium, strontium, barium, radium. See Table 16.

Alkaline sodium picrate An *Etchant* for *Steel*. Typically 25 g of sodium hydroxide plus 2 g of picric acid dissolved in 100 cm^3 of water (hot). Used hot it will slowly darken cementite.

Alligator(ed) surface Same as Orange peel effect and Pebbling.

Alligatoring Cracking of plate or bar material in a plane parallel with the rolled surface.

Allomeric Different substances having the same crystallographic structure are termed allomeric.

Allomorphous Having different crystalline forms for a given composition.

Allotropic Occurring in two or more solid forms having differing physical characteristics and where the change is reversible. If the change is not reversible, polymorphism is the usual term.

Allowed (energy) bands The band of energy in which the valence electrons of a metal crystal are allowed to exist. Individual bands may be empty, partially or completely filled and they are separated by Forbidden bands in which the electrons can not normally exist other than to jump across. See also Band theory and semiconductor.

Alloy Any combination of a *Metal* with one or more further metals or non-metals where an intimate, fairly homogeneous mixture has been achieved by a process such as melting. The term implies that the additional element has been introduced deliberately with the intention of improving some characteristic of the material. *Brass, Bronze, Steel* and *Sterling silver* are examples of alloys. *Steel* is an alloy of iron with a small but vital carbon content so the term 'Alloy steel' implies the addition of further elements such as nickel, chromium and molybdenum to improve specific properties. The term is used colloquially, as in 'alloy wheels' when referring to aluminium probably as a corruption of the long-established casual conversational use in some industries of 'alley' referring to aluminium in either its pure or alloy form. The potential for confusion is obvious.

Alloy plating The *Electrodeposition* of a coating comprising two metals such as copper and zinc to produce a *Brass* plate. The two metals may be supplied from a single-alloy *Anode* or from individual anodes.

Alloy steel A steel containing significant amounts of alloying elements in addition to, or instead of, carbon. See Steel.

Alluvial tin (or other material) Material deposited on the beds of rivers in areas of slow flow, having been transported by the stream from locations where it was eroded from the rocks.

Alnico See Alcomax.

Alpha α The first letter of the Greek alphabet. The various *Phases* occurring in alloy systems are designated by Greek letters, α, β, γ, etc., respectively alpha, beta, gamma, etc. Alpha usually refers to the primary *Solid solution* in any system. See Table 20.

Alpha beta brasses The *Brasses* which contain about 40% zinc. They are duplex, alpha plus beta, and are readily hot worked but will accept only a small amount of cold work. See Table 8.

Alpha brasses The *Brasses* which contain up to about 37% zinc. They are single, alpha, phase and are readily cold worked following initial hot working. See Table 8.

Alpha particles Sub-*Atomic* particles comprising two protons and two neutrons, i.e. the helium nucleus, and hence positively charged. They have relatively low penetrating power.

Alumel An alloy of nickel with about 2.5% manganese, 2% aluminium and 1% silicon widely used in conjunction with *Chromel* for *Thermocouples*. See Table 10.

Alumina Aluminiun oxide Al_2O_3, also termed **Corundum**. It forms naturally and very rapidly on aluminium exposed to the atmosphere. Ruby is alumina with small quantities of chromium oxide; sapphire is alumina with small quantities of cobalt or titanium.

Aluminium A metallic element in common use in a (fairly) pure form or with a few per-cent of alloying elements. In its commercially pure form it has high electrical and thermal conductivity with good resistance to corrosion but low strength. It is highly reactive with oxygen but derives its corrosion resistance from its ability to form a thin impervious surface oxide film which prevents further reaction (also see Anodizing). Various elements can be added to form aluminium alloys with improved mechanical properties but, usually, reduced corrosion resistance. Some additions, such as copper, silicon and magnesium in suitable proportions, allow the alloy to be hardened by a heat treatment cycle referred to as *Precipitation hardening* or *Age hardening*. Commercially pure aluminium and other alloys that cannot be hardened by this form of heat treatment are referred to as non-heat treatable. However, as with other metals, they can be hardened by working, i.e. rolling, drawing, etc. and, subsequently, they can be softened by *Annealing*, which is, of course, a heat treatment process. Various levels of hardness and corresponding tensile strengths can be achieved by selecting the severity of the working process. The various strength levels of work-hardened aluminium are sometimes referred to by a variety of terms such as Half hard condition or Quarter hard temper or by temper designations such as T1, T4, T6, etc. Table 7 summarizes the alpha-numeric system for identifying aluminium alloy composition and condition and gives an indication of the range of properties achievable.

Aluminium brass Alloy of copper with 22% zinc, 2% aluminium and, usually, 0.4% arsenic. The addition of aluminium to the brass improves resistance to *Impingement attack* and the arsenic inhibits *Dezincification*. See Table 8.

Aluminium bronze Alloy of copper with up to 10% aluminium and possibly other elements such as iron, manganese or nickel in small quantities. It is corrosion resistant with good mechanical properties. See Table 8.

Aluminothermic welding Same as Thermit welding.

Aluminizing Any treatment which causes aluminium to diffuse into the surface of a component (usually steel) to improve corrosion resistance, particularly resistance to oxidation at temperature up to about 900 °C. Processes include dipping in molten metal, metal spraying followed by a diffusion heat treatment or exposure of the steel at elevated temperature to a gaseous or liquid environment capable of releasing aluminium into the metal surface. The common feature is that diffusion and the formation of an intermetallic alloy layer ensures good adhesion in contrast to metal spraying which merely applies a mechanically adherent coating.

Amalgam   Any alloy of mercury. Amalgams are commonly made from a mixture of liquid mercury and finely divided metal powder. The mixture is pressed to squeeze out the excess mercury liquid and the remaining paste solidifies by a *Diffusion* process.

Amalgamation process   A process for the initial extraction of gold and silver. The crushed ore is mixed with liquid mercury which amalgamates with the metals leaving behind the impurities. The amalgam is then heated to drive off the mercury leaving impure gold and silver.

Amorphous   Non-*Crystalline* solid, i.e. the atoms are not arranged in any regular pattern.

Amorphous silicon   A non-crystalline form of silicon or silicon hydride. It is formed as a thin film by chemical vapour deposition and is used as a *Semiconductor*.

Anaerobic   Environments containing little or no oxygen.

Anatomical materials   Metals and other materials used for long-term insertion into the human body as prostheses, bone reinforcement, etc.

Andrade's Law   An equation for calculating *Creep* strain in metals and other materials:

$$\varepsilon_{cr} = ct^{1/3}$$

where ε_{cr} is creep strain, t is time and c is a constant for that material.

Anelasticity   Deformation in which *Strain* is time dependent and not proportional to *Stress* although on removal of the stress the component will return eventually to its original unloaded shape.

Angle of contact/bite/nip   See Rolling angle.

Angle of repose   The steepest angle of slope that particulate matter can achieve. It develops naturally when powders etc. are poured from a single fixed source to form a heap.

Angle iron   Wrought metal 'L' section. Rolled mild steel is usually implied and the two legs of the section are usually equal.

Ångström unit 10^{-8} cm.

Anion   A negatively charged *Ion*. It carries an excess electron and is drawn to the anode. See Electrochemistry.

Anisotropic   Having a variation in physical or mechanical properties in different directions within the material.

Annealing   Heating a metal to a temperature at which it becomes fully softened. It is usually implicit that the softening will involve *Recrystallization* and, in some cases such as ferritic *Steels*, that cooling will be at a controlled slow rate. In addition to softening, the process will allow *Recovery*, eliminate *Residual stresses* and may also allow considerable *Diffusion* producing a more *Homogeneous* material. Some *Grain growth* will occur if the time at temperature is protracted. Usually this is undesirable although there are exceptions, for example austenitic steels for *Creep*-resisting applications. See also Steel, Stress relief and Sub-critical annealing.

Annealing twin   Twins formed during *Recrystallization*. See Twins.

Anode   The pole, or the material forming the pole, in an electrolytic cell towards which anions travel. The usual site of corrosion. See Electrochemistry.

Anode efficiency   The observed rate of anode dissolution as a percentage of the theoretical rate predicted by *Faraday's laws* of electrolysis.

Anodic cleaning/pickling Cleaning of a metal in an *Electrolytic* cell in which the component forms the *Anode* receiving an electrical current from an external source.

Anodic coating A surface coating produced by *Anodizing*.

Anodic corrosion The corrosion at the anode of an *Electrolytic* cell, the normal site in wet environments.

Anodic oxidation See Anodizing.

Anodic protection The application of a sufficiently high anodic current to render the surface passive in an aqueous environment. Normally, anodic areas would be expected to corrode.

Anodizing The process of forming an oxide film of greater thickness than normal by making the component, usually aluminium, the anode in an electrolytic cell carrying an electric current from an external source. See Electrochemistry. A freshly exposed surface of aluminium will normally develop a thin, hard, tenacious oxide film which resists further corrosion. The thicker film produced by anodizing, typically 0.01 mm, has improved resistance to abrasion and corrosion and can be dyed various colours.

Anticlastic deformation The deformation experienced in the transverse plane when a beam is bent in the longitudinal plane. Bending induces compressive stresses at the exterior of the bend and compression at the interior. In the case of a simple rectangular beam the exterior face of the longitudinal curve tends to become concave and the interior face to become convex while the sides taper out towards the inner face.

Anti-fouling Coatings applied to the immersed surfaces of vessels and other structures to inhibit adhesion by marine organisms. The term usually implies materials that have a toxic component that is continuously released to poison or at least deter organisms.

Anti-friction alloys Tin-, lead-, aluminium- and zinc-base alloys having a low coefficients of friction in contact with, particularly, steel.

Anti-friction bearing An imprecise term often indicating only that the bearing option under discussion is better than the previous practice. In very general terms friction reduces in the sequence–metal to metal, plain bearing such as *White metal, rolling element bearing, air bearing*, magnetic levitation in air and magnetic levitation in vacuum, the first three being vastly improved by lubrication.

Antimonial alloys Any alloys based on, or containing large quantities of, antimony. The two classes of alloy best meriting the description are the very hard and relatively brittle antimony-based bearing alloys with about 25% copper or 15% tin and 5% copper and the lead-based alloys with up to about 30% antimony for chemical plant. The tin-based *Babbit metals* also contain antimony but in lesser amounts.

Antimony A metallic element having limited application in the pure form or as the primary constituent in an alloy (see Antimonial alloys). In small quantities it is a common addition to tin- and lead-based alloys, as in lead battery alloys, tin and tin–lead *Babbit metals* for bearings and printers' type and in decorative alloys such as pewter. See Tables 11, 15, and 16.

Anti-pitting compound/agent An agent to solutions used for *Pickling, Electroplating*, etc. to prevent pitting of the component being treated.

Anvil The stationary surface on which a component is supported to be struck by a hammer. Alternatively, a pair of surfaces which both move to retain or manipulate a component.

Apatite A mineral, calcium phosphate and calcium fluoride, $CA_3(PO)_2.CaF$. It is a fertilizer and is No. 5 on *Mohs'* hardness scale.

Apparent density The density of powder measured under defined conditions of filling, settlement, etc.

Apparent modulus The same as Secant modulus or, less commonly, Tangent modulus.

Aqua regia A mixture of 25% nitric acid and 75% hydrochloric acid by volume.

Aqueous Relating to water, for example a water environment or water-based solution.

Aramid fibre A high-strength polymer fibre often used as the strengthening component of composites.

Arbitration bar A test bar produced with a component and intended for destructive examination as a check on the component quality. In the case of castings the bar will be from the same melt as the component. In other cases a bar taken from the same stock accompanies the component through treatment cycles such as hardening or coating.

Arbor (1) The rotating shaft on which a grinding wheel or machining cutter is mounted.

Arbor (2) The system of metal reinforcement bars in a large sand mould. This material does not contact the casting, unlike *Chills*.

Arc The passage of an electric current across the gap between two electrodes in vacuum, air or liquid. The electrical transfer is accompanied by radiation of heat and light.

Arc air cutting/gouging Processes in which an electric arc melts the material which is then ejected by a high-velocity air stream. The arc may be struck between the component and an electrode or between two electrodes.

Arc blow (in welding) The undesirable deflection of the electric arc by magnetic fields usually arising from the welding currents.

Arc brazing Techniques of *Brazing* in which the heat is provided by an electric arc struck between two non-consumable electrodes.

Arc cutting Any process in which an electric arc melts the material to be cut, the metal being removed by a combination of melting and oxidation. The arc may be struck between the component and an electrode or between two electrodes. See also Arc air cutting.

Arc energy The rate of heat release in an electric arc. See Weldability.

Arc furnace A vessel in which metal is melted by an electric arc struck between two electrodes or between one electrode and the *Charge*.

Arc melting Melting of a material by an electric arc. The term normally refers to melting in a furnace rather than in processes such as arc welding.

Arc plasma welding Same as Plasma arc welding.

Arc strike (in electric arc welding) The unintentional striking of an arc on parent material or previously deposited weld and the damage caused thereby. Such damage may range from superficial blemishes to serious cracking.

Arc welding Any welding in which the heat input is derived from an electric arc. See Electric arc welding.

Argon An inert gas used to provide a protective environment for various processes such as Argon arc welding.

Argon arc welding *Electric arc welding* in which the *Weld zone* is protected by a shroud of Argon gas.

Argonaut welding A proprietary form of *Metal inert gas welding*.

Armco A proprietary name used as a colloquial term referring to the *Steel* barriers at the edge of, and between, the carriageways of roads.

Armco iron A proprietary grade of commercially pure iron with low hardness, high ductility and good corrosion resistance compared with carbon steel.

Armour plate An imprecise term usually indicating *Steel* with a strength of about 1000 MPa. It may be through-hardened, face hardened, carburized on one face or composite, i.e. in multilayer form with other materials. The term is also applied to non-ferrous materials such as higher-strength aluminium alloys when these are used as the skin of armoured vehicles.

Arrest marks Concentric marks on the surface of a *Fatigue* fracture. They are visible to the unaided eye and mark the crack front at irregularities in the load cycling such as interruptions in cycling or an abnormally high load. Also termed Conchoidal marks, Shell marks or Beach marks. Compare with Striations.

Arrest points The inflections observed on the curves plotted in the *Thermal analysis* of metals and alloys. They reflect changes in the rate of heat release or absorption and hence identify the temperature of *Phase* changes.

Arrhenius equation A semi-empirical formula applicable to many rate-controlled processes such as diffusion, some stages of creep and some corrosion processes. It is usually presented as:

$$\text{Rate} = A\,e^{-Ea/RT}$$

where A is the Arrhenius constant, Ea is the activation energy, R is Boltzmann's constant and T is temperature absolute.

Arsenic A toxic, allotropic, semi-metallic element. It has little commercial application in the pure form but, as an alloying element, it improves oxidation resistance and high-temperature strength in copper, inhibits *Dezincification* in some *Brass* and hardens lead.

Arsenical copper Alloys of copper with up to 0.5% arsenic to improve high-temperature properties including *Creep* strength and enhance resistance to atmospheric oxidation. See Table 15.

Artefact Generally a man-made, as opposed to a naturally occurring, structure. In its metallurgical sense a feature not arising naturally in the material or microstructure in question, for example an inadvertent scratch on a specimen under the microscope which could be mistaken for a structural feature.

Artificial ageing An ageing process deliberately induced by heating a component. See Precipitation hardening.

Asbestos Various magnesium silicate minerals which are non-flammable with high melting point and low thermal conductivity. They have a fibrous structure which is capable of progressively shred-

ding to finer and finer sizes. Airborne fibres cause lung damage, sometimes long after ingestion.

As-cast/as-welded etc. The condition reached upon completion of the indicated process but prior to heat treatment or other processing.

ASME American Society of Mechanical Engineers.

Aspect ratio Depending upon context, various ratios such as length to diameter (fibres), length to width (crack), depth to diameter (corrosion pit), etc.

Asperities High spots or protrusions on a surface.

Assay value The measure of the metal content of an ore of precious metal. It may be expressed as a weight percentage or, traditionally, as troy ounces of metal per avoirdupois ton of ore.

Assaying Analysis of metals and ores. The term is usually used in the context of the noble metals intended for retail sale particularly jewellery, or of ores being analysed and assessed for exploitation.

Assel mill A three-roll piercing mill. See Tube making.

ASTM American Society for Testing Metals.

Aston process A obsolete process for producing *Wrought iron*. The impure *Pig iron* is melted in a *Cupola* furnace and then refined in a *Bessemer converter* prior to being mixed in a ladle with a molten slag of iron oxide and silica. The semi-solidified paste is then immediately forged as described under the *Puddling process*.

Athermal transformation A *Phase* transformation occurring without the need for a temperature change. It is usually implicit that no diffusion is involved and the rate of transformation can be very rapid.

Atom The smallest particle of an element that can exist and retain its chemical characteristics. Further division of the atom, termed **Fission** or, colloquially, splitting the atom, reduces it to its constituent, sub-atomic, particles:

protons, electrons, neutrons, etc. See Atomic structure.

Atomic arrangement The way atoms are arranged in a *Crystal structure*.

Atomic hydrogen (or other element) Hydrogen (or other element) in a state in which the *Atoms* are not bonded to each other. Normally hydrogen atoms bond strongly together in pairs to form *Molecules*. If the hydrogen molecule, H_2, is raised to a high energy level by heating it dissociates into two individual H atoms. When the atoms recombine a large amount of heat is released. In the **Atomic hydrogen (welding) process** molecular hydrogen is dissociated in an electric arc between two electrodes, usually tungsten. The high-energy gas is directed on the component to be welded where it produces very high temperatures as it reverts to the molecular form.

Atomic mass Strictly, Relative atomic mass and previously termed atomic weight. The mass of one atom of an element relative to the mass of a reference element. Originally the reference element was hydrogen, atomic mass of 1.0, then oxygen of atomic mass 16.0 or carbon of atomic mass 12.

Atomic number The number of protons in the atomic nucleus of the element in question. See Atomic structure.

Atomic % The composition of an alloy or mixture in terms of the number of atoms of each constituent per hundred of mixture.

Atomic radius Half the distance between the centres of two nuclei of the element that are at equilibrium separation but not bonded.

Atomic structure (1) The internal structural of atoms. An atom can be visualized by the non-specialist as a central nucleus, composed of protons and neutrons, with electrons in orbit around it. The **Nucleus** is small, about 10^{-12} of the volume of the atom. **Protons** carry a positive electrical charge, **Electrons** an equal but negative charge and **Neutrons** no charge. Protons and neutrons have the same mass, 1.675×10^{-27} kg; electrons

are much less massive, 9.11×10^{-31} kg. All atoms of an element contain a specific number of protons and the same number of electrons so the electrical charges balance. The number of protons and electrons defines the **Atomic number** of the element, i.e. hydrogen, atomic number 1, has one proton and one electron, aluminium, atomic number 13, has thirteen of each. Although, simplistically, the electrons can be considered to be in orbits around the nucleus they do not wander freely. They are constrained to specific energy levels usually termed **Shells** which can be considered as concentric spheres. Hydrogen with a single electron obviously has a single shell as does helium with two electrons in the same shell. However, this innermost, or No. 1, shell is **Filled** when it contains two electrons. Consequently, with larger numbers of electrons more shells are formed so lithium has two electrons in the 'filled' inner shell 1 plus one in shell 2, beryllium has two electrons in both shells 1 and 2 while boron haas two in shell 1 plus three in shell 2. This buildup of shell 2 continues up to element No. 10, neon, with eight electrons in shell 2. At this stage, shell 2 with eight electrons is now also 'filled' so further shells have to form to accommodate larger numbers of electrons. This buildup progresses as more electrons necessitate more shells, and, with large numbers of electrons, the structure becomes more complex with some inner shells accommodating more than eight electrons and also forming sub-shells. These subshells are designated with a letter suffix following the shell number, for example 2s or 3p. However, the essential feature is that the maximum number of electrons in the outermost shell is always eight (apart from shell 1 with a maximum two electrons). The electrons in the outer shell are termed the **Valence** electrons as they are the only ones available for interaction with other atoms and hence they dominate the chemical characteristics of the element. As the number of electrons available for interaction is limited to eight there is a distinct pattern in the chemical behaviour of the elements. Elements having similar numbers of electrons in their outer shells react in similar but not identical ways, for example chlorine, fluorine, bromine and iodine all have seven valence electrons and react similarly with other elements. Equally, neon, argon, krypton, xenon and radon, all with eight electrons filling their outer shells, plus helium with its only shell filled by two electrons, are all inert gases. This distinct pattern, or **Periodicity**, to the chemical behaviour of the elements gives rise to the **Periodic Table** in which all elements are presented in columns and rows. The table can be developed in various forms, one being shown in Table 16, but the common basic features are that the columns comprise **Groups** of elements having the same number of valence electrons and the rows comprise **Periods** of elements having similar numbers and arrangements of inner shell electrons. The complexity imposed on the table by the **Transition elements** arises because some sub-shells start to fill before earlier shells have 'filled'. The groups of elements are numbered, often in Latin numerals, i.e. Groups I to VIII, and may also have names, for example Group I is the alkali metals.

Neutrons do not affect Atomic number or chemical properties and their number, in the atoms of a given element, may vary within limits. As noted above, neutrons have the same mass as protons so they make a major contribution to the **Atomic mass** of the element. Versions of any particular element having different numbers of neutrons are termed **Isotopes**. For example, hydrogen has one proton and one electron but a small proportion of its atoms will also have a neutron, or even two. The various isotopes have identical chemical properties but the atoms with the additional neutrons have a higher density, hence these isotopes are commonly termed **Heavy**

hydrogen. Also in the case of hydrogen, the isotopes have recognized scientific names, **Deuterium** for the two-neutron form and **Tritium** for the three-neutron forms. In the case of all elements, isotopes are usually indicated by a numeral suffix identifying the total of protons plus neutrons in the nucleus. For example, Uranium-238 and Uranium-235 each contain 92 protons but one has 146 neutrons the other 143. See also Band theory.

Atomic structure (2) Occasionally this term is used to refer to the manner in which atoms in bulk solids are arranged with respect to each other. A better term for this usage is *Crystal structure*.

Atomic weight Obsolete term for *Atomic mass*.

Atomicity The number of atoms in a molecule of the element.

Atomize Processes for producing metals as fine particles or powder by causing a jet of liquid or air to impinge on a stream of the molten metal. The particles produced may be small but they are far bigger than an atom.

Attenuation The reduction in a signal, for example radio or ultrasonic, as a result of effects such as beam spread, scattering or absorption.

Attrition Continuous abrasion. An **Attrition mill** is a device for breaking metals or minerals into small particles.

Auger analysis A technique for analysing very shallow surface features.

Auger effect The ejection of an outer shell electron by an atom which has been ionized by the ejection of an inner shell electron (see Atomic structure). The resultant energy emission spectrum is characteristic of the element and hence the effect offers an analytical tool. See previous entry.

Ausforming Mechanical working of a suitable steel that has been rapidly cooled to retain the austenitic phase below the lower critical temperature, 723 °C in carbon steels. It is implicit that the steel composition is such that the austenite is unstable below this temperature and it will eventually transform to martensite or bainite. Such treatment, which effectively combines hardening by working with hardening by transformation, can produce very high-strength components. See entries on Steel and Isothermal transformation. Note that the term is not appropriate for steels that will remain austenitic.

Austemper A heat treatment process in which steel is rapidly cooled from the austenitic region to a temperature just above the martensite start temperature. This suppresses the normal pearlite transformation allowing isothermal transformation to bainite. With appropriate steels transformation is delayed until temperature gradients and associated stresses are reduced. The absence of stress together with transformation to bainite rather than martensite reduce the risk of cracking. After it has fully transformed the steel is finally allowed to cool to ambient temperature. *Martempering* is rather similar except that after an initial hold at the intermediate temperature the component is cooled to form martensite. See entries on Steel and Isothermal transformation.

Austenite The *Phase* in which iron or steel has a face centred cubic *Crystal structure*. See Steel.

Austenite bay The recessed area on an *Isothermal transformation diagram* where there is the maximum delay in transformation to pearlite or bainite.

Austenitic grain size Generally, the grain size of steel when in the austenitic state (see Steel). The term is also used in the sense of an inherent austenitic grain size, established by the original steelmaking practice, which influences the steel properties even after a number of subsequent working and heat treatment operations. The term **Prior austenite grain size** usually refers to the grain size when the steel was last in the austenitic condition.

Austenitic stainless steel Any steel containing sufficient chromium to render it 'stainless' and sufficient nickel to render it austenitic. (See Steel.) Typically, these steels contain at least 18% chromium and at least 8% nickel, hence the common designation '18/8 stainless'.

Austenitic steel Any steel that, following slow cooling, is austenitic at ambient temperature. This usually implies a high content of nickel or manganese (see Hadfield's manganese steel). The steel will be non-magnetic but will be 'stainless' only if it contains more than about 12% chromium.

Austenitizing Heating a steel to a temperature at which it transforms to austenite. See Steel.

Autoclave A vessel in which reactions or tests are performed under pressure.

Autofrettage A process in which tubular components, prior to service, are deliberately expanded by internal pressure to a level exceeding their *Yield strength*, so leaving favourable compressive *Residual stresses* at the bore. Internal pressurization of a cylinder induces tensile stresses in the longitudinal and hoop (circumferential) directions, the hoop stresses being approximately double the longitudinal. In addition, the hoop stresses vary through the thickness with the bore developing the maximum levels, an effect which increases as wall thickness increases. Thus service performance of thick-wall pressure vessels, gun barrels, etc. can be dominated by the peak tensile hoop stresses at the bore. In the autofrettage process the cylinder is pressurized to a level which is above the service pressure and is selected to induce yielding of the material at and close to the bore. After removal of the pressure the vessel is slightly larger in diameter than originally and has high levels of compressive stress at the bore balanced by moderate tensile stresses in the exterior area. When the cylinder is subsequently pressurized in service the pattern of tensile stress will be more evenly distributed across the wall thickness, in particular the high tensile peak at the bore will be reduced, and no further yielding will occur.

Autogenous weld A weld made by the fusion of the parent components without the addition of *Filler* material.

Automatic welding Processes in which welding is completed by the equipment without any operator involvement apart from initial adjustment of settings.

Autoradiography A radiographic technique which relies on radiation emitted by the component under test rather than radiation from an external source.

Avogadro's constant/number The number of atoms in the *Atomic mass* in grams of any element, $6.022\,52 \times 10^{23}$.

B

b Symbol for barn $= 10^{-28} \text{m}^2$.

BA British Association, usually in the context of BA threads, a series of standard thread dimensions and profile.

Babbit metals Various alloys based on tin, antimony (5–15%) and, usually, copper (2–5%) used for plain bearings and printers' type. Lead may substitute some of the tin in cheaper alloys. The common feature of all the alloys is the formation of a hard, tin–antimony intermetallic compound with a cubic form. The resultant microstructure of hard particles embedded in a soft compliant matrix provides excellent bearing properties. The cuboids are of low density compared with the matrix and hence, during solidification, tend to rise towards the surface. The function of the copper is to produce, early in the solidification process, a tin–copper intermetallic compound, also hard, which precipitates in an interlocking needle formation inhibiting movement of the cuboids. These alloys are also termed *White metals*. See Table 11.

Back annealing Partial *Annealing* of work-hardened material to some predetermined strength level.

Back end defect See Extrusion defect.

Back face See Fire side.

Back gouging (of a weld) The machining, grinding etc. of the underside of the first *Run* of a weld to allow a good-quality run to be deposited on that side.

Back reflection X-ray technique A technique for studying *Crystal structure* in which a narrow, collimated beam of monochromatic X-rays is directed at the metal through a hole in a photographic film. Crystallographic planes in the crystal structure reflect X-rays back onto the film producing dot patterns which can be interpreted to indicate grain size, cold work, etc. Sometimes termed **von Laue** analysis after the pioneer in such work.

Back scatter techniques/gauge The measurement of radiation, usually gamma or beta, reflected from some source by the component under test. Within limits, the amount of reflected radiation increases as a direct function of the thickness of the sample so the technique is used for applications such as the continuous measurement of sheet during production.

Back side See Fire side.

Back step sequence (of a weld) A welding technique, normally by hand, in which a long run of weld is built up as a sequence of short deposits. After the first short run is deposited the welder moves back beyond the start of the first run and deposits a second short run terminating at the initiation point of the first run. The third run then terminates at the start of the second and the sequence is repeated until the required length is complete. See Figure 49, located at entry on Welding terminology.

Back weld The weld, usually a single run, made on the rear, *Root*, side of a weld previously commenced on the front side.

Backhand welding The technique of gas welding in which the flame points back towards previously formed weld as it moves forwards towards unwelded joint. Any filler rod points into the unwelded joint. Also termed **Backward(s) welding** and **Rightwards welding** although the latter may lead to some confusion in the case of left-handed operators.

Backing pass (of a weld) An initial pass of weld to form a surface on which the main weld is deposited. It may remain untouched or be partly machined away and, possibly, a *Back weld* deposited.

Backing plate/ring, strip, etc. (of a weld) Material placed behind the *Root* of the weld to provide a surface onto which the first weld run is deposited. A **Temporary backing plate** etc. is one which undergoes little or no fusion and is subsequently removed. A **Permanent backing plate** etc. is one which is partly fused by the weld and may remain in position or be partly removed after welding. See also Fusible insert.

Backward(s) welding Same as Backhand welding.

Bacterial attack/corrosion Any corrosion caused, or assisted, by the presence of microbes or bacteria. Such attack usually arises from the reaction between the metal and the aggressive environment produced by the bacteria.

Bag filter Systems for filtering particulate material from gas streams originally by using fabric bags but now often extended to other forms of fabric and fibre pads, packs, etc. Such systems are often located in **Bag houses** which may be a separate building or merely a room or compartment.

Bainite A range of microstructures in steel comprising fine carbide particles in an acicular ferrite matrix. Bainitic structures normally result from cooling, from the austenitic temperature range, at a rate too rapid to form pearlite but not fast enough to form martensite. The precise structure is a function of the temperature at which the carbide precipitates. See Steel and Isothermal transformation diagram.

Bainite nose See Isothermal transformation diagram.

Bainite shelf See Isothermal transformation diagram.

Bake A term applied to various heating processes involving modest temperatures and no change to the primary properties of the component. For example, components being vitreous *Enamelled* require baking to fuse the coating, welding electrodes are baked to drive off moisture from the coating and components containing high levels of *Hydrogen*, introduced by treatments such as electroplating or welding, are baked to allow the potentially damaging gas to diffuse out.

Bakelite Phenol formaldehyde resin.

Ball mill A process in which balls are used to crush ores or otherwise treat components. The balls may tumble freely in a cylinder or barrel as in the case of mills used for *Peening* or some pulverizing processes. Alternatively, in the case of some crushing mills, the balls may run in grooved tracks in a pair of rings, hence the term ring mill. Typically the free-tumbling pulverizing ball would be small, say about 50 mm diameter when charged, while a track ball would be much larger, say about 400 mm, although there are wide variations.

Ball pein hammer The hammer typically carried by an engineering fitter. It has one smooth hemispherical face and, usually, a flat square face. The pein end is the likely origin of the term *Peening*.

Ball sizing/ball broaching Processes in which a ball or possibly a round nose bar is forced through a hole. Depending upon the application various benefits ensue—burnishing the bore, developing an accurate size and introducing a favourable pattern of *Residual stresses*. No metal is removed, unlike *Broaching*.

Balling (1) Spheroidization.

Balling (2) The aggregation of a pasty mass of impure iron and slag in the *Puddling* process.

Band theory The valence electrons of a metallic crystal are not tied to individual atoms but are shared as a general cloud.

The atoms collectively impose a periodic electropotential function which causes the valence electrons to be confined to specific ranges, or bands, of energy level. Individual bands may be partially or completely filled or empty and they are separated by *Forbidden bands* in which the electrons cannot normally exist other than to jump across. The electrons in a partly filled band can move allowing conduction, hence the terms **Conduction electrons** and **Conduction band**. See also Atomic structure and Semiconductor.

Banding/banded structure   A structure in which non metallic inclusions or other forms of *Segregation* are aligned in bands. The effect arises when structural variations from casting are elongated by subsequent processes such as *Rolling* or *Drawing*. *Steels* containing relatively high phosphorus (0.03–0.05%) are most susceptible.

Bar   Material produced in long lengths of constant cross-section and, in most contexts, with the width and height approximately similar as opposed to strip. The cross-section of material produced by *Rolling* or *Drawing* is usually simple but that produced by *Extrusion* may be extremely complex.

Bar drawing   *Tube drawing* using a *Mandrel* which travels with the tube.

Bar magnet   A straight magnet with a north pole at one end and a south pole at the other.

Bar mill (1)   A *Rolling mill* for producing *Bar*.

Bar mill (2)   A barrel or similar container in which components are contained with short lengths of bar and possibly an abrasive. The container is rotated and the bar burnishes or abrades the components.

Bare wire/electrode welding   *Electric arc welding* in which the wire or electrode does not have any coating to form a protective shield for the molten metal. It is normally implicit that no alternative protection, such as an inert gas or powder covering, is applied.

Barffing   Heating iron or steel in steam at about 750 °C to form a protective and decorative coating of black magnetite (Fe_3O_4). Any non-protective red Fe_2O_3 is reduced by injecting carbon monoxide at a final stage in the process. The term usually also implies subsequent oiling of the surface to enhance appearance and protection.

Barium   A metallic element, one of the alkaline earth group. It is highly reactive, being spontaneously flammable in moist air and hence is used as a *Getter*, usually alloyed with other metals such as magnesium or aluminium. Alloys with nickel are used as filaments in some thermionic valves and alloys with lead are used for some bearing applications. Some compounds are useful for their fluorescence and barium titanate has strong piezoelectric characteristics. See Table 15 for physical properties.

Bark   The rough textured surface beneath the scale on metal, particularly steel, that has been heated at high temperature in a non-protective atmosphere.

Barlow's (thin wall) formula   A simplistic formula for calculating the *Hoop stress* in an internally pressurized tube.

$$S = \frac{PD}{2t}$$

where S = stress, P = pressure, D = diameter (mean), t = wall thickness. All pressurized tubes develop a tensile hoop stress which increases towards the bore but the effect becomes more pronounced, the thicker the wall. Barlow's formula is most accurate for tubes with a thin wall relative to the diameter.

Barn   A unit at the nuclear level, symbol b ($= 10^{-28}$ m^2).

Barrel plating/finishing/cleaning etc.   Any process in which components are placed in a rotating barrel usually with some other material to effect the plating, polishing, etc. The barrel may be self-contained or it may be perforated and rotate in a bath of plating solution, etc.

Barret effect   See Magnetostriction.

Barstock Bar intended for subsequent manipulation or machining into some finished component.

Basal plane The plane perpendicular to the principal axis in hexagonal *Crystal structures*.

Basalt An igneous mineral containing large quantities of iron and magnesium silicates, typically dark-green to black/ brown, often as columnar strata.

Base bullion A base metal, usually lead, which contains sufficient noble material, particularly gold or silver, to be profitably extracted.

Base material (of a weld etc.) Parent material to be welded etc.

Base metal (1) Metals not classified as *Precious* or *Noble*.

Base metal (2) Material undergoing some action or process.

Base metal (3) In a welding context, the unmelted parent material as opposed to metal that has fused.

Base units See SI system.

Basic In a metallurgical context, processes in which the environmental conditions including the furnace *Refractory* lining and the *Slag* are alkaline or basic as opposed to *Acidic*.

Basic electrode (for welding) See Electrode (welding).

Basic open-hearth process The largely obsolete *Open-hearth* process utilizing a basic lining with a lime *Flux* to deal with *Pig irons* having higher phosphorus contents than could be dealt with in the Acid open hearth.

Basic oxygen steelmaking A widely used modern steelmaking process utilizing an unfired, vertical pear-shaped converter vessel with a *Basic* lining comprising *Magnesite* brick covered with *Dolomite* and having a capacity up to about 400 tonnes. Initially, scrap is charged followed by molten *Pig iron*. An oxygen lance, inserted through the top, directs a jet at the molten surface to oxidize the impurities which then form a slag with the lime, fluospar and mill-scale *Flux*. The process is rapid and after less than an hour's treatment the

vessel is rotated, first to discharge the slag and then to pour the steel.

Basic refractory A refractory material for furnace linings and similar applications which, at operating temperature has a basic, i.e. alkaline as opposed to acid, reaction with the *Charge*. Such refractories include dolomite (calcium magnesium carbonate, $Ca.Mg(Co_3)_2$) and magnesite (mainly magnesium carbonates and silicates).

Basic slag The slag from *Basic* steelmaking processes. It contains about 25–30% phosphates which may be recovered or the crushed slag may be used untreated as a fertilizer.

Basic steel A steel produced by one of the *Basic* steelmaking processes.

Bastard Usually, anything abnormal, unintended, or otherwise undesirable but sometimes merely a mixture or hybrid. The usage is usually vulgar but not always, for example, a **Bastard file** is a recognized name for a mechanic's file having a specific cut of medium coarseness.

Batt A loosely pressed pad or sheet of fibre such as is used for reinforcing plastic composite materials.

Battery (1) An *Accumulator* for storing electricity.

Battery (2) A factory containing a number of *Forges*.

Battery (3) Multiple devices operating in concert.

Bauschinger effect The phenomenon by which stressing beyond *Yield* in one direction reduces the *Yield strength* when the material is subsequently stressed in the opposite direction.

Bauxite A mineral, mainly hydrated alumina (Al_2O_3. $2H_2O$), the primary ore source of aluminium.

BCC Body centred cubic. See Crystal structure.

Beach marks Concentric marks on the surface of a *Fatigue* fracture. They are visible to the unaided eye and mark the crack front at irregularities in the load cycling such as interruptions in cycling or an abnormal load. Also termed

Conchoidal marks, Shell marks or *Arrest marks.* Compare with *Striations.*

Bead (of a weld)   A single run of weld metal laid without interruption.

Beam (1)   A length of material of constant cross-section, usually wrought but occasionally cast.

Beam (2)   A length of material carrying a bending load.

Beam (3)   A projected stream of radiation.

Bearing (1)   A support carrying some component or structure. The term usually implies a small support area and a capacity to allow relative movement of the component being carried, for example a bearing carrying a rotating shaft. The term encompasses plain bearings, rotating element bearings (balls, rollers) air bearings, etc.

Bearing (2)   In terms such as 'niobium-bearing steel' it means a steel containing niobium.

Bearing bronzes   Various copper–tin alloys having applications as plain bearings. Alloys include tin–bronzes, leaded tin–bronzes and phosphor–tin bronzes. See Table 8.

Bearing metal/material   Any metal/material having characteristics making it suitable for use as a plain bearing (as opposed to a ball or roller bearing) supporting a shaft or similar component in sliding contact. Most bearings are lubricated, usually by oil or grease, but occasionally by other materials including air. Consequently, if they operate within the design limits, there is theoretically no metal-to-metal contact and hence no wear. In practice, however, some transient contact is almost inevitable during starts, running-in or because of interruption of lubricant. A frequent principle of bearings, therefore, is that ultimately they are sacrificial; that is they, rather than the shaft, should fail in the event of oil starvation or overload. Apart from this often overlooked aspect, various desirable characteristics can be recognized. Usually resistance to deformation is important but also some com-pliance to accommodate misalignment may be of value. A hardness lower than the shaft material is a normal requirement to minimize damage during transient contact but many bearings contain hard particles to support the shaft in multiple-point contact. Corrosion resistance is important but also the development of a strong oxide film may be necessary to provide a low-friction, galling-resistant surface if metal-to-metal contact occurs. The tin base *White metals* are often quoted as classic bearing materials since they comprise hard particles in soft matrix. The hard particles provide the multiple-point contact while the soft matrix provides general support with some compliance, and it wears to slightly below the surface of the hard particles to provide a reservoir of oil for limited periods of oil starvation. The tin – and phosphor – *Bronzes* have similar duplex structures. However, compared with the white metals, they have a higher bulk strength and greater load-carrying capacity but less tolerance of misalignment. *Sintered* bearings comprise metal particles, often bronze, sintered to form a porous mass. The sinter is often pre-impregnated with PTFE, grease or even lead or it may be installed to become impregnated by the machine lubricant. The metal sinter particles provide the load-bearing capacity and the other component is a source of long-term or transient lubrication. Solid plastics such as PTFE or nylon have very low coefficients of friction but limited strength so are often reinforced with metal particles. Such plastics are useful where oil lubrication is not possible.

Beilby layer   The deformed surface layer produced on *Metallographic* specimens by mechanical polishing. It can be removed by *Etching.*

Bell (butt) joint   Tubular joints, particularly for *Braze welding* and *Soldering*, in which one of the tube ends is expanded to a bell into which the other tube is inserted to form a *Fillet* for the deposition of molten *Filler*. Where diameters vary significantly the smaller is bell expanded to match the bore of the larger. This may be termed a **Diminishing bell (butt) joint**.

Bell furnace A furnace in the shape of an inverted bowl, i.e. a bell. The bell is placed over the *Charge* on a fixed hearth and the required heating carried out. The bell can then be transferred to a second and subsequent hearths while the charge on the first hearth cools and is removed. Occasionally, a simple unheated inner bell is left in place to minimize atmospheric contamination and draughts. A **Top hat** furnace is similar.

Bell metal Alloy of copper with 20–25% tin, i.e. a high tin-content tin bronze. It is hard and corrosion resistant with a low damping capacity, and is used in the cast state for bells, plain bearings, slide valves etc.

Bellows A thin-wall tube with circumferential corrugations that allow repeated longitudinal flexing.

Bend test Any test in which material, usually in the form of a purpose-machined testpiece, is bent and its performance monitored. In simpler tests, for example of general quality or *Ductility*, the testpiece is bent in a vice or over an anvil of specified radius. Often the loading is not measured and the testpiece is merely bent until it cracks, bent to a specified angle or, in a **Reversed bend**, bent to a prescribed angle and then bent back. Acceptance criteria depend on the particular test but may range from simply surviving intact to complete freedom from cracking. More complex tests where loading is measured include **Three-point bend testing** in which the testpiece is supported at its extremities and the load applied at a single central point. This induces a peak load at the line on the surface opposite the central loading point. For **Four-point bend testing** the testpiece is supported at its extremities and loaded via two points symmetrically located between the support points. This induces an even stress over the length between the two loading points. In tests of *Fracture toughness* the testpiece is pre-cracked and the properties of interest include the stress required to extend the crack, as indicated by *Potential drop* techniques, and the activity at the crack tip as reflected by *Crack opening displacement*.

Bending brake A machine for forming sharp-edged folds and corners on sheet material.

Bending moment The algebraic sum of all the couples resulting from all the forces on one side of the section.

Bending rolls An assembly of, usually, three rolls having their axes parallel but the centre one offset from the line joining the other two. Bar or sheet passed through the appropriately adjusted rolls is bent to a curve or cylinder.

Benefication Various physical processes applied to ores to separate the desirable constituents from the *Gangue*. Processes include magnetic techniques, flotation and centrifuging.

Bentonite A clay mineral, mainly hydrated calcium and magnesium silicates, capable of absorbing five times its weight of water.

Berkelium One of the man-made elements, No. 97.

Beryllia Beryllium oxide, BeO. It is chemically stable but toxic.

Berylliosis A severe illness related to pneumonia resulting from the inhalation of beryllium.

Beryllium A metallic element with an exceptionally low thermal neutron capture cross-section and good corrosion and oxidation resistance in dry conditions, hence its suitability for nuclear canning applications. Like its oxide *Beryllia* it is toxic, causing *Berylliosis*, a form of pneumonia, if it enters the lungs or ulceration, etc. if it enters body tissue. See Table 15 for physical properties.

Beryllium bronze/copper Alloys of copper with up to 3% beryllium. They can be precipitation hardened to a high strength. See Table 8.

Bessemer converter/process A largely obsolete process for converting impure *Iron* into *Steel*. Molten pig iron is poured into the vertical cylindrical converter vessel and air or oxygen, possibly with a steam addition, is injected through the perforated bottom. The oxygen in the injected gas oxidizes the impurities, releasing sufficient heat to maintain the increasingly pure metal in the molten state. When the reaction is complete the metal is poured out by tilting the vessel. The Thomas converter is similar.

Beta *β* One of the Greek letters designating *Phases*. See Table 20.

Beta brass Alloys of copper with zinc in the range 46-49% which at ambient temperature comprise a Beta intermetallic compound.

Beta compounds See Electron compounds.

Beta iron See Steel.

Beta particles Particles with a mass similar to that of an electron. They may be positively or negatively charged with an energy up to 4-MeV, a velocity approaching that of light but limited penetrating capacity. They are emitted by a radioactive source undergoing Beta decay.

Beta phase See Alpha.

Beta radiation Emissions of *Beta particles*.

Beta structures This term can indicate either phases with a *Body centred cubic* crystal lattice similar to the Beta phase in *Brass*, or an *Electron compound* having three valence electrons to two atoms.

BG *Birmingham Wire Gauge.*

BHN *Brinell* hardness number.

Biaxial Applying to or acting on two axes, for example stresses.

Biaxiality The ratio of the two stresses acting in a biaxial stress field.

Bifurcated Forked. A **Bifurcated rivet** has a solid head and a forked shank which is inserted through the sheet ma-

terials being joined and the prongs of the fork bent apart.

Billet A bar, slab, etc. of simple section that has undergone some hot-shaping operation following casting but which requires further working.

Billion Originally a million million, now, usually, following US practice, a thousand million.

Bimetallic Involving two metals.

Bimetallic corrosion The corrosion which occurs between different metals in electrical contact in a wet environment because of the difference in *Electrochemical* potential between the two.

Bimetallic strip A strip which comprises a layer of one metal bonded to another. The two metals are chosen to have a large difference in thermal expansion coefficient so that as the temperature rises one will expand considerably more than the other, causing the strip to bend.

Binary Having two components. The term is often used casually in a metallurgical context in reference to binary *Phase* diagrams.

Binder A material added to hold together and fill the gaps between solid particles to form an aggregate.

Binding energy The increment of energy maintaining components in a bond or in position. The **Binding energy of an electron** is the energy that must be introduced to release a valence electron. The **Binding energy of the nucleus** is the difference between the total energy in the system with the nucleons (protons and neutrons) combined and the greater energy with the nucleons in the free state. See Atomic structure.

Bio-composite (1) A naturally occurring material such as bone which has the characteristics of a composite, i.e. a bulk material with a reinforcing structure.

Bio-composite (2) A man-made composite suitable for insertion in the human body as a prosthesis etc.

Bio-degradable Capable of being broken down by sunlight or natural organisms such as fungi, bacteria or their life cycle products. It is usually implicit that the breakdown products will be environmentally acceptable. The term is sometimes extended to cover materials that break down under natural ultraviolet radiation.

Bio-engineering The development of materials and devices that are compatible with the human body to replace, support or assist defective parts.

Birmingham platinum An alloy of zinc with 20-45% copper (no platinum). It is hard and brittle with mainly decorative applications.

Birmingham Wire Gauge An early numerical system of standard wire diameters.

Bismuth A metallic element with limited commercial application. Its low melting point allows it to be included in alloys specifically designed to melt at low temperatures, such as fusible plugs in sprinklers and pressure vessels. See Table 15.

Bit A rotating tool with cutting edges at its tip for drilling holes.

Bite In *Rolling*, the reduction in thickness on a single pass through the rolls.

Bite angle See Roll angle.

Bitter pattern A technique for revealing the magnetic domains at an iron or steel surface. The surface is covered with a suspension of fine magnetic powder in a suitable carrier such as white spirit and the component is then magnetized. The magnetic powder then delineates the domain boundaries. The technique is essentially the same as that used for non-destructive crack detection.

Black annealing The *Annealing* of a component in an enclosed container to minimize contamination from the furnace environment. The container may contain some material to assist the protection.

Black body The theoretical surface that absorbs all radiation falling upon it and emits a continuous spectrum when incandescent.

Black body radiation The radiation emitted by a theoretically perfect *Black body*.

Black bolt See Bolt.

Black copper Copper prior to final refinement.

Black lead Graphite, one of the crystalline forms of *Carbon*.

Black plague Nickel sulphide formed by severe corrosion of nickel base alloys in a high-temperature sulphur-rich environment.

Black spot(ting) Casting porosity filled with carbon-rich material. The term is usually used when the defects are exposed by machining.

Blackheart (cast) iron A malleable cast iron. See Cast iron.

Blacking Graphite, plumbago, etc. as fine powder or in liquid suspension used for coating moulds to improve surface finish and assist parting of the casting from the mould.

Blacking holes/spots Same as Black spot.

Blacksmith welding See Forge welding.

Blank (1) Material in a partly formed state, cut to size or otherwise prepared for the final major shaping process. Examples include discs stamped from sheet for *Stamping* or *Impact extrusion* and cut lengths of bar to be machined or *Forged*.

Blank (2) In various test procedures, especially chemical analysis, a sample that does not contain the material being investigated or analysed but which is subject to the full test procedure to identify any experimental error or contamination.

Blank carburizing/nitriding Submitting a component to the *Carburizing* or *Nitriding* heating cycle but without exposing it to the carburizing or nitriding environment.

Blanking press A machine for stamping out *Blanks (1)*.

Blast furnace A vertical shaft smelting furnace for producing impure pig iron from iron ore (see Figure 1). The *Charge*, loaded at the top, comprises iron ore, limestone as *Flux*, coke as fuel and reducing agent. The pre-heated air blast injected around the periphery at

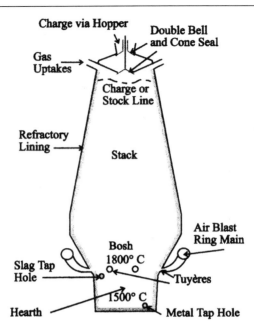

Charge via Hopper

Double Bell
and Cone Seal

Gas
Uptakes

Charge or
Stock Line

Refractory
Lining

Stack

Air Blast
Ring Main

Bosh
1800° C

Slag Tap
Hole

Tuyères

1500° C

Hearth

Metal Tap Hole

Figure 1   The blast furnace for iron production

about quarter height increases the temperature derived from burning the coke to about 1800 °C. Carbon monoxide formed by the coke reduces the iron oxide to molten iron. The limestone combines with the contaminants to produce a fluid *Slag*. The slag and impure *Pig iron* are tapped off separately close to the bottom. Similar processes are used for some other metals.

Blasting (of a surface)   See Shot blasting.

Blebs, blebbing   Growths of material on the exterior of ingots or castings as a result of inverse *Segregation*. As they tend to contain a disproportionate amount of alloy and impurity they may be relatively hard and brittle, causing problems in subsequent operations.

Bleeding (1)   Exudation of metal, usually high in impurities, from the newly solidified surface of a cooling ingot. See Solidification.

Bleeding (2)   The trail of red iron oxide seeping from an interface suffering from *Fretting*.

Blending   Any process where two or more components are mixed to produce a homogeneous material.

Blind hole   A hole that does not penetrate to the far surface

Blind rivet   A rivet which requires access to one side only for insertion and securing. For example, see Pop rivet.

Blister   A planar defect just beneath the surface which results from the release of dissolved gas, almost invariably hydrogen, entrapped during casting. The defect usually lies just beneath and parallel with the surface forming an unsightly hump. Alternatively, it may be revealed only by machining operations. Also see Hydrogen damage.

Blister copper   Part-refined copper ingot which has a severely blistered surface resulting from the release of gas during cooling immediately following casting.

Blister steel   An early form of steel produced by recarburizing a good-quality *Wrought iron* by heating in charcoal.

Block sequence (of weld)   A technique

in which a long multi-layer weld is made in a series of short blocks. Each block comprises multiple layers which are deposited before the next block is commenced. Adjacent blocks may be made in sequence or by leaving gaps which are subsequently block filled. See Figure 49, located at entry on Weld sequences.

Blocking A first stage of forging in some *Closed die* forging operations. The procedure helps ensure a true shape particularly with more complex forms.

Blocky structure See Free machining.

Blood See Bleeding (2).

Bloom (1) A bar of simple section that has undergone some hot-shaping operation following casting but which requires further working.

Bloom (2) A surface film. It may be applied as in the case of the coating on lenses or it may be an inadvertent tarnish.

Bloom (3) A sudden growth of algae.

Blooming mill A *Rolling mill* for the first-stage rolling of ingots to produce *Blooms*.

Blow (as in 'the blow' and 'after-blow') The periods in which air or other gas is being injected in the *Besse-mer* and other air or oxygen injection steelmaking processes.

Blow hole An internal void in cast metal, including weld metal, resulting from the entrapment of gas during solidification. Such holes may occur throughout the casting, within grains and at the boundaries. The term usually implies voids of substantial size and, in the case of weld metal in particular, voids smaller than about 1.5 mm diameter are usually termed 'gas pores'.

Blow moulding Processes in which a hollow body is formed by inflating the material, typically plastics, in a mould by air injection.

Blown metal Steel manufactured by the *Bessemer* and other injection processes.

Blowpipe (1) In welding, cutting, etc. a device in which fuel gas and oxygen are mixed and, at the exit nozzle, burned in a controlled manner forming a flame, **Torch** is the same.

Blowpipe (2) Any pipe by which air is injected into some system by mouth or other means.

Blue anneal The *Annealing* of steel without any protection from the environment at a temperature of about 750–800 °C. This avoids the severe oxidation associated with full annealing or normalizing and the resultant adherent dark blue oxide film may be a desirable feature.

Blue brittleness A reduction in the *Ductility* of carbon *Steels* due to heating at about 200–300 °C. The term results from the light-blue oxide produced at this temperature so it has been applied by various writers with reference to both *Strain age embrittlement* and to *Temper Brittleness*, so some caution must be applied in its interpretation.

Blueing Any process for developing a thin blue oxide on a steel surface, usually at about 300 °C, in air or steam. Such oxidation may be incidental to a *Tempering* process but the term 'blueing' usually implies a deliberate intent as such oxides have an attractive appearance and offer some limited corrosion resistance particularly when oiled. See Barffing.

BNF British Non-Ferrous Metals Research Association.

BNF jet impingement test A test to measure resistance to *Impingement* attack in which a jet of aerated sea water is directed at the material under test immersed in sea water.

BNF jet test A measure of plating thickness in which a jet of appropriate fluid is directed at the surface and the time to achieve penetration is noted.

Body centred cubic A crystal structure with an *Atom* at the centre and corners of the unit cube. See Crystal structure.

Body centred tetragonal A distorted *Body centred cubic* structure.

Bohr atom This model regards an atom as a central nucleus with electrons in fixed orbits around it.

Boiler   A vessel or system of vessels and pipework in which water is boiled to produce steam. In many cases the system will be under pressure thereby raising the boiling point considerably; 350 °C is not uncommon for steam for power plant. The steam produced may be heated further, termed 'superheating'.

Boiler plate   An imprecise term usually indicating a good-quality mild *Steel*, typically 0.12–0.2% carbon, less than 1% manganese, and probably in the normalized condition.

Boiler scale   An imprecise term, usually referring to the deposits of calcium sulphate, possibly intermixed with magnetite, found on the *Water side* of boiler circuits evaporating poor-quality water. The term has occasionally been used regarding the scale and deposits on the *Fire side* of tubes.

Boiler tubes   The tubes separating the water and the heat source in a boiler. The water evaporates to form steam whereas in *Superheater* tubes steam is further heated above the boiling temperature. In a **Fire tube boiler** the flame and combustion products pass through the tubes which are immersed in the water. In a **Water tube boiler** the water is contained in the tubes which pass through the combustion chamber or form the wall of the combustion chamber; hence the term **Water wall tubes**.

Boiling   The vigorous evolution of bubbles as a fluid changes to a gas.

Bolster   Generally, a support or a containing reinforcement for some moving or removable component.

Bolt   A threaded fastener having a head at one end, a threaded length at the other and a plain **Shank** between. **Bright bolts** are normally machined all over. The term **Black bolt** originally referred to bolts that were at least partially covered with a black scale; nowadays the term usually indicates not colour but a bolt with wider machining tolerances than a bright bolt.

Boltzmann's constant   The gas constant divided by Avagadro's number, 1.380×10^{-23} J K^{-1}.

Bond   The joining material, energy or other link holding components together.

Bond strength   The force or energy required to separate two components.

Bonderizing   A proprietary process for producing phosphate *Conversion* coatings on steel components by immersion in a hot acid solution of zinc and manganese phosphates. The coatings improve corrosion resistance and paint adhesion and provide lubrication during *Drawing*.

Book mould   A two-piece mould hinged along one edge to open like a book.

Borax   Sodium diborate, $Na_2B_4O_7$. Often used in powder form or solution as a *Flux* in soldering, brazing and welding, it forms a hard glassy deposit on cooling.

Bore (1)   The hollow space within a tube or similar vessel.

Bore (2)   The internal diameter as a dimension, or the internal surface of a tube, gun, etc.

Bore (3)   The action of cutting a central axial hole by drilling, trepanning, etc.

Bored rotor   See Bottle bored.

Boron   A semi-metallic element occasionally added to steels in small quantities, up to 0.03%, as a deoxidizer etc. See Table 15.

Boron carbide B_4C. A very hard, chemically resistant man-made mineral used as an abrasive.

Boron nitride   BN. A man-made mineral, very hard and oxidation resistant, used as a refractory and structural material up to about 1900 °C.

Bosh   The tapered section of the *Blast furnace* above the hearth from the tuyere zone up to the plane of maximum diameter at the bottom of the stack.

Boss   A raised area or reinforcement. It might, for example, act as a location, support point or be drilled to act as a bearing.

Bottle   A partly pierced billet. See Tube making.

Bottle bored   A bored forging with a chamber at some position along the bore. Large forgings for high-speed rotors for turbine generators and similar applications are bored axially along their centre line to remove the poor-quality material remaining from the original ingot. In some cases defects extend deeper than the (typically 100 mm diameter) bore. In such cases a short length is machined deeper to cut out the defects forming a **Chamber** or bottle, provided that the design can accommodate the variation.

Boundary lubrication The thin film of lubricant, possibly only a molecule thick, remaining when supply is interrupted. It may be sufficient to avoid metal-to-metal contact for a short period.

Box anneal *Annealing* within a sealed container with the intention of minimizing reaction with the furnace environment.

Bower-barf process See Barffing.

Bragg's law A formula concerned with the diffraction of X-rays by a crystal lattice:

$$n\lambda = 2d \sin \theta$$

where n = any integer, λ = the wavelength, d = the lattice spacing and θ = angle between the lattice planes and the incident X-ray beam.

Brale A diamond indentor, of conical form with a hemispherical tip, used on the *Rockwell hardness* testing machine for testing hard meterials.

Brass Alloys of copper and zinc. The binary copper–zinc *Equilibrium* diagram covering the range of commercial brasses is shown in Figure 2. The single-phase alpha brasses containing up to about 38% zinc are highly ductile and hence are particularly suitable for high surface quality, precision cold forming following initial hot working. Maximum ductility is provided with about 30%

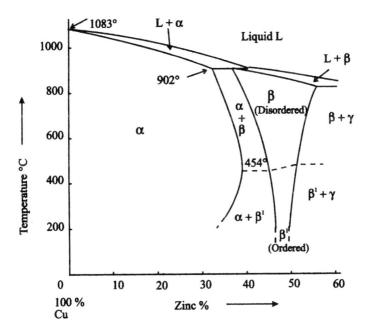

Figure 2 The portion of the copper–zinc equilibrium diagram covering the commercial brasses

zinc, maximum economy with alloys as close as practical to the 38% limit. The duplex, alpha plus beta alloys, with about 39–46% zinc have poor ductility at ambient temperatures because of the brittle beta phase and hence are formed by hot working in the 650–740 °C range. The beta alloys contain about 46–49% zinc. Many alloys mentioned are also suitable for castings. Other elements, including aluminium, tin, iron and manganese, can be added to improve specific properties. Various named alloys are defined at their respective entries and properties of some of the more important alloys are listed in Table 8.

Bravais lattice A system of symbols, letters and numbers for defining the locations of atoms on a crystal lattice.

Braze A joining process in which a metal is melted and introduced to the interface between two components which are not melted in the process. It is implicit that the braze material has a composition different from, and a melting point lower than, those of the components being joined. The braze metal, sometimes termed the **Filler** (metal), melts and bonds to the two components producing a strong joint. Use of the term is normally confined to joints where the interface to be joined is narrow but of large area relative to the cross-section and where the molten filler is either drawn into the interface by capillary action or introduced as a pre-placed strip or sleeve, etc. prior to heating. If this restriction is recognized then joints made with bulky deposits of lower melting point filler are termed **Braze welds**. Originally brazing referred to the use of brass to join copper or steel and it is still widely used in this narrow sense. However, the term is now applied to any joint meeting the above general description where the filler material melting range exceeds about 450 °C. For example, aluminium may be brazed with a lower melting point aluminium–silicon alloy. Where the filler melting range is below about 450 °C the process is normally termed 'soldering' or 'soft soldering' (see entry on this topic). In most brazing processes a *Flux* or other means of cleaning the component surfaces is required. Many processes are used, including gas or flame brazing, furnace brazing, bath or dip brazing, vacuum and induction brazing, the various terms indicating the heat source or the environment. The term **Hard soldering** means brazing or *Silver soldering*.

Braze filler The low melting point material introduced into a brazed joint (see Braze) to form the bond between the two components. The composition of the filler depends on the materials being joined but for steels and copper alloys, typical fillers include phosphor–copper, various plain or complex *Brasses* and *Silver solders*.

Braze welding See Braze.

Brazing temperature The temperature to which components are heated to melt the filler and allow bonding to, and between, the parent materials.

Break The process of parting into two or more pieces usually by some form of loading but excluding processes such as cutting. In metallurgical contexts there are ill-defined subtleties in the choice between the terms 'break', 'crack', 'cracking', 'fracture' and 'rupture'. This may reflect the fact that the first three terms, with their Anglo-Saxon roots, are possibly seen as common usage whereas the latter two with their Latin origins may sound appropriately scientific to some ears or merely confusing jargon to others. See also the various terms mentioned.

Break-away oxidation A form of oxidation in which the oxide repeatedly falls off. During the time that the oxide remains on the surface the rate of oxidation falls progressively but after a layer has *Spalled* off the rate immediately increases only to fall again as the oxide thickness builds again. As a result, over a long period, a continuing and fairly constant rate of attack is observed.

Break(ing) down The initial hot-working process applied to an *Ingot* to break down the coarse cast structure.

Breaking length That length at which a freely suspended rope, cable, etc. would break under its own weight.

Break(ing) out Removal of a casting from its mould.

Breeze Furnace ash of substantial form as opposed to powder or fly ash. Such material can be bonded by cement to form building blocks.

Bridge die See Extrusion.

Bridging (1) The premature *Solidification* of metal across the top of an *Ingot*. The underlying molten metal then solidifies with associated contraction leaving a large void beneath the bridge.

Bridging (2) The development of a blockage in a furnace or other equipment with down-flowing contents. This causes voids and interruptions in the flow and, if the bridge collapses, damage to the furnace or associated plant.

Bright anneal *Annealing* in a protective atmosphere or a vacuum to avoid oxidation and tarnishing.

Bright bolt See Bolt.

Bright dip Immersion in a solution to produce a clean, lustrous surface.

Brightener Material added to electroplating baths, etc., with the intention of producing a bright lustrous surface.

Brinell hardness test A test in which a steel or tungsten carbide ball is pressed into the surface in question by a known load. The resultant indentation diameter is then measured and, by comparison with established tables, a hardness determined. See Hardness.

Brinelling Indentation due to local overload, particularly on the race surface of ball or roller bearings. See also False brinelling.

Britannia metal An alloy of tin with, typically, about 7% antimony and 2% copper, although there are variations. It is one of the range of *Pewters* with mainly decorative applications.

British Association (threads) A series of thread sizes of standard profile used mainly in electrical and electronic application rather than mechanical engineering.

British Standard Whitworth/Fine Two series of threads of standard profile. The Whitworth (**BSW**) are relatively coarse pitch compared with Fine (**BSF**).

BSI British Standards Institution. The body responsible for standards and specifications in the UK.

British thermal unit The quantity of heat required to raise the temperature of one pound of water by one degree Fahrenheit.

Brittle Failing without significant *Ductility*. Sometimes the term is used to imply susceptibility to failure under impact loading but materials can fail in a brittle manner even under static loads.

Brittle boundary technique Same as Brittle edge technique.

Brittle edge technique A technique for *Case hardening* the side surfaces of notched *Impact* and *Bend* test specimens to prevent the initiation of *Shear lips*.

Brittle fracture Generally, cracking without associated *Ductility* and, usually, with only a small energy input. In a narrower sense the term is associated with the change from ductile to brittle behaviour that some metals, particularly iron and other metals with a *Body centred cubic* structure, undergo as the temperature falls. In the case of many *Steels* this occurs at about ambient temperature. Above the **Brittle to ductile transition** the steel will accept deformation, i.e. absorb energy and act in a ductile manner, before failing by overloading. Such behaviour makes the metal tolerant of minor cracks and other *Stress raising* features, reduces sensitivity to impact and may give warning of imminent failure. Below the transition the steel will be sensitive to cracks and impact loads and may fail without warning even under apparently static loads. Above the transition the fracture mode is *Shear*, i.e. planes of atoms in the *Crystal structure* sliding over each other, while at lower temperatures the mode changes to *Cleavage* in which the planes peel apart. Although the fractures are both transgranular the surfaces produced by the two modes are quite distinct, the shear being a dull grey and the cleavage bright and crystalline, so the change from one to the other is readily observed on a series of test pieces broken over a range of temperatures. The percentage crystallinity of the test pieces broken at the various temperatures is noted to determine the *Fracture appearance transition temperatures (FATT)*, i.e., the temperature at which the fracture is 50% ductile and 50% cleavage. The test-pieces are often broken in an *Impact test* and the energy required to break the specimens is then plotted against temperature (see Figure 20, located at entry on Impact test). The large step change in

toughness, as measured by the energy absorbed, takes place over a limited range, typically about 50 °C. Above and below this range any change is gradual. Consequently, the terms **Upper shelf** and **Lower shelf** are used with reference to the respective parts of the curve and for behaviour which is, respectively, fully ductile or fully brittle. Brittle fracture has been responsible for many major and dramatic failures of ships, bridges, pressure vessels and pipelines, etc.

Brittleness   A lack of *Ductility* with sensitivity to cracks and low absorption of energy during crack propagation. See Brittle fracture.

Broaching   A process in which a shaped cutting tool, often with multiple cutting edges, is forced through an orifice or aperture to develop its final, usually high-precision, shape.

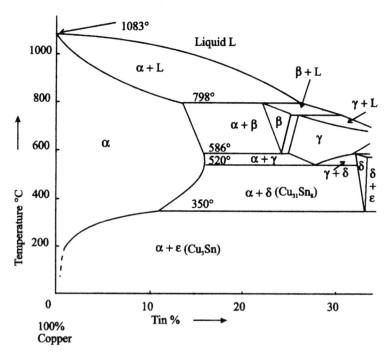

Figure 3   The portion of the copper–tin equilibrium diagram covering the commercial tin bronzes

Bromine   A non-metallic element, one of the halogen group, liquid at ambient temperature.

Bronze   Historically, this term indicated various alloys based on copper and tin (5–15%), possibly with other elements, which combined good strength with corrosion resistance plus, in some alloys, good casting characteristics and bearing properties. The binary copper–tin *Equilibrium diagram* (Figure 3) has to be interpreted with some caution. The alpha-plus-liquid field is very broad and deep and diffusion rates in these alloys are very low. Consequently, bronze castings are prone to severe *Coring* and are capable of producing phases much richer in tin than the nominal composition. Also, because diffusion rates become very low below about 400 °C, extremely slow cooling rates are required to allow the indicated transformations below this temperature. As a result, commercial products rarely develop the epsilon phase. The term 'bronze' is now also commonly applied to other copper-based alloys in which tin is absent, for example, aluminium–bronze (copper with up to 10% aluminium) or silicon–bronze (copper with up to 5% silicon). These alloys generally have excellent corrosion characteristics comparable with, or superior to, the tin bronzes and in most cases considerably superior to the *Brasses*. Unfortunately, the term 'bronze' has occasionally been coupled with a trade or brand name for some alloys having properties inferior to what might be inferred from the name. See also Gunmetal, Phosphor bronze and Cold cast bronze. See Table 8 for typical properties.

Bronze welding   Processes for making fusion joints with brazing *Fillers*, particularly but not exclusively, using *Arc welding* techniques but forming large bulky deposits rather than the thin capillary film characteristic of true brazing. See Brazing.

Brush plating   An electroplating technique in which the plating solution is carried in, or fed through, a brush or pad which acts as the anode.

BSF and **BSW**   See British Standard Whitworth.

Buckling   Lateral deflection of a column, etc. under a compressive stress.

Buckminsterfullerines   The *Allotropic* forms of *Carbon* in which large numbers of carbon atoms are arranged in three-dimensional structures reminiscent of the geodesic frame developed by the architect R. Buckminster Fuller. The carbon atom has three valence electrons so it bonds with three other carbon atoms. Where the atoms combine to form only hexagons the structure is confined to a flat plate but if pentagons are introduced the structure curves. An assembly of 60 atoms arranged as 20 hexagons and 12 pentagons forms a highly stable sphere. The shape is technically described as a truncated icosahedron cage but is more readily recognized in the form of a modern soccer ball assembled from hexagonal and pentagonal panels on which the carbon atoms would be located at the points where three panels meet. The terms may be used specifically of this 60-atom arrangement as it is the most common and probably offers the best prospect for commercial exploitation. However, other combinations of hexagons and pentagons are possible, for example the 70-atom cluster is also very stable and clusters of 76, 78 and 84 have been confirmed. All may be described as Buckminsterfullerines, **Buckyballs** or **Fullerines**.

Buckyballs   See Buckminsterfullerines.

BUE   See Built-up edge.

Buffing   polishing of a surface usually by a high-speed rotating soft fabric wheel, the **Bob**, carrying some polishing compound. The term may imply a process of surface smearing rather than a clean-cutting action.

Build-up (1)   In *Electroplating*, the excessive thickness that can develop on projections and corners.

Build-up (2)   In welding, the sequence of

deposition of multiple layers or beads to form a bulk weld.

Building-up   The deposition of weld metal, electroplate or metal spray, etc. to increase or restore dimensions. Compare with *Surfacing*. Also the same as Build-up.

Bulb section   Rolled or extruded bar having one edge much thicker than the other, the transition being fairly abrupt rather than a continuous taper.

Bulging   A process in which a tube end is expanded without deliberate change of the wall thickness. See Figure 45, located at entry on Tube manipulation for comparison with related processes.

Built-up edge (BUE)   A deposit, on the edge of a cutting tool, formed by the passage and adhesion of material being cut.

Bulk density   The weight per unit volume of a material.

Bulk modulus   The ratio of shear stress to shear strain. Same as Shear modulus.

Bull block   A powered drum, effectively a winch, onto which wires or small-diameter tubes are coiled as they are pulled through the *Drawing* die.

Bullion   Precious metal, particularly gold and silver, in bulk, i.e. as *Ingots* or bar rather than a finished product although the term sometimes includes coinage.

Burden (1)   The total material charged into a furnace or other reaction vessel.

Burden (2)   In submerged arc welding, the total amount of fused and unfused flux over the weld metal.

Burgers vector   The total lattice translation resulting from the passage of a dislocation. See Dislocation.

Burn off rate   The rate at which a welding electrode or other consumable item is melted or otherwise consumed.

Burn through (of a weld)   A hole completely through a welded joint due to local collapse of the weld pool.

Burning (1)   Permanent, irreversible damage caused by heating at an excessive temperature or for an excessive period. The various forms of damage include severe *Grain growth*, surface

cracking, grain boundary *Liquation, Decarburization*, gross oxidation and penetration of oxide intrusions down grain boundaries.

Burning (2)   An alternative term for *Flow* welding techniques where bulk molten metal is poured into a substantial gap between the two faces to be joined. Sufficient heat input to melt the joint faces is ensured by the molten metal being *Superheated* or by being allowed to flow to waste across the joint. **Lead burning** usually refers to the joining of lead components, by *autogenous* welding, i.e. without the addition of *Solder* or any other *Filler*.

Burning (3)   Colloquially, *Oxy-fuel* gas cutting.

Burning on   A *Flow* weld technique for localized repair in which new material is cast into place in a temporary mould formed round the damaged area of the component. In practice molten metal is usually allowed to flow through the mould to waste until the parent metal edges have fused. It is essentially the same as Burning (2) but 'burning on' is particularly associated with marine repairs such as to propellers.

Burnishing   The development of a smooth, bright surface by a surface smearing action rather than a cutting and polishing action. It can be effected by rubbing with hard, but non-cutting, tools or by *Tumbling* with steel balls.

Burnt   See Burning (1).

Burnt deposit   In electroplating the nodules and scabs formed by faulty conditions such as excess current density.

Burnt weld   A weld which has suffered significant oxidation particularly penetration at *Grain* boundaries. It usually results from heating for excessive periods or at excessive temperatures, particularly in the presence of higher than normal levels of oxygen.

Burr (1)   The turned-over edge, usually thin, ragged and sharp, formed by various machining operations.

Burr (2)   A cutting tool in the form of a rotary file.

Bus (bar) A metal conductor for electricity. Usually a high current-carrying capacity and no insulation is implied.

Bush(ing) A guide or plain bearing, for a shaft etc., usually carried in a housing.

Butt weld/butt joint These terms are used loosely as synonyms. More strictly, the 'joint' term refers to the parent component geometry so 'butt joint' indicates that the two components are welded edge to edge, usually defined as being at an angle of 135–180°. The 'weld' term refers to the weld profile so 'butt weld' indicates that the weld metal extends across the end face of at least one of the components. Usually there is no confusion but see Figures 47(a) and 47(b), located at entry on Welding terminology, for examples of butt welds of butt joints, fillet welds of butt joints and butt welds of T joints. See also Strap joint.

Butt welded tube Tube formed by rolling or otherwise curving strip to form a near-circle and then welding the butting edges together by various welding processes.

Butterfly extrusion A technique for pro-ducing deeply recessed sections by extruding a section with an open, flatter profile and then folding the product to produce the deep profile.

Buttering The practice of laying a weld deposit over a parent metal surface and then depositing the main weld on the buttered layer. The composition of the butter usually differs from those of the main weld and the parents. The technique is used, for example, where there is a need for a progressive transition between differing parent materials or where some characteristic of the main weld metal is incompatible with the parent.

Butting A process in which a tube end is formed to a thicker wall at constant outside diameter. The thicker section can accommodate the local increased stresses due to joining processes such as brazing, welding, screwing or tapping. See Figure 45, located at entry on Tube manipulation, for sketches of this and related processes.

Button With reference to *Spot welds*, the central core left behind on one surface when the weld is torn apart.

BWG *Birmingham Wire Gauge.*

C

C Curve See Isothermal transformation diagram.

CCT diagram See Isothermal transformation diagram.

Cable (1) Rope or chain particularly in marine contexts.

Cable (2) Flexible electrical conductor, usually multistranded.

CAD Computer Aided (or Assisted) Design.

Cadmium A medium-density metallic element having little application in the bulk form but used as a *Sacrificial* protective coating on steel and copper, particularly for components to be soldered. Small quantities added to copper enhance its strength with only a moderate decrease in electrical conductivity. Its application as an alloying element in *Silver solders* is reducing because its toxic fume presents a hazard. It is not toxic in normal handling but is not used in contact with foodstuffs, etc. See Table 15 for properties.

Cadmium copper An alloy of copper with up to 1% cadmium for electrical conductors. The cadmium addition increases the tensile strength by about 50% but only slightly reduces the conductivity.

Caesium A highly reactive metallic element. It has limited commercial application but is used in photoelectric devices. See Table 15 for physical properties.

Cal Symbol for *Calorie*.

Calcareous scale The scale deposited by hard water, mainly calcium carbonate and magnesium hydroxide.

Calcination Various processes in which material is heated to effect a physical change such as driving off water, particularly in the combined form, rather than a chemical change.

Calcite A natural mineral, calcium carbonate, $CaCO_3$. In its pure form it is transparent and capable of polarizing light. It is reference material No. 3 on *Mohs* scale of hardness.

Calcium A metallic element having little commercial application in metallic form. About 0.1% may be added to lead to improve its strength for cable sheathing or battery plates. It is also widely used as a reducing and deoxidizing agent in various extraction processes. See Table 15 for physical properties.

Calcium silicide A mineral used in powder form for *Inoculating* cast iron.

Calomel electrode A reference electrode against which metals are compared to determine their position in the *Electrochemical* series.

Calorie The quantity of heat required to raise the temperature of one gram of water by one degree Celsius. The exact value of this depends on the water temperature. The calorie (International Table) is 4.186 Joules. The dietician's calorie, sometimes termed the large calorie or the kilocalorie, is 1000 Cal.

Calorimetry The measurement, in a calorimeter, of heat–energy relationships.

Calorizing A process in which steel is coated with aluminium powder and heated to about 1000 °C to form an aluminium oxide coating over an aluminium iron intermetallic layer. The coating offers good oxidation resistance up to about 950 °C.

CAM   Computer Aided (or Assisted) Machining (or Manufacture).

Camber   Slope or deviation from straight or level. Roll camber is a deliberate double tapering of rolls with a maximum diameter at mid-width. The normal deflection under load then results in the contact face flexing to a straight line.

Campaign   The period in which a furnace or other equipment is operated between major overhaul such as the replacement of the refractory lining.

Canada balsam   The thin resin produced by Canadian firs. It has a refractive index similar to glass and hence is used as an adhesive for multi-element lenses.

Canning (1)   The containment of foodstuffs and other materials in sealed cans made from tin-plated steel, aluminium or other materials.

Canning (2)   Enclosing a reactive material in a container of more corrosion-resistant or innocuous material to facilitate working, manipulation, etc. The materials used for canning nuclear fuel elements are selected for low neutron capture cross-section, short half-life, good thermal conductivity, compatibility with the working environment, adequate mechanical properties and forming characteristics.

Cantilever   A loaded beam supported at one end only.

Capillary action   The force which tends to draw liquids into a narrow gap. It is a consequence of surface tension and depends critically on the liquid *Wetting* the solid surface.

Capped steel   A steel cast in a process in which a solid cap inserted in the top of narrow moulds encourages some solidification from the top downwards and builds up internal pressure. This can improve the surface quality of the ingot.

Capture cross-section   The measure of the capability of a material to absorb neutrons without undergoing fission. See Atomic structure.

Carat (1)   Units of purity of gold, 24 carat being pure.

Carat (2)   Metric carat, for gemstones etc. = 200 mg.

Carbide   A compound of carbon with some other element(s).

Carbide   tip/tool   Intermetallic compounds between carbon and metals such as tungsten and titanium, used for severe cutting applications. Usually the carbides are in the form of relatively small particles carried in a matrix of a metal or alloy such as cobalt or nickel.

Carbon   A non-metallic element occurring naturally in the long-recognized *Allotropic* forms of graphite and diamond and the more recently identified form, *Buckminsterfullerines*. Carbon is the primary alloying addition to iron to form *Steel* and larger quantities are present in *Cast iron*. See Table 15 for properties.

Carbon arc cutting   Processes utilizing the heating action of an electric arc struck between a carbon electrode and the workpiece to melt the metal along the line of cut.

Carbon arc welding   Processes in which the electric arc is struck between two carbon electrodes or between one carbon electrode and the component being welded. Usually no shielding gas is introduced into the weld zone although gases produced by the burning of the carbon may offer protection. A *Filler* material may be used.

Carbon dioxide CO_2. The gas produced by the complete combustion of carbon, i.e. with sufficient, or an excess of, oxygen, and hence non-flammable. A gas at normal temperature and pressure, it can exist as a solid, colloquially termed *Dry ice*, below its sublimation temperature of minus 73 °C. It is exhaled in our breath and is non-toxic. It provides the protective gas shroud in some *MIG* welding applications for *Steel*.

Carbon dioxide welding   A *MIG* welding process for steel in which carbon dioxide is fed into the arc vicinity to protect the weld zone from atmospheric contamination. Carbon dioxide is effectively inert in these circumstances and its choice, rather than argon, arises from its favourable effect on arc characteristics as well as the economic factor.

Carbon equivalent The measure of the effect of one or more alloying elements in *Cast iron* or *Steel* expressed in terms of the amount of carbon that would have the same effect. For example, the alloying elements in steel affect characteristics such as *Hardenability*, and hence *Weldability*, or *Microstructure* in a way which reinforces or detracts from the effect of carbon. The element can therefore be ascribed a factor reflecting its potency relative to carbon, taken as unity. This factor is termed the carbon equivalent of the element. In any iron or steel, individual alloy contents multiplied by their respective carbon equivalents can be added together to provide a total carbon equivalent for the material. One of a number of formulae for predicting weldability of low-alloy steels is:

Carbon equivalent

$$= C + (Mn \div 20)$$

$$+ (Ni \div 15)$$

$$+ (Cr, Mo, V/all \div 10)$$

Also see Schaeffler diagram for an example of carbon equivalents related to microstructure.

Carbon extraction replica See Extraction replica.

Carbon fibre A filament composed of nearly pure carbon typically up to 10 μm diameter used for reinforcing epoxy or other plastics to form **Carbon fibre composite** materials.

Carbon monoxide The gas produced by the incomplete combustion of carbon, i.e. combustion with inadequate oxygen, and hence capable of burning. It is toxic even in small concentrations.

Carbon steel Any steel with a significant carbon content, usually in the 0.1–1% range although occasionally up about

2%, but no other deliberate alloying addition. See Steel.

Carbon tetrachloride A volatile, non-flammable liquid, CCl_4. It is a good degreasing agent and has been used as a fire extinguisher but is less used now because of its toxicity.

Carbonize, carbonization (1) The conversion of carbon-rich organic material to elementary carbon by a process of destructive distillation, i.e. heating without access to oxygen.

Carbonize, carbonization (2) Same as Carburising.

Carbonizing flame (in gas welding) Same as Carburising flame.

Carbonitriding See Case hardening.

Carbonyle process A process in which a metal carbonyl gas is heated to form a purified metal powder.

Carborundum Silicon carbide particularly when used as an abrasive.

Carburizing The process by which carbon diffuses into *Steel*. The process can be deliberately induced by appropriate treatments (see Case Hardening) or it can occur accidentally during some manufacturing processes with potentially adverse consequences for the component.

Carburizing flame (in gas welding) A flame having an excess of carbonaceous fuel gas relative to the oxygen supply, which can have a carburizing effect on the workpiece. Such flames have a long feathery inner *Cone* and a relatively low temperature.

Cartridge brass An alloy of 70% copper and 30% zinc having very high *Ductility* and hence suitable for manufacturing cartridges. See Table 8.

Cascading, cascade failure A term applied where one event or failure triggers another, which in turn triggers the next and so on.

Case The outer surface layers particularly in the context of *Case hardening.*

Case hardening Any process for increasing the surface hardness of metals, usually steel or cast iron. Some components require a combination of a very hard exterior surface with a relatively soft but *Tough* core. This is achieved either by applying a heat treatment selectively to the surface or by modifying the surface composition by some diffusion technique. Case hardening by localized surface heating is applied to steel containing sufficient carbon and perhaps other alloy additions and may be termed **Surface hardening**. After initial heat treatment to give the required core properties the surface is rapidly heated by various techniques such as oxy-gas **Flame hardening**, electrical **Induction case hardening**, or **Laser hardening**. Subsequent rapid cooling by the underlying cool core, possibly assisted by water sprays, causes hardening of the layer that was heated above the transformation range. The most common technique employing surface diffusion is **Carburizing** which involves heating the component in a carbon-rich environment so that it absorbs carbon into the surface. The technique is applicable to low-, medium-carbon and alloy steels and the temperatures consequently vary but are normally about 900 °C, i.e. in the *Austenitic* range. If the carburizing environment is solid, i.e. charcoal, etc., the process is termed **Pack carburizing**; if liquid it may be termed **Bath carburizing** or **Cyaniding** in recognition of the molten cyanide salts commonly used. **Gas carburizing** is carried out in a methane- or similarly carbon-rich gas. Depending upon the process and circumstances carburized case depths of up to 3 or 4 millimetres are practical. After these treatments the steel will have a surface layer of high carbon content and a core with the original lower carbon content. Heat treatment, as described in the entry for Steel, will then give the required combination of properties. The higher-carbon case will have a lower austenitizing temperature than the core so, in the case of high-quality components, a double heat treatment may be employed (see Core refining). In an alternative diffusion process termed

Nitriding appropriate alloy steels, usually containing combinations of a per-cent or two of aluminium, chromium and molybdenum, are heated to about 550 °C in an ammonia atmosphere. The ammonia dissociates, releasing nitrogen in atomic, rather than molecular, form at the metal surface. The nitrogen diffuses into the surface and combines with the alloying elements and iron to form very hard nitrides. No further heat treatment is required. Case depths are usually limited to about one millimetre maximum, often less, and hence components are usually finished to final size prior to nitriding. Compared with carburizing, nitriding offers a case that is shallower but harder, more fatigue resistant, more resistant to softening up to about 500 °C and with a measure of corrosion resistance provided that it is not ground or polished. **Carbonitriding** is carburizing in a gas to which a small amount of ammonia is added. The nitrogen is beneficial mainly in improving the *Hardenability* and wear resistance.

Cast (1)   The practice of pouring molten metal into a mould and allowing it to solidify. If the resultant item is of approximately the final shape, requiring only final machining and other treatments that do not involve significant deformation, it will be described as a *Casting*. If the cast item is to be wrought to shape by rolling, forging, etc., it will usually be described as an *Ingot*. See also Continuous casting.

Cast (2)   The natural curved form that a wire takes as it is drawn from a reel and allowed to lie unrestrained. Its measure is the diameter of the circle formed when a sufficient length is allowed to lie freely on a horizontal surface.

Cast iron   Alloys of *Iron* containing, at the time they are cast, carbon in the range from about 2% to 4.5% and usually with about 1–3% silicon. This contrasts with *Steel* which is also an iron–carbon alloy but which contains considerably less carbon. It is common for the term 'cast' to be omitted in descriptions such as 'grey (cast) iron', 'malleable (cast) iron' or even '(cast) irons'. Iron containing about 3% carbon solidifies over a temperature range from about 1350 °C to 1130 °C, i.e. significantly lower than the 1535 °C or so melting point of pure iron. This wide freezing range coupled with a high fluidity ensures that complex shapes are readily cast. Other favourable characteristics of the simple irons include good corrosion resistance (relative to carbon steel) and erosion resistance, a high *Damping* capacity and good compressive strength. Cast irons can develop, under appropriate cooling conditions and by subsequent heat treatment, the structures described in the entry on Steel. However, the very high carbon content has a significant effect and, in the as-cast condition, the structure will contain either soft, weak, free carbon, i.e. carbon not combined with the iron, or large quantities of hard, brittle cementite (iron carbide, Fe_3C). Factors such as precise composition and cooling rate during casting influence the structures that initially develop. In many irons the free carbon will exist as graphite flakes. This usually forms as an *Eutectic* with the austenite at 1130 °C but in adverse circumstances a coarse form, termed **Kish** graphite, can be precipitated at an earlier stage of solidification. However, even normal graphite flakes have a pronounced weakening effect although some improvement can be obtained by **Inoculation** which involves adding a finely particulate refractory powder, such as calcium silicide, to the molten iron. The particles provide a large number of nucleation sites for precipitation thereby refining the graphite. When irons containing graphite flakes are broken the fracture follows the weak flakes producing a characteristic grey fracture surface. Historically, irons have been categorized on the appearance of their fracture surface and hence such irons are termed **Grey irons**. Apart from the graphite the microstructure will com-

prise a matrix of other phases such as *Pearlite* or *Ferrite* hence terms such as **Pearlitic grey iron** and **Ferritic grey iron**. Where the composition and cooling rate are such that no graphite is formed and the structure comprises cementite and pearlite the material will be very hard and brittle with a bright fracture surface, hence the term **White iron**. The mechanical properties, particularly *Ductility* and *Malleability*, of white irons can be improved by various heat treatments to form **Malleable irons**. One process involves heating a white iron of fairly low carbon content, about 2.2–2.9%, at about 850 to 950 °C in a non-reactive, i.e. non-decarburizing, environment for a long period (up to a week). This causes the cementite to decompose, producing a microstructure of ferrite with **Temper carbon**, i.e. graphite in a nodular or rosette form which is not so damaging as flake. These carbon nodules are more prolific towards the centre of the section so the appearance of the fracture surface leads to the term **Blackheart (malleable) iron**. A variation on this process requires a second heating to 900 °C allowing some of the carbon to redissolve in the austenite so that pearlite is produced on cooling, hence the term **Pearlitic malleable iron**. Such irons can be heat treated as described for steel. An alternative malleabilizing treatment involves heating white iron, of relatively high carbon content, about 3–4%, at about 950 °C in an oxidizing environment for up to 100 hours. This allows most of the carbon to diffuse to the surface where it reacts with oxygen, forming carbon dioxide which is released into the atmosphere. The resultant structure is largely ferritic or pearlitic, with a little temper carbon at the centre (more in large castings), hence the name **Whiteheart (malleable) iron**. This non-homogeneous structure usually results in inferior mechanical properties compared with the blackheart grade. An alternative approach to producing ductile irons is to modify the form of the graphite in the as-cast state, usually by small additions of magnesium or cerium to the molten iron. These promote the formation of graphite in the nodular or spheroidal form which, compared with the flake form, has only a slightly adverse effect on the mechanical properties. These materials are referred to by various terms such as **Ductile irons, Nodular irons, Spheroidal graphite irons (SG irons)**. The descriptions may also be extended by referring to other microstructural features as in terms such as Pearlitic SG iron and so on. See Table 6.

Cast structure The structure of a material in the as-cast condition; in particular the term refers to various features which can be eliminated or at least improved by subsequent *Working* process. Such features include grains that are very coarse, possibly *Cored* or *Dendritic*, extensive columnar grains at the exterior and large quantities of internal voids.

Casting copper Impure copper, typically 99.4% pure.

Casting cracks Generally, any cracking developing during solidification, cooling or immediately afterwards. The term is sometimes used loosely to refer to any planar defect since their cause is not always obvious to a casual inspection. In its more restricted sense the term refers to cracks caused by *Casting stresses*.

Casting shrinkage The decrease in volume of a metal between the molten state and the fully cooled state. The total shrinkage comprises, first, contraction during cooling of the molten metal after the first solid has formed, second, contraction during the continued solidification and, finally, contraction during cooling down to ambient temperature.

Casting stresses *Residual stresses* resulting from restraint of the thermal contraction during solidification and cooling following casting.

Casting wheel A wheel carrying, at its edge, a number of moulds which are filled as they pass beneath the molten metal stream. As the wheel rotates, the moulds are replaced or emptied if sufficiently cooled.

Catalyst A material which promotes, assists or accelerates a chemical reaction without itself being permanently affected by the reaction.

Catenary (1) The drooping curve adopted by a wire, beam or shaft, etc., supported at its ends.

Catenary (2) A wire supported at its ends which in turn supports at multiple positions a further wire, for example the overhead conductor wire of an electrified railway.

Cathode The pole in a cell towards which cations travel. See Electrochemistry.

Cathode copper Copper in slab form produced by *Electrolytic* refining. It is subsequently remelted to produce high-conductivity grades such as Electrolytic tough pitch high conductivity copper and Oxygen free high conductivity copper. It is also the basis for high-conductivity copper castings and high-conductivity alloys.

Cathode ray tube An electronic device in which a stream of electrons, i.e. the negatively charged cathode ray, is formed and directed at a *Fluorescent* screen. The path of the ray is deflected by electromagnetic coils to produce the screen display.

Cathodic corrosion The corrosion in the vicinity of the cathode resulting from attack by products of the electrolytic reaction, usually alkaline in nature.

Cathodic protection The protection of a metal from corrosion by making it the cathode of an electrolytic cell. Electrical currents can develop in metal components exposed to wet environments and corrosion will occur at the anodic areas (see Electrochemistry). However, attack may be prevented by inducing a reverse current that causes the component to be cathodic with respect to some external anode. One technique is to impose a DC electrical current from an external power source—termed **Impressed current** cathodic protection. Alternatively, the component may be protected by attaching to it, ensuring electrical contact, another metal that is more anodic. For example, steel is protected by zinc either in the form of plating (**Galvanizing**) or as large blocks of zinc bolted to, or buried close to, and in electrical circuit with, the component. These anodes will be progressively consumed by corrosion, hence the term **Sacrificial** cathodic protection, but the rate of anode loss, in practical cases, is acceptably economic and the main component will be preserved.

Cathodoluminescence Fluorescence induced by irradiation in an electron beam.

Cation A positively charged ion. See Electrochemistry.

Cat's tongue (surface) A rough surface texture composed of sharp spikes and ridges separated by deep tapered pits. The texture often has a 'lay', i.e. it lies at an angle to the original surface and hence feels very rough when stroked in one direction but relatively smooth in the opposite direction, as does a cat's tongue. As the damage progresses, the texture becomes increasingly coarse although, in terms of weight loss, the rate of damage may fall. The effect is characteristic of damage caused by repeated impacts by particles including *Water droplets*. The damage mechanism is not a form of cutting *Abrasion* but results from a fatigue action by the repeated impacts producing surface mechanical damage.

Cauliflower A growth of material on the exterior, particularly the top, of an ingot as a result of inverse *Segregation*.

Caulking The process of sealing joints such as those in butting or overlapping plates. In some cases a sealing material or **Caulk** is pushed or hammered into the gap. In other cases the metal close to the interface is deformed by a **Caulking tool**, similar to a chisel, which is struck by a hammer. This produces a line indentation forcing the edge against the mating surface to provide a mechanical

seal. Such joints are notoriously prone to seepage which, together with the effects of severe local work hardening, has led to major failures of riveted pressure vessels by *Caustic embrittlement*.

Caustic Material capable of corroding organic materials.

Caustic attack Generally, any corrosion caused by caustic alkali. The term often refers specifically to the attack on the water side of steam-generating tubes in power plant. The cause is excessive sodium hydroxide resulting either from the introduction of quantities well in excess of that required for normal feed water treatment or from local concentration effects. Factors promoting concentration are very high levels of heat input and bore irregularities such as deposits or the crevice produced when the internal projection of a flash butt weld is smeared along the bore rather than being cleanly cut away. When such factors are present the tube bore facing towards the fire develops a persistent **Steam blanket** rather than the normal bubbles of steam which can be swept away. The caustic alkali concentrates in this area and the normally protective magnetite film is disrupted, leading to severe corrosion and in some cases to Hydrogen embrittlement (see entry on this topic for further comment but particularly as use of the term in this context is not always accepted). In some cases the corrosive attack produces deep grooves along the tube leading to terms such as **Caustic grooving** or **Gouging**.

Caustic dip Immersion in caustic alkaline solution for cleaning purposes, etc.

Caustic embrittlement/cracking Intergranular *Stress corrosion cracking* of steel in caustic, alkaline, solutions. It usually only occurs above about 70 °C and in fairly highly alkaline conditions. However, it can be a serious problem in cases where the solution is nominally too weak to have any significant effect but where an inadvertent concentration mechanism is active. For example, a slight leak can allow a concentration to develop in the leak path interface as water evaporates at the exterior face. Historically, this mechanism caused catastrophic failure of a number of riveted boiler drums. See also Caulking.

Caustic grooving/gouging See Caustic attack.

Caustic lime Quick lime, calcium oxide.

Caustic potash Potassium hydroxide.

Caustic soda Sodium hydroxide.

Cavitation The production of voids in liquids as a result of severe pressure fluctuations (see also next entry).

Cavitation erosion/damage The removal of metal from an immersed surface as a result of local severe pressure fluctuations associated with turbulent flow. Water subjected to a sudden pressure drop will form cavities, i.e. bubbles. These rapidly collapse, producing shock waves which damage adjacent surfaces by various mechanisms including a direct mechanical action, a *Fatigue* action or by disrupting protective films. The development of local pits exacerbates cavity quantity and size, so metal loss is accelerated. Pumps and propellers are commonly affected.

CCT diagram Continuous cooling transformation diagram. See Isothermal transformation diagram.

CCP Cubic close packed crystal structure. Same as Face centred cubic. See Crystal structure.

Ceiling The maximum that should not be exceeded in the process in question.

Cell A local system of activity, in particular an electrochemical cell. See Electrochemistry.

Cellular A network comprising a large number of associated cells of broadly similar characteristics.

Cellulosic electrode (for welding) See Electrode (welding).

Celsius The *SI* recommended scale of temperature based on reference points including 0° as the freezing point of water and 100° as the boiling point of water. It is essentially identical to Centigrade.

Cementation Any process by which any element diffuses into a surface of another material; usually a high temperature is involved. See Carburizing.

Cementation process An obsolete process in which *Wrought iron* packed in charcoal was heated at about 900 °C for a few days allowing carbon to diffuse into the low-carbon iron to produce *Steel.*

Cemented carbide Material composed of hard metallic carbides of tungsten, vanadium, etc., held together by a relatively *Ductile* binder such as cobalt. Such materials are widely used for the tips of cutting tools and similar applications. The term also may refer to *Cementite.*

Cementite Iron carbide, Fe_3C. Prolonged heating at high temperature will cause it to decompose to iron and graphite. See Steel.

Centigrade The original metric measure of temperature. Identical to *Celsius* on the *SI* system.

Centre-spun, centri-spun, centrifugal casting Casting processes in which molten metal, poured into a spinning cylindrical mould, centrifuges to the periphery where it solidifies, forming a hollow component.

Centrifuge A mechanism in which a material is spun either by having a swirl induced in it or by being contained in a vessel that is spun at high speed. In either case the intention is to apply high gravitational forces that separate materials of different densities. Very high-speed machines termed 'ultracentrifuges' are capable of developing greater than half a million g.

Ceramic A solid compound formed between a metal or metalloid and a nonmetal or a metalloid. Examples are aluminium oxide, silica (silicon oxide) or magnesium silicate.

Cerium A metallic element, one of the *Rare earth* group. It is a strong reducing agent used as a *Getter* in thermionic valves and in a number of refining processes usually in the form of **Mischmetal**, an imprecise mixture of about

50% cerium with other rare earths. With 50% iron, it forms a pyrophoric alloy for products such as cigarette lighter 'flints'. Small additions are made to aluminium and magnesium to improve their strength and *Ductility* and to cast irons to promote the formation of nodular graphite (see entry on Cast iron). See Table 15 for the physical properties of cerium.

Cermet   A solid mixture in which *Ceramic* particles are bonded by a metal, such as cobalt, with a high melting point. The material is usually formed by *Powder metal* processes and the bond may be chemical or mechanical. Ceramics tend to have excellent high-temperature strength but are sensitive to notches, impact and thermal shock. The metal matrix of the cermet is intended to introduce a sufficient improvement in *Ductility* to remedy these adverse features without significantly impairing the strength.

CFRP   Carbon fibre reinforced plastics.

CGS system   Centimetre, gram, second system, the original metric system of units, now superseded by the *SI System*.

Chafing   Local *Abrasion* or sometimes an alternative term for *Fretting*.

Chain (1)   A length of interlinked loops intended to carry tensile loads and, sometimes, transmit drive between toothed **Chain wheels**.

Chain (2)   A string of linked atoms forming, usually, an organic molecule.

Chain intermittent weld   A weld made intermittently along the two sides of a joint, for example a T fillet weld, with the welds on the two sides lying opposite each other. Where the welds on one side lie opposite the gaps on the other, the joint is termed a **Staggered intermittent weld**. See Figure 49, located at entry on Welding terminology.

Chain reaction   Any continuing process occurring in repeating steps where each step initiates the next. The term often refers to *Nuclear* reactions where *Fission* of one *Atom* releases neutrons which cause fission in further atoms and so on.

Chambered bore   See Bottle bore.

Chaplet   Metal spacers used to position *Cores* in the *Mould* during sand casting. They are made of similar metal to the casting and are intended to fuse completely into it.

Characteristic radiation   The monochromatic radiation, specific to the element, which is emitted by an element excited by an electron beam or other radiation.

Charcoal   The high-carbon, usually 75% or more, material produced by heating organic material in low-oxygen conditions.

Charge (1)   All solid materials fed into a furnace. In the case of smelting processes this includes *Ore*, *Flux* and (solid) fuel but not liquid fuel or air. In the case of secondary production processes it includes the components being treated and usually any individual support systems but not usually items such as bogies or removable hearths that are a basic part of the furnace.

Charge (2)   Any process of adding material to a body or system, of imposing a static electric field or of introducing a quantity of electricity to a battery or similar equipment.

Charpy test   See Impact test.

Chatter   Generally, any vibration-induced relative movement between contacting surfaces that causes noise or local surface damage. More specifically vibration and judder of a cutting tool at the point of metal removal. This causes poor surface quality termed **Chatter marks**.

Checks, checking   Shallow surface cracking, often extensive, observed after treatment or service at elevated temperature. They are the result of *Thermal shock* arising from the stresses associated with temperature transients.

Chelant/chelating agent   A substance, typically an organic acid, that can combine with metal ions to form a **Chelate** or **Chelated complex**. For example, a chelant EDTA (ethylenediaminetetraacetic acid) is sometimes added to the water circuit of boilers to retain iron in solution as a chelated complex rather than forming deposits.

Chelant/chelate corrosion/attack Corrosion on the water side of steam-generating tubes resulting from excessive quantities, generally or locally, of *Chelating agent* in the same manner as *Caustic attack*. Corrosion may also be promoted by excess oxygen, turbulence and high water velocities.

Chemical attack An imprecise term implying any adverse effect resulting from a chemical reaction between a material and its environment. Such effects include chemical dissolution or combination with a component of the environment resulting in oxidation, scaling, rusting, pitting, etc. The term usually does not include damage mechanisms with a non-chemical component such as *Stress corrosion cracking* or *Corrosion fatigue cracking.*

Chemical bonding See Inter-atomic bonding.

Chemical cleaning The removal of surface films, rust and other contamination by the immersion in, or application of, appropriate chemicals.

Chemical compound A substance that comprises two or more elements joined in an *Inter-atomic bond* in fixed weight ratios.

Chemical deposition The deposition of a coating on an immersed surface by a chemical reaction between constituents of the solution or between the constituents and the surface.

Chemical equivalent (of an element) The atomic weight divided by the valence.

Chemical lead Lead, usually as sheet, of high purity, 99.9% or better, for chemical reaction vessels or linings.

Chemical machining/milling The removal of material in a controlled manner by the use of chemicals as solvents.

Chemical polishing The polishing of a metal surface by chemical dissolution without the assistance of abrasives or an external electric current.

Chevrons Vee- or arrow-shaped lines on a *Brittle* (crystalline) fracture surface. They point to the fracture origin.

Chill A metal plate inserted into the surface of *Sand casting* moulds. It causes a local increase in cooling rate and so modifies the solidification characteristics in its vicinity. It does not become incorporated into the casting. In cast irons, chills can induces local hardening, the hardened area being referred to as chill, chilled metal or similar terms.

Chill casting Any casting technique which promotes rapid cooling but particularly casting into metal moulds. See also Splat casting.

Chill crystals The first fine *Crystals* formed on the faces of castings as a result of solidification initiating at the large number of nuclei that result from the local undercooling, i.e. chilling, by the cold mould surface.

Chill plate (in welding) A substantial piece of material held in good thermal contact with a more flimsy component to protect that component from overheating during welding.

Chilled iron Cast iron poured in metal moulds to produce a white outer case over a grey core. See Cast iron.

Chinese script Various intermetallic *Phases* observed in the metallographic examination of some aluminium alloys, so called because their complex shapes are reminiscent of written Chinese characters.

Chip (1) Small metal particles cut away during machining.

Chip (2) A integrated circuit comprising a series of interconnected electronic devices such as transistors, resistors etc. A large number of chips are formed together on a wafer, i.e. a sheet of, usually, silicon.

Chip breaker (1) Features in the microstructure, particularly inclusions, which provide a plane of weakness causing break up the material removed during machining operations. See Free Machining.

Chip breaker (2) Details on the profile

of a machining tool that promote material removal as chips rather than continuous strands.

Chisel steel Any steel used for chisels and similar tools. Hand-held carpenters' chisels are typically high-carbon steels, larger chisels for more onerous duty may be of various low-alloy steels. They will be hardened and tempered as appropriate to the duty. See Steel.

Chlorination Any process in which chlorine is used to treat a material. The most common is chlorination of water to kill bacteria. In a metallurgical context chlorine may be injected into molten magnesium as a deoxidizing and degassing agent.

Chlorine A toxic gaseous element, one of the halogen group.

Chromadizing / chromodizing / chrome pickling (and similar terms or names, some of which may be proprietary) The treatment of aluminium or magnesium alloys in a chromic acid solution to improve paint adhesion.

Chromating The development of a metal/chromium *Conversion coating* by immersion in a solution containing chromium compounds. The coating enhances corrosion resistance and improves paint adhesion.

Chrome Abbreviation of chromium.

Chromel A proprietary series of nickel–chromium alloys, the most well known having 90% nickel and 10% chromium which, with *Alumel*, is widely used for *Thermocouples*.

Chromium A metallic element having a strong affinity for oxygen but forming a thin impervious oxide coating highly resistant to further attack. It has little application in the pure, bulk form but is widely used as an electroplate either for decorative and reclamation applications or, because of the very high plating hardness (up to 900 HV), for wear-resisting deposits. On steel the chromium deposit is not completely impervious so it is often preceded by copper and/or nickel layers. In addition to being more impervious and so enhan-

cing corrosion resistance, these layers improve the adhesion of the plating. Chromium is widely used as an alloying element. In *Steel* chromium (often in conjunction with other elements) up to about 5% improves *Hardenability* and mechanical properties. Amounts up to about 17% stabilize carbides and enhance high-temperature properties for tooling and creep-resisting applications. More than about 12% renders steel and cast iron 'stainless'. The austenitic stainless steels contain chromium in the range 12–30% and nickel in the range 8–20%, possibly with further additions. See Steel. A few per-cent in *Chromium copper* enhances creep strength. Nickel–chromium alloys, sometimes including other elements (see Nimonic), are extensively used for their good creep resistance and other higher-temperature characteristics. See Table 15 for physical properties of chromium.

Chromium copper An alloy of copper with up to 1% of chromium. It is *Precipitation hardening* with good tensile and creep properties.

Chromium plating See Chromium.

Chromizing A treatment in which components (usually steel) are exposed to a chromium rich environment, usually at elevated temperature, which causes chromium to diffuse into their surface to improve corrosion resistance.

Ciment fondu Aluminous cement used as a *Refractory*.

Cire perdu Essentially the same as the *Lost wax process*. The term tends to be favoured by the artistic and antiques fraternities whereas the other alternative term, investment casting, tends to be used in engineering contexts.

Civil transformation A transformation from one solid *Phase* to another in which the *Atoms* at the advancing interface realign themselves to the new *Crystal lattice* in an uncoordinated manner and without regard for the original grain boundaries. Most transformations are of this form. The opposite is *Military transformation*.

Clad, cladding (1) The attachment of sheet to a structural framework.

Cladding (2) The application of a substantial coating of one type of metal to a substrate of another. For example, a strong, corrosion-prone aluminium alloy plate can be clad with a corrosion-resistant but weak 'pure' aluminium. Such cladding is accomplished by *Rolling* either a pair of the materials or a sandwich comprising a slab of the alloy between two slabs of 'pure' material. Rolled gold is another example.

Cladding (3) *Canning.*

Clamping load See Preload.

Clay Finely particulate mineral, such as hydrated alumino silicates, with a minimal amount of water forming a plastic bulk solid. Despite the water content, suitably prepared layers can be virtually impervious. As the moisture content reduces, the bulk material becomes less plastic. Firing at high temperature forms a hard, brittle ceramic.

Cleavage Cracking along specific planes of the *Crystal structure* producing a low-*Ductility* fracture. The effect can be visualized as layers of atoms peeling apart in contrast to ductile shear failure, where the layers slide across one another. The relevant atomic planes are referred to as cleavage planes and the resultant cracks as *Brittle, Flat,* or cleavage fractures.

Clenching/clinching The final process of tightening a mechanical joint by, as examples, tightening a bolt, bending over the projecting point of a nail or closing over the shank of a rivet.

Clink, clinking The noise of a crack occurring in metals, usually steel, during heating or cooling. The cracks themselves may be termed 'clinks'. They result from restraint of thermal expansion or contraction ·or, during cooling, from *Hydrogen damage.*

Clinker Coarse bulk material remaining from the combustion of coal, as opposed to fine ash.

Close(d) annealing *Annealing* in a sealed box to minimize reaction with air. In the case of steel the term may also imply sub-critical annealing as opposed to full annealing.

Close packed hexagonal See Crystal structure.

Close(d) joint A joint in which the component faces to be welded are in contact prior to welding.

Closed die forging See Forge.

Closed pass A *Rolling* operation in which one profiled roll face is inset into the other. The product completely fills the roll gap with minimal flash or fin at the interface position.

Closed pore/cell *Sintered* material that contains a significant amount of porosity but in which the individual voids are not continuously interconnected. In contrast, open-pore material allows lubricant or other fluids to permeate the material.

Cluster mill A *Rolling mill* in which each of the two working rolls is supported by two or more large backing rolls.

CMOD See Crack opening displacement and Fracture toughness.

CO₂ Carbon dioxide.

CO₂ flux welding process See CO₂ welding.

CO₂ process A process for producing strong sand moulds for casting. The sand is mixed first with a quantity of 3–5% sodium silicate solution just sufficient to lightly coat the grains but allow normal moulding. After the mould has been formed CO_2 is blown through to react with the silicate and strongly bind the sand. Also termed the **Sodium silicate** or the **Silicate process**.

CO₂ welding A form of *Metal inert gas* (MIG) welding in which the electric arc is struck between the component to be welded and a continuously fed bare wire filler electrode. Carbon dioxide (CO_2) is delivered, usually via the wire feed system, to provide a protective gas shroud for the weld zone. The process is commonly used for steel with which CO_2 is effectively inert. **CO₂ flux welding** is similar except that the consumable electrode is a *Flux*-coated wire or a flux-filled tube.

Coalescence (1)   Any process of separate components, usually in some surrounding matrix, coming together to form a larger entity. Examples include the coalescence of voids during *Creep* or the coalescence of carbides in steel where a large number of fine particles can, by a process of *Diffusion*, coalesce to a small number of coarse particles.

Coalescence (2)   The bonding that results when powders are *Sintered*.

Coalescence (3)   In the context of welded or brazed joints the term indicates a satisfactory bond between components.

Coarsening   The increase in *Grain* size by the process of *Grain growth* or, in steel, the increase in size, with associated reduction in number, of carbide particles or pearlite plates. See Steel.

Coated electrode   (for welding)   See Electrode (welding).

Coating (1)   Any process of applying a surface layer to a component, or the layer itself. The term is sometimes used in contrast to *Cladding* to imply a relatively thin surface layer, less than about 1 millimetre.

Coating (2)   The layer of *Flux* and other materials on some welding *Electrodes*.

Coaxing   A technique for increasing the *Fatigue* strength and life of a material by subjecting it to a series of cycles at rising stress ranges.

Cobalt   A ferromagnetic metallic element. It is the primary element in many high-strength, high-temperature alloys and is an alloying addition to superior grades of *High-speed steels*. It is used as a binder or matrix for some types of *Carbide* tool materials. See Table 15 for properties.

Cocoa   The cocoa-like powder of red iron oxide, Fe_2O_3, formed by *Fretting* of iron and steel.

Cockling/Cockles   The wavy edge of rolled sheet and plate.

COD   Crack opening displacement. See Fracture toughness.

Coefficient   A factor defining the relationship between two components. For example, the **Coefficient of thermal**

expansion is a measure of the change in dimensions resulting from a unit change in temperature.

Coercive force   The reverse magnetic force necessary to eliminate the remanence resulting from a previous magnetic force and thereby return the magnetic flux density to zero. See Magnetic.

Co-extruded   An extrusion process in which two materials are extruded simultaneously so that one forms a surface layer around the other. The process is, for example, applied to tubes which require a combination of strength and corrosion resistance not obtainable in a single material.

Cogging   The first *Rolling* operation on steel ingots in a **Cogging mill**, also called a **Blooming mill**.

Coherency   The state in which the lattices of a precipitate and its surrounding parent matrix remain substantially in alignment but are distorted in the vicinity of the interface.

Coherent (precipitate)   An intermediate stage of precipitation, preceding formation of a distinct and separate phase, in which the lattices of the solution and the emerging precipitate are still in alignment but distorted.

Cohesion   The bond between atoms, molecules or particulate matter.

Cohesive strength   The theoretical strength developed by interatomic cohesion assuming failure without plastic deformation.

Coinage metals   Metals used for currency coinage because of their availability, durability and perceived value, originally gold, silver, copper and bronze, now presumably including aluminium, nickel, etc.

Coining   A *Forging* operation, usually cold, in which the component being formed just fills the *Closed die* ensuring a crisp, clean imprint.

Coke   The solid carbonaceous material produced by heating coal in the absence of air.

Coke oven   A furnace in which coal is

tightly packed and heated with a limited air access to form coke.

Coke oven gas The gas produced in the manufacture of coke. It is a mixture of carbon monoxide, methane and hydrogen plus numerous other volatile constituents.

Cold anodizing A process for increasing the thickness of the normal oxide film on aluminium by immersing it in cold 50% nitric acid for a few minutes. Also see Anodizing.

Cold bend test A test in which a bar of metal is bent to some specified degree as a test of **Ductility** or, especially in the case of tests that include a weld zone, to confirm freedom from cracking. The specimen may be plain or notched. See also Bend test.

Cold cast A term used, dubiously, for a technique where a metal powder such as *Bronze* is mixed with liquid resin and poured into a mould to solidify. The resultant item is then described as **Cold cast bronze**.

Cold chisel A chisel of relatively short length, substantial cross-section and relatively obtuse angled cutting edge for cutting materials such as metal or brick as opposed to timber. It has no handle and the flat, bevelled edge, head is struck with a hammer.

Cold cracks Cracks formed in a cold component following some heating cycle, particularly steel that has been hardened or welded. Such crack are caused by *Residual stresses* and, often, high levels of dissolved hydrogen. See Hydrogen damage.

Cold Extrusion/forming/drawing etc. See Cold work, Extrusion and Drawing, etc. In some uses of the terms there is an implication of close dimensional control of the product.

Cold galvanizing The application of zinc-rich paint by brush or immersion in an attempt to obtain the protection offered by *Hot dip* or electro *Galvanizing*. The coating must have good adhesion and provide electrical continuity between zinc particles and the underlying

steel if any *Cathodic protection* is to be obtained. The term has also been used of electro-galvanizing.

Cold hardening Natural ageing. See Precipitation hardening.

Cold heading Cold forging the heads of bolts by an *Upsetting* process.

Cold hearth melting The use of a furnace with a water-cooled hearth, usually copper, on which to melt materials to avoid contamination by refractories.

Cold joint (soldered or brazed) A joint with inadequate bonding resulting from insufficient heating of the parent components. See also Dry joint.

Cold junction The cold end datum point of the pair of wires forming a *Thermocouple*.

Cold lap Same as Cold shut.

Cold pressing Pressure compaction of powder at a temperature too low to allow *Sintering*.

Cold rolling (1) *Cold working*, by rolling, of sheet, bar, etc., to effect a size change.

Cold rolling (2) Local rolling, of surface features such as crankshaft radii, with hardened steel wheels or rollers to induce surface hardening with beneficial *Residual stresses*.

Cold short Having poor ductility and hence a susceptibility to cracking during cold working operations.

Cold shut Casting defects characterized by a thin oxide film at the interface between neighbouring areas solidifying in isolation. This occurs when the flow of metal is interrupted, allowing surface solidification that fails to fuse with subsequently poured metal or when molten metal flows in from two directions and the interface is too oxidized to allow the flows to merge.

Cold welding (1) Welding, without any heating, between two surfaces in intimate contact. In practice, high pressures and very clean, oxide free surfaces are required. Also termed 'adhesion', 'galling' and 'seizure', particularly when these consequences are undesirable.

Cold welding (2) The application of special procedures allowing fusion welding, in particular arc welding, without *Preheat* and/or *Post heat*, in circumstances where they would normally be expected, for example alloy steels in restrained thick sections.

Cold work Any process of plastic deformation in which the component does not *Recrystallize* but becomes progressively harder and stronger but less ductile up to some limit. The temperature will normally be at about ambient. If it is deliberately and significantly higher, but still below the recrystallization temperature, the process may be termed **Warm working**.

Colloid A finely particulate material that in a fluid remains fully dispersed as a suspension rather than settling.

Columbium The American name for niobium.

Columnar grains/structure Elongated grains lying parallel with each other, particularly those perpendicular to the surface of a casting. They occur when initial solidification commences at a limited number of nuclei at the surface. As heat is extracted from the surface, lateral growth of grains is limited by the neighbouring grains but inwards growth continues as the solidification front progresses.

Combination mill A rolling mill in which the initial roughing, or *Cogging*, mill is followed immediately by a further series of rolls to complete the hot rolling process.

Combined carbon The carbon in a *Cast iron* that is combined with, or in solution in, the iron as opposed to free carbon (graphite).

Comminution The production of fine powder by various processes including pulverizing, attrition, chemical or electrochemical techniques.

Common brass Alloy of 63% copper and 37% zinc. The highest zinc level and hence the cheapest *Brass* that remains single alpha phase allowing it to be cold worked.

Compact The action of compressing powder in a die to form a briquette or the briquette so formed.

Compact tension specimen (CTS) See Fracture toughness.

Compact(ed) graphite cast iron A *Cast iron* with a graphite structure midway between the flake and spheroidal forms.

Compensating leads The electrical conductor wires between a *Thermocouple* and the instrument measuring the emf. The wire materials are selected to have the same thermoelectric characteristics as their respective thermocouple wires over the temperature range that the lead experiences. The benefits offered are that the lead can be stranded to improve flexibility and it can be of cheaper material than the thermocouple.

Compliance The measure of the ability of a shaft or similar body to flex elastically.

Component (1) Generally, any item being manufactured or under consideration.

Component (2) In the context of the *Phase rule*, one of the minimum number of elements that allow a system to be fully defined.

Composite (material) Any combination of two or more individual materials, including metals, ceramics, polymers, etc., in particle, fibre, sheet or bulk form. There may be some bonding, diffusion or alloying between components but they will be individually identifiable under the microscope. It is implicit that the composite will combine the favourable features of its components.

Compression ratio The ratio of the volumes of gas, powder, etc. prior to and after compression.

Compression test Any test in which a compressive load is progressively applied and the component deflection monitored. Failure modes include cracking, buckling and general collapse.

Compressive strength The maximum *Stress* that can develop in a material under compressive loading. Unless the material fails by some distinct cracking mechanism at a readily measured stress the compressive strength has to be defined in terms of the degree of deformation regarded as failure.

Computer Aided (Assisted) Design (CAD) Design techniques employing computer-based mathematical and graphics programs to assist in matters such as stress determination, component layout and product appearance.

Computer Aided (Assisted) Manufacture (Machining) (CAM) Processes employing computers to assist in matters such as monitoring of product dimensions, rectification of deviations, etc.

ConCast Continuously cast material or the *Continuous casting* process.

Concave root An indentation along the line of the weld *Root*. Same as Root concavity (see Figure 52, located at entry on Welding terminology).

Concave weld A weld including *Butt* and *Fillet* in which the final weld surface falls below a straight line joining the points where the weld metal meets the parent metal (see Figure 52, located at entry on Welding terminology).

Concentration (1) A local increase in some feature or element.

Concentration (2) The quantity of one substance contained in another, usually as a percentage.

Concentration cell The *Electrolytic* cell produced by the variation, from point to point, in concentration of the electrolyte. The anode of the cell is likely to corrode.

Concentration gradient A variation, from point to point, in the proportion of two or more elements. The gradient may be a sharp step, as at the boundary between two *Phases*, or it may be a shallow slope as in *Coring*.

Conchoidal markings Concentric marks on the surface of a *Fatigue* fracture. They are visible to the unaided eye and mark the crack front at irregularities in the load cycling such as interruptions in cycling or an abnormal load. Also termed *Beach marks*, *Shell marks* or *Arrest marks*. Compare with *Striations*.

Condensate Liquid produced from the gaseous phase by *Condensation*. The term usually implies bulk liquid such as would be produced in a *Condenser*.

Condensation The process of liquid precipitating from the gaseous phase or the liquid so produced. In the latter case the term may imply films or small volumes of liquid.

Condenser (1) A heat exchanger, in which a gas, often steam, is cooled causing it to revert to the liquid state. In a **Tube condenser** cooling water is circulated through tubes to condense the steam exhausting from a steam turbine or other cycle. In a **Jet condenser** the cooling water is injected into the exhaust steam; this requires high-quality cooling water assuming that the condensate is to be re-used.

Condenser (2) An electrical capacitor.

Condenser foil Metal foil, usually aluminium or tin base, used for electrical capacitors, i.e. condensers. The foil may be varnished or, in the case of aluminium, anodized, to provide an insulating surface. This avoids the need for an nonconducting interleaf when rolled to form the capacitor.

Conduction band/Electrons See Band theory.

Conductivity The (measure of the) ability of a material to transfer heat or electricity. See also International Annealed Copper Standard.

Conductor A material capable of transmitting heat or electricity. See also Semiconductor.

Cone (of oxy-fuel gas flame) The innermost cone at the exit of the burner.

Constantan Alloy of 60% copper with 40% nickel having high electrical resistivity, about 48 $\mu\Omega$-cm, a low-temperature coefficient of resistivity up to about 450 °C and good corrosion resistance. It is used for various electrical purposes, such as resistors and thermocouples.

Constitution diagram Synonymous with *Phase* diagram.

Constraint (of deformation) Localized restriction of deformation resulting from the *Poisson* effect. The Poisson effect causes material being extended along the longitudinal axis to contract across the transverse axes. At notches and other stress concentrations the material at the highly stressed notch tip is constrained from contraction in the transverse axes by the surrounding large volume of lower stressed material. This gives rise to complex triaxial stresses at the notch tip. The constraint is usually confined to small volumes of material while all stresses are in the elastic range, but large volumes may be involved if plastic deformation has occurred.

Constricted arc   See Plasma arc.

Consumable insert (of weld)   A preplaced *Filler* having specific dimensions that is located in the *Root* of a joint to be made from one side. It is intended to be fully fused to become an integral part of the joint and is not normally machined following welding. Same as Fusible insert. See Figure 47(a), located at entry on Welding terminology.

Contact angle (1)   See Roll angle.

Contact angle (2)   The angle formed at the junction between a solid and a liquid, measured through the liquid. If the liquid 'wets' the solid as with water on clean glass the angle is acute, if it does not wet as with water droplets on a waxed car body the angle is obtuse.

Contact corrosion   The corrosion in the vicinity of the point of contact between dissimilar metals in an *Electrolyte*.

Contact electrode (for welding)   An *Electrode*, the coating of which during the welding operation maintains a cup profile projecting slightly beyond the tip of the metal electrode. The edge of the cup is rested against the component being welded to maintain a fixed arc length.

Contact fatigue   Cracking and pitting damage resulting from the high-tensile stresses, termed Hertzian stresses, immediately beneath the surface at locations carrying point and line contact loads. It is common on the races of highly loaded ball and roller bearings. Also see Fatigue.

Contact resistance   The resistance to the passage of electricity at the contact interface between two components. It reflects aspects such as contact area, roughness, interfacial pressure and contamination.

Contact tinning/tin plating   Tin plating of components of steel, cast iron, copper and other metals by dipping them into a molten bath of tin covered by a *Flux*.

Continuous casting   Any system using a mould open at both ends such that solid metal is steadily drawn out of the bottom while molten metal continues to enter the top. The mould can be in various forms including a simple die or rolls and is usually water cooled. Often, equipment beyond the mould cuts the ingot, bar, tube, etc. into lengths allowing the process to continue virtually indefinitely. If the process is occasionally interrupted to allow removal of the ingot the process may be termed **Semi-continuous** casting.

Continuous cooling transformation diagram (CCT diagram)   A diagram depicting the *Phase* changes that occur in a metal cooled continuously at various rates, particularly steel cooled from the austenitic state. See Isothermal transformation diagram.

Continuous covered electrode   Welding electrode *Filler* wire, supplied as coils, comprising a central main filler wire carrying an external coating and helical wire(s) which reinforce the coating and act as current pick-up.

Continuous furnace   A furnace which operates continually with materials being charged at one end and discharged at the other. A variation is the **Rotary hearth** where the charge and discharge points are a short distance apart around the radius.

Continuous mill   A series of powered *Roll* stands, arranged in tandem, such that the material leaving the first roll stand immediately enters the second and so on. The speeds of the various rolls are adjusted to match or exceed the increase in speed of the product that results from the reduction in gauge. See Figure 33, located at entry on Rolling mill.

Contraction (and related terms) See Shrinkage.

Controlled cooling Any cooling at a prescribed rate but it is usually implicit that the rate is slower than would occur naturally, as opposed to *Quenching*. It is imposed after processes such as heat treatment, welding or hot working for various reasons including avoiding *Hardening*, minimizing *Residual stresses* or allowing dissolved *Hydrogen* to escape.

Controlled rolling The practice of continuing a sequence of hot *Rolling* operations on steel to a temperature lower than normal. This produces a fine grain size with a slight increase in tensile strength and a beneficial reduction in *Brittle fracture* transition temperature.

Controlled thermal severity test (CTS) A test of the weldability of steels, in particular the relationship between cooling rate and susceptibility to cracking. For the test, a 75 mm (originally 3-inch) square steel plate is bolted to a much larger and thicker steel plate and substantial *Fillet* welds deposited along two opposite sides. Two test welds are then deposited, in sequence, on the remaining sides. The concept is that when the first test weld is deposited the cooling rate is controlled by the conduction of heat through the two assembly welds while the cooling rate for the second test weld is controlled by conduction through the previous three welds. The severity of the test is defined in terms of **Thermal Severity Number (TSN)**, which is the number of conduction paths multiplied by the plate thickness in units of 6 mm (originally 0.25 inch). For example, with 6 mm (quarter-inch plate) the TSNs of the test welds would be 2 for the first test weld and 3 for the second test weld, and with

12 mm (half-inch) plate they would be 4 and 6.

Conventional stress The stress calculated on the original cross-section area, as opposed to the real stress. See Tensile test.

Conversion coatings A coating developed on a metal surface by chemical or electrochemical treatments that form a compound between the applied material and the base metal rather than merely depositing a layer. *Anodizing*, *Chromizing* and *Phosphating* are examples.

Converter A furnace such as that in the *Bessemer* process in which air is blown through or across the charge of molten metal to oxidize impurities. The heat released by the reaction is sufficient to maintain, or even increase, the metal temperature.

Coordinate bonding See Interatomic bonding.

Coordination number The number of equidistant closest neighbours that surround each atom in a *Crystal structure*. It is 12 for the close packed hexagonal and face centred cubic structures and eight for body centred cubic structures.

Cope The top section of a two-piece moulding box split horizontally. The lower part is the **Drag**.

Copper A metallic element having good electrical and thermal conductivity and readily formed by a range of manufacturing operations. It is extensively used in its pure, or near-pure, condition to take advantage of these characteristics. Various production techniques are available for removing undesirable contaminants in the manufacture of nominally pure coppers and in many cases the resultant descriptions used in specifications are self-explanatory, for example oxygen-free high-conductivity copper or **Phosphorus deoxidized copper**. The common description, **Tough pitch copper**, refers to copper in which the significant oxygen content has been rendered non-damaging to mechanical or physical properties by allowing it to form fine, well-dispersed copper oxide particles (although see Gassing). The description arises from the use in the industry of the term 'Pitch' to refer to the oxygen content of copper. Copper is the basis of many alloys including *Brasses*, *Bronzes* and *Cupronickels* and is a minor constituent in many other alloys. See Tables 8 and 15.

Copper–constantan   A pair of metals used for *Thermocouples*. See Constantan.

Copper loss   The power loss resulting from the resistance in an electrical circuit.

Copper matte   An intermediate stage in the production of copper from sulphide ores; it contains about two-thirds copper.

Copper shortening   See Shortening.

Core (1)   The shaped former set in a mould to establish the internal shape of a hollow *Casting*.

Core (2)   The inner material beneath a *Case hardened* or other surface layer.

Core (3)   The central magnetic mass of a transformer around which the conducting windings are arranged.

Core plug   A plug inserted into the wall of a hollow casting to close a hole left by the core support or through which a sand core was extracted.

Core refining   The initial stage of hardening a carburized steel component. The carburizing treatment produces an undesirably large grain size in the component. Good-quality work requires a fine grain size in both the low-carbon core and the high-carbon case. Initially the component is heated to and quenched from about 880 °C, which is just within the austenitic region for the typical lower-carbon core. This refines, hardens and toughens the core. Subsequently, the component is heated to and quenched from about 750 °C, which is just within the austenitic region for the high-carbon case. This hardens the case. See Steel and Case hardening.

Core wire (for welding)   The central main wire of a *Continuous covered electrode*.

Cored hole   A hole or cavity in a casting formed by a *Core* rather than being machined subsequently.

Cored solder   *Soft solder* containing *Flux* in one or more longitudinal cavities.

Coring   A variation in composition from centre to edge of individual grains of an alloy solidifying as a solid solution. It is most pronounced where there is a large difference between the liquidus and the solidus and arises during casting because the first material to solidify is relatively rich in one element and the remaining molten material becomes progressively richer in the second element. The change in composition may be so extreme that a non-equilibrium phase precipitates in the latter stages of solidification. See Phase, Lever rule and Solidification.

Corner joint (welded)   A joint made between the edges of sheet or plate components that meet at, very approximately, right angles, usually between 30° and 135°.

Corrosion   Any process by which a metal reacts chemically or electrochemically with another element in the surrounding environment. It is usually implicit that the consequences will be damaging as opposed to processes such as *Anodizing*. Damage includes loss of bulk section, pitting, rusting and scaling. See also Electrochemistry and Corrosion resistance.

Corrosion embrittlement   Deeply penetrating corrosion damage, essentially cracking, often along grain boundaries. The damage is usually not obvious to a visual inspection but causes a major reduction in load carrying capacity, not merely a reduction in *Toughness*. See also Hydrogen damage.

Corrosion–erosion   A combination of corrosion and erosion. There are considerable similarities with *Impingement attack* and some authorities recognize only one mechanism. Where the two are differentiated, and this may not be easy in some cases, impingement is regarded as having a major electrochemical component arising from the potential difference between uncorroded areas and bare metal exposed by erosion. The term 'corrosion–erosion' would then be confined to cases where the erosion merely serves to remove the products of corrosion which, if undisturbed, could impede further attacks. See also Electrochemical and Cavitation.

Corrosion fatigue   A cracking mechanism in which *Corrosion* and *Fatigue* act concurrently to reduce significantly the total life that would be expected if either fatigue loading or a corrosive environment had acted alone or if the two had acted consecutively. Purists sometimes argue that any fatigue, other than in vacuum or perhaps an inert gas, is corrosion fatigue. A more practical approach is to differentiate on the basis of the major difference in life seen in a clean dry environment compared with that in a moist and contaminated one.

Corrosion potential   The electrical potential, in an electrolyte, of a metal relative to another metal or reference electrode. See Electrochemistry. In a more casual sense, the general susceptibility of a metal to corrosion.

Corrosion resistance   The ability of a metal to avoid unacceptable reactions with its environment. Generalizing, metals can be divided into three categories. Some metals such as gold and platinum have **Inherent resistance**, that is, they do not react to any significant extent with the atmosphere and many other environments. Other metals such as aluminium or titanium react very rapidly when exposed to the atmosphere but the oxide film is thin, tightly bonded to the metal and forms an impervious barrier to further attack. This is often termed **Acquired resistance**. Finally, metals such as iron and steel in moist environments form a bulky, loosely adherent and porous oxide layer. This does not significantly slow the rate of attack and it may even detach, exposing bare metal

for further attack. See also Stainless steel.

Corundum   The natural, impure form of alumina, aluminium oxide, Al_2O_3. Gemstones such as ruby, emerald and sapphire are also natural forms having specific individual impurities which confer the respective colours.

Cosmetic   In a technical context, actions such as weld repairs performed solely to improve appearances. The work may actually be damaging and the term often has a pejorative note.

Cosworth process   A process for producing high-quality precision castings. Quality is assured by extreme efforts to avoid entrainment of oxide or other contaminants, in particular by the use of extended periods of holding the molten metal followed by pumped, up-hill feeding. Dimensional precision is achieved by the use of moulds of zircon sand having a low expansion coefficient.

Coulomb   The SI unit of quantity of electricity, the electrical charge transported by one ampere flowing for one second.

Counter-current principle/flow   Maximum efficiency in, for example, a heat exchanger is achieved if the two flows pass parallel to each other but in opposite directions. The fluid being heated reaches the maximum possible temperature because, immediately before it leaves the exchanger, it is exposed to the incoming heat source at its maximum temperature. The converse applies at the outlet of the heat releasing fluid ensuring maximum efficiency. The principle holds true for many processes and reactions.

Counter-forge/hammer   A forging process in which the component is forged between two forging heads moving together in unison.

Couple   (1)   The ability of dissimilar metals in electrical contact to produce electrical currents. Similar effects can occur in a single component because of variations, from point to point, in the composition of the material or of the environment. See Thermoelectric effects.

Couple (2)   In mechanical systems, a pair of forces of equal size but acting in opposite, parallel directions.

Coupling   A device or mechanism joining two components or structures, usually for some specific function such as transmitting a fluid, traction or power.

Coupon   A sample for test purposes that accompanies a component during some production process, treatment or merely any sample subjected to testing or process simulation.

Covalent bonding   See Interatomic bonding.

Cover pass   The final layer of a multilayer weld. The term often implies a deposit having a good appearance.

Covered electrode   (for welding)   See Electrode (welding).

Covering power   The ability of an *Electroplating* bath to apply a coating to all surfaces including recesses. **Throwing power** is the ability to apply coatings of even thickness to such irregular surfaces.

CPH   Close packed hexagonal crystal structure. See Crystal structure.

Crack   A planar defect that has developed in an area that was substantially defect free. There is often an implication of limited *Ductility* and 'crack' as opposed to 'break' tends to be favoured where the component has not parted completely. See also Break, Rupture and Fracture.

Crack arrest temperature   The temperature at which a running *Brittle fracture* in *Steel* will cease to propagate. The term may be used in connection with some particular laboratory test such as the *Robertson test*. More generally the term refers to the lowest service temperature of a steel component at which a brittle fracture will not propagate further.

Crack arrester   Any feature, intended or otherwise, that stops a crack from propagating further. Such features may be on a microscopic scale, for example

non-metallic inclusions, or on a macro scale, for example reinforcement plates, ductile materials or even holes placed across the path of an advancing crack.

Crack initiator   Any feature, such as a pre-existing crack, section change or inclusion, which acts as a local *Stress raiser* and hence assists the initiation of cracking by any mechanism.

Crack opening displacement (COD)   The distance that the faces of a crack move apart at the tip during loading. More specific terms are **Crack tip opening displacement (CTOD)** and **Crack mouth opening displacement (CMOD)**. See Fracture toughness and Fracture mechanics.

Cracking   One or, more often, multiple *Cracks*. Also see Break.

Crackled   Either a craze cracked surface or a severely wrinkled paint coating.

Crater/crater pipe   (of a weld)   A depression or hole at the termination of a weld *Bead* resulting from local shrinkage during solidification.

Crater crack   A crack in or from the *Crater* of a weld.

Crazing   Multiple intersecting surface cracks, usually shallow, arising from various causes but particularly *Thermal shock*.

Creep   Permanent, progressive deformation, occurring over a period of time, at loads below the *Yield stress*. Most metals at ambient temperature will carry a *Stress* which is less than the yield stress for, effectively, an indefinite period. When the stress is removed the component will revert to its original dimensions. However, if the temperature of the component is raised above a critical level any significant stress will, over a period of time, cause progressive deformation that remains after the stress has been removed. If the load is maintained for a sufficient period internal voids and cracks will develop and ultimately the component will fail. A typical, perhaps idealized, creep test has the characteristic three stage, strain to time relationship, termed a **Creep curve**, shown in

Figure 4. In the usually brief **Primary** or **Transient** stage, deformation is rapid as it is dominated by mechanisms such as *Dislocation* movement and *Grain boundary* slide, i.e. the two neighbouring grain faces rotate across each other. This high initial rate then slows as work hardening plays an increasing role. In the longer **Secondary**, or **Steady state**, stage, deformation continues at a lower uniform rate arising from the balance between work hardening and thermal softening. Towards the end of this stage damage, in the form of grain boundary voids, starts to become visible under the microscope. In the final **Tertiary** or **Runaway** stage, deformation accelerates as voids join up, gross grain boundary cracks develop and the specimen *Necks* prior to failure. This conventional view of creep being accompanied by major deformation is often confirmed by service experience but it is not universal. Some metals fail by creep with negligible reduction in area and only a fraction of one per-cent elongation. Such behaviour may be termed **Creep brittle**. Creep failure is often termed **Creep rupture** but this does not imply any special characteristic of failure; it could equally be termed 'creep fracture' or even 'creep breakage' although these

phrases tend to sound inappropriate to most metallurgists. The temperature at which creep commences is broadly related to the melting point of the metal, it often being suggested that creep of pure metals becomes significant above half their absolute temperature melting point. For example, lead will creep, at least over decades or centuries, at normal ambient temperature and mild steel will creep at a significant rate above about 400 °C. The rate of damage, above the critical temperature, rises with increasing stress and temperature. Creep deformation characteristics, often termed the **Stress rupture properties**, are determined by testing specimens of the material over ranges of stresses and temperatures to establish the time to fail or to reach a specified percentage creep *Strain*. Unfortunately, the results obtained from testing identical specimens under identical conditions are rarely the same. Hence, the results of numerous tests plotted on a graph do not form a neat, smooth curve but tend to be scattered over a fairly wide band. This gives rise to such terms as creep **Scatter band** so designs may be made on the basis of **Lower scatter band** (i.e. the lower line enclosing all test results) or **90% scatter band** (the line enclosing 90% of all

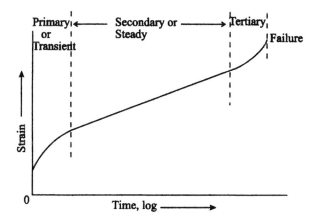

Figure 4   The three stages of creep

results) or **Average stress rupture** properties. Stress rupture properties are often presented in diagrams such as Figure 5 but also see the entry on the Larson–Miller parameter for another example of creep data presentation.

Creep ductility   The *Elongation* at failure of a component failing by creep. See Creep.

Creep limit   This term has a range of interpretations and some potential for confusion. The most basic definition is the maximum *Stress* that can be allowed without causing unacceptable damage. Damage, in this context, may be defined as failure or as some level of *Strain*, in which cases it is necessary also to specify the service temperature and life required. Damage may also be specified as a rate of deformation in which case only the temperature need be specified. See Creep.

Creep relaxation   The reduction of imposed load as a result of deformation by *Creep*, i.e. by the conversion of elastic strain to plastic strain by a creep mechanism. The phenomenon occurs in assemblies where the loading mechanism is rigid and does not move sufficiently to maintain the stress. One example is high-temperature bolts which are tightened to induce a large amount of tensile elastic extension while the flange through which they pass, being of much larger cross-section, experiences negligible elastic compression. If the bolt creeps, assuming it does not fail, the stress in the bolt and hence the joint-closure force will progressively fall. This contrasts with the case of, for example, a boiler tube which experiences a constant internal pressure. If the tube distends the internal pressure remains the same but is carried by a reducing

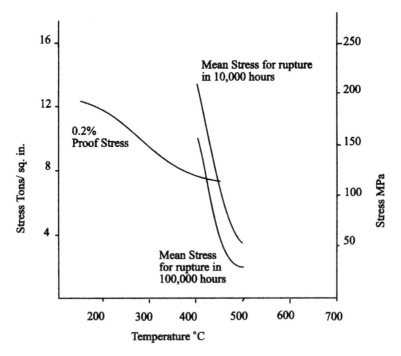

Figure 5   High-temperature tensile and creep rupture properties for a 0.2% carbon steel

wall thickness so the stress rises. See also Preload and Displacement control.

Creep rupture Failure by *Creep*. See also Larson–Miller parameter.

Creep strength Generally, the strength of a material in its *Creep* range. More specifically, the same as Creep limit.

Crevice corrosion The corrosion, occurring within a narrow gap, which is severe compared with that on a plain surface exposed to the same environment. The mechanism usually involves the development of variations in environment composition from point to point leading to *Electochemical* corrosion. The variation may be a concentration effect, for example of dissolved salts, or it may be a depletion effect, for example a reduction in oxygen within the crevice.

Crimping (1) The formation of circumferential corrugations around one side of a tube to induce a bend, more usually termed *Wrinkle bending*.

Crimping (2) The deliberate kinking of a folded *Lap* joint to lock the joint.

Crimping (3) The introduction of angular ridges in a sheet to improve rigidity.

Crimping (4) The introduction of multiple transverse indentations in bar to improve grip or to increase its length.

Critical cooling rate The cooling rate that has to be exceeded to achieve or avoid a particular effect, for example to develop a fully martensitic structure in steels. See Steels.

Critical humidity The atmospheric humidity above which the rate of corrosion increases dramatically.

Critical frequency/speed The vibration frequency or rotational speed which coincides with a natural resonance frequency of the component. At this critical condition, perturbations which would normally be innocuous can rapidly increase in severity leading to damage or failure. Examples include gross instability of shafts, collapse of bridges and rapid fatigue failure of turbine blades. Design and operating practices

aim to avoid completely, or to pass rapidly through, the critical stage.

Critical mass The minimum amount of fissile material required to sustain nuclear *Fission*.

Critical point The temperature (and pressure) at which some change, in particular a *Phase* change, occurs.

Critical range In steels it is the temperature range between the upper critical and lower critical temperatures. See Steel.

Critical strain The minimum amount of plastic *Strain*, i.e. deformation, required to induce *Recrystallization*. Some metals, such as iron, can develop a completely new grain structure by heating then through a *Phase* transformation. However, in the absence of such a transformation, a metal cannot form a new structure by heating below the melting point, unless it has been plastically deformed by more than a critical amount, the critical strain, typically a few percent. This process of developing a new grain structure with associated reduction in the hardness is termed *Annealing* and, generalizing, the more severe the deformation, the finer the subsequent grain size. Consequently, components deformed only slightly more than the critical amount can form structures with exceptionally large grain sizes when annealed. This effect is termed **Critical strain grain growth**. It arises because individual grains grow from nuclei and, at the critical strain, only a few nuclei are available.

Critical strain energy release rate The *Toughness* as measured by the energy input required for a crack to continue propagating.

Critical stress intensity See Fracture toughness.

Critical temperature The temperature at which some change, in particular a *Phase* change, occurs.

Criticality (of a nuclear reactor) The state at which the fission reaction is exactly in balance with the number of neutrons released being just sufficient to maintain the *Chain reaction*.

Cropping   The practice of cutting off and discarding the ends of bars, ingots, etc., as these are the site of defects.

Cross-country mill   See Rolling mill.

Cross-rolling   The technique of rolling plate or sheet in which the rolling direction of each pass is at 90° to the previous pass.

Cross-section (1)   Of a bar, extrusion, etc., it is the shape and dimensions of the face formed by a full transverse cut.

Cross-section (2)   In the context of nuclear interactions it is a function of the probability that a particle entering the material in question will meet a particle with which to react. The SI unit is the barn.

Cross-slip   The extension of a *Dislocation* from one slip plane to an adjacent plane.

Crucible   A vessel in which metal is melted. Usually, it will be of a size that can be handled manually or with limited mechanical assistance.

Crucible (cast) steel (1)   An obsolete process in which steel produced by the *Cementation* process was melted in crucibles and recast. This retained the high carbon of the cementation steel but allowed most of the slag impurities to be removed.

Crucible steel (2)   The term is sometimes still used to refer to steels made in small quantities and, possibly, claimed to have superior characteristics.

Cruciform   Having the geometry of a cross. A **Cruciform weld** joint is formed when two bars or plates are welded in the same plane on either side of, and perpendicular to, another plate. See Figure 47(b), located at entry on Weld assemblies. A **Cruciform test** is one in which such a joint is made, inspected and, usually, tested to destruction.

Crud   Originally a American term for solid particulate contamination in the boiler water of steam power plant. The term is now widely used for any solid suspension or deposition in any liquid.

Cryogenic   Low temperature, usually taken as below minus 100 °C.

Cryostat   A reaction chamber or test vessel working at low temperature.

Crystal   A volume of solid material with all of its atoms aligned on a lattice (see Crystal structure). 'Crystal' is often used interchangeably with 'grain'. Unfortunately, there are no widely accepted rules to guide usage but, generalizing, 'crystal' tends to be favoured in the context of features on an atomic scale, for example crystal lattice, whereas 'grain' tends to be favoured for larger-scale matters, for example grain size. Note, however, '**Crystalline**' is used, when describing a fracture surface, to indicate a bright, angular, faceted appearance resulting from cleavage along specific crystallographic planes, characteristic of *Brittle* fracture.

Crystal lattice   See Crystal structure.

Crystal structure   The arrangement of atoms in crystals including virtually all solid metals. Atoms in crystals are arranged in simple, repeating, three-dimensional patterns rather than in the random jumble found in amorphous, i.e. non-crystalline, materials. Considering, first, pure metals, the basic unit of the pattern can be visualized as comprising atoms located at specific points on a simple geometric body, for example the corners of a cube. This basic unit is repeated for very large distances, measured on an atomic scale, and the atoms can then be regarded as lying in layers or on **Atomic planes**. The three-dimensional arrangement is termed a **Crystal lattice**. An individual **Crystal** or **Grain** has all its atoms and atomic planes in continuous alignment. All crystals of the metal will have the same pattern but the planes of neighbouring crystals will not be in alignment. The interfaces between grains are referred to as the **Crystal boundaries** or **Grain boundaries**. The various crystal structures can be illustrated as sphere models or wireframe models. **Sphere models** represent atoms as solid balls in surface contact, as shown in Figure 6, which illustrates three common structures. The **Close packed**

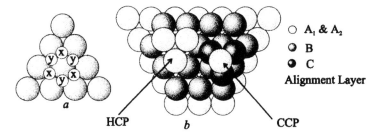

○ A₁ & A₂
◑ B
● C
Alignment Layer

In the close packed structures any atom is in contact with 12 others, six on its own plane, three each on the planes above and below. These sets of three, for the centre atom of sketch a, lie in the hollows of the first plane and can be at either the x positions or the y positions. Consequently, when layers build up, as in sketch b, there is an option as the third layer atoms take position on the second layer, B, atoms. If these third layer atoms lie above the first, A₁, layer atoms then the alignment sequence repeats as A₁ B₁ A₂ B₂ ... This is the Hexagonal Close Packed structure (HCP). Alternatively, if the third layer atoms do not lie above those on the first layer then a C alignment is formed. Alignment with the A₁ atoms is then re-established at the fourth layer so the alignment sequence repeats as A₁ B₁ C₁ A₂ B₂ C₂..... This is the Cubic Close Packed (or Face Centred Cubic) structure (CCP or FCC).

In the Body Centred Cubic structures (BCC) any atom is in contact with 8 others, four each on the planes above and below. The packing can not vary. The third layer atoms align with those on the first as in sketch c.

Figure 6   Sphere models of common crystal structures

structures are so called because the spheres are at maximum packing density, 74% of the equivalent solid volume, and each atom is in contact with twelve others, i.e. a **Coordination number** of 12, again the maximum possible. Each atom is in contact with six others on its own, 'A', plane and three others on the planes above and below. The atoms on the planes above and below nestle in hollows in the 'A' plane but, as shown in sketch a of Figure 6, there are six hollows so the trios of atoms can lie in either of two positions as the second, 'B', plane builds up. Consequently, when the third layer builds it has the option of coinciding or not coinciding with the 'A' layer. If it does coincide all atoms in the crystal will align with layer A or layer B

atoms, developing in the sequence: A B A B A B .... This is termed the **Hexagonal close packed (HCP)** or **Close packed hexagonal (CPH)** structure. However, if the third layer atoms do not align with the 'A' layer then a third alignment, 'C', is formed and the layer sequence develops as A B C A B C A B C .... This is termed the **Cubic close packed (CCP)** or **Face centred cubic (FCC)** structure. Also illustrated in Figure 6 is the **Body centred cubic (BCC)** structure. This is not close packed; each atom is in contact with eight others, a coordination number of 8, giving a density of 68% of the equivalent solid volume. **Wireframe models** represent atoms as lying at the intersections of a three-dimensional grid, as in Figure 7

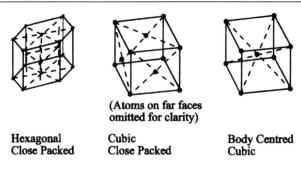

(Atoms on far faces
omitted for clarity)

Hexagonal   Cubic   Body Centred
Close Packed   Close Packed   Cubic

Above: '3-dimensional' models of
common crystal structures.

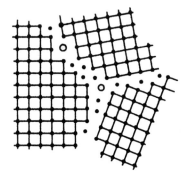

Left: '2-dimensional' model of
grains with unaligned and foreign
atoms in the grain boundaries.

Figure 7   Wire frame models of common crystal structures with the atoms at mesh
intersections

which illustrates the same three structures described above. Such models can
be more easily understood when extended and so are more useful than the
sphere models for describing long-range
features such as grain boundaries or dislocations. Where two or more metals
form an alloy the general principals for
individual grains are similar to those for
pure metals except that the atoms of the
second element can be located either in
place of the primary metal atoms or they
can be arranged in the gaps between
them. In most cases the sizes and other
characteristics of the atoms of the two
metals will differ considerably so there
is usually a limit to the number of secondary atoms which can be accommo-
dated on, or in, the lattice of the primary

metal. When this limit is exceeded a
second *Phase* will form. Also see Solid
solution, Ordering and Dislocations.

Crystalline (1)   Having the characteristics of a *Crystal*.

Crystalline (2)   A bright, angular faceted
fracture surface. Each facet corresponds
to the fracture path following a specific
crystallographic plane across each crystal. Such surfaces are normally characteristic of low *Ductility*, *Brittle fracture*.
See Crystal.

Crystallite   An imprecise term which
may mean no more than a very small
Crystal. Alternatively, the term may refer to sub-crystals within a grain or to
small crystals or parts of crystals which
do not exhibit all the characteristics of
the full crystal.

Crystallography The study of *Crystal structure.*

Crystobalite An *Allotrope* of quartz. See Quartz.

CTOD See Crack opening displacement.

CTS (1) Compact tension specimen. See Fracture toughness.

CTS (2) Controlled thermal severity (test), see entry on this topic.

Cubic close packed See Crystal structures.

Cubic structures Crystal structures having atoms located at the corners of a cube. Other atoms may be placed elsewhere in the cube. See Crystal structure.

Cumulative damage The total amount of damage resulting from a number of damage mechanisms or from a number of exposures to a mechanism. The term is commonly used in the case of *Fatigue*, reflecting the variation in stress range or other variables that a component might experience. The basic concept in all cases is that each exposure consumes a fraction of the fatigue life and failure will occur when these fractions total unity.

Cup and cone fracture The pair of fracture surfaces produced by *Tensile* failure of round (or nearly so) specimens of ductile material. One surface, the cup, comprises a flat central area and steep sides rising to a thin circular edge, often termed the *Shear lip.* The other surface, the cone, has a corresponding mating surface.

Cupel A thick base, shallow dished refractory vessel, usually of bone ash, on which *Cupellation* is undertaken.

Cupellation A technique for extracting gold and silver from a lead-rich alloy. The alloy is heated on a *Cupel*, oxidizing the lead which is absorbed into the cupel leaving the precious elements on the surface. The rate of oxidation is usually increased by an air jet.

Cupola A vertical shaft, air blast furnace, lined with refractory. It is used for remelting metals, in particular *Cast iron.* The charge of iron with coke or charcoal

as fuel and a *Flux* are introduced at the top; the molten iron and *Slag* are tapped at the bottom.

Cupped/cuppy The term applied to wire and rod with large numbers of curved, transverse, sub-surface cracks. They result from effects such as over-drawing or excessive impurities.

Cupping test A test in which sheet metal is deeply indented in a suitable die and plunger arrangement to measure its *Deep drawing* capability. The *Erichson* is one such test.

Cupronickel Copper-based alloys containing significant amounts of nickel. These alloys have good mechanical properties and good resistance to atmospheric, aqueous and marine environments. See Table 8.

Curie One Curie is the quantity of radionuclide in which the number of disintegrations in one second is 3.7×10^{10}, i.e., 1 Curie (Ci) $= 3.7 \times 10^{10}$ s^{-1}.

Curie point/temperature For a metal it is the temperature below which it is *Magnetic* (more strictly, ferromagnetic) and above which it is nonmagnetic (more strictly, paramagnetic). For iron it is 770 °C. It is also the temperature at which *Piezoelectric* materials lose their electromechanical characteristics.

Current density In *Electrochemistry*, the current, in amps, per unit area of electrode.

Current efficiency In *Electrochemistry*, the ratio of the quantity of material actually deposited or dissolved relative to the theoretical quantity predicted by Faraday's law.

Cut Apart from obvious meanings, it is the depth of material removed by a single pass of the cutting tool.

Cutlery steel Originally, approximately 1% carbon steel used for domestic cutlery and other blades but now usually 12% chromium, 0.3% carbon (martensitic stainless) steel.

Cutoff wheel A thin disc, coated or impregnated with abrasive, rotating at high speed and often water cooled, for cutting purposes. It can produce a narrow, clean cut with minimal surface damage even in hard materials. Also variously termed 'cutting wheel', 'slitting wheel', 'elastic wheel', etc.

Cutting compound/fluid/paste A substance applied at the cutting edge during machining. Where it is a stream of fluid its primary function is cooling but all these substances act as a lubricant and limit adhesion to the tool edge. They

also contribute to surface quality, tool life and precision.

Cutting speed The relative linear speed between the tool and the workpiece.

Cyaniding Heating steel components in a molten cyanide bath at about 900 °C to cause carbon and nitrogen to diffuse into the component. See Case hardening.

Cycle A series of events starting and finishing at an equivalent position or state.

D

Dalic process A portable electroplating system in which the plating solution is delivered to the workpiece through suitable conducting pads or brushes.

Dalton's atomic theory In essence this stated that: 'Matter is composed of atoms which are discrete, indivisible particles. All atoms of an element are the same. Chemical reactions are the result of atoms combining in simple proportions.' Although it has subsequently been demonstrated that atoms can be divided into smaller particles and *Isotopes* have been discovered, the general principles are the basis of chemistry.

Damascening The repeated folding and forge welding of layers of steel. Individual layers have varying carbon contents, deliberately or otherwise, which show as light and dark areas after polishing and *Etching* to provide a decorative finish. The word derives from Damascus, where the technique was used for high-quality swords and other blades. Generalizing, the greater the number of repetitions of folding and forging, the finer becomes the decorative effect and the stronger and tougher becomes the steel.

Damping Reducing the severity or duration of vibration.

Damping capacity The ability of a material to absorb vibration-induced strain energy and dissipate it internally. Materials having a low damping capacity 'ring' when struck, those with a high capacity sound dead and dull.

Dative covalent bonding See Interatomic bonding.

DBTT Ductile/brittle transition temperature. See Impact test.

DC Direct (electrical) current.

DCPD Direct current, potential drop, a technique for measuring crack propagation. See Potential drop.

Dead soft Fully annealed, having the lowest hardness that can be achieved in the material in question. In the case of steel the term also implies a low-carbon steel.

De-aluminification A form of corrosion affecting aluminium bronzes (see Table 8) in which the aluminium is selectively removed leaving a weak, porous, copper-rich material. Similar to *Dezincification* of brass.

Debye–Scherrer technique An X-Ray diffraction technique for analysing crystalline materials in powder form.

Decalescence The absorption of heat as a material undergoes an isothermal *Phase* transformation during heating.

Decarburization Loss of carbon from the surface of (ferrous) metals, by diffusion into, or reaction with, the environment. It is usually undesirable except in the case of certain treatments of *Cast iron*.

Decrepitation The crackling noise emitted during the *Decrepitation process*.

Decrepitation process The rapid heating of small particles so that the differential stress resulting from the severe temperature gradients causes the material to shatter into smaller particles.

Deep drawing Pressing operations producing cups or similar shapes having a large cup depth relative to the diameter.

Defect Generally, any deviation from perfection. However, in a metallurgical context it is usually implicit that the defect has significant consequences in the context as all metals contain defects if examined with sufficient rigour.

Defect lattice A *Crystal lattice* in which *Vacant sites* act in effect as foreign atoms and hence significantly influence the material properties and characteristics.

Deflagration Rapid combustion with noise and flames.

Deflection Flexure, as opposed to permanent deformation, under load.

Deformation A change in shape.

Deformation bands Same as Lüders lines.

Deformation twins Same as Mechanical twins.

Degassing Any process for removing gas from a solid or liquid. Heating allows hydrogen to diffuse from solid metals. Liquid metals may be degassed by, for example, adding powder to act as deoxidants or by bubbling through another gas which chemically reacts with or physically scavenges the unwanted gas.

Degaussing The elimination of all magnetic fields either by demagnetizing the material or by superimposing another field.

Degrees of freedom The independent variables, such as temperature and pressure, that can alter without causing a *Phase* change.

Dehydration The loss of water by processes such as natural evaporation or heating.

De-ionized water Water which has been passed through columns of anion and cation ion exchange resins to remove ionic contaminants. Such water will have low conductivity but may contain non-ionic material such as air, sand or bacteria.

Delamination The parting of layers.

Delayed elasticity Same as Elastic after-effect.

Delayed fracture The fracture, after a significant period of time, of a compo-

nent subject to a bulk stress less than its *Tensile strength*. The stress may be *Residual* or externally imposed. The usual cause is *Hydrogen embrittlement*.

Delayed hydrogen cracking Any cracking induced by the presence of *Hydrogen* remaining from some process but particularly after welding. See Weldability.

Delayed yield The period of a few milliseconds between the rapid imposition of a load exceeding *Yield* and the onset of yielding. It is the result of the time taken for *Dislocations* to disengage from lattice features such as *Interstitials*.

Deliquescence The absorption, by a solid material, of water to form a liquid.

Delta δ. One of the Greek letters used to refer to particular phases in various alloy systems. See Table 20.

Delta ferrite See Iron and Steel.

Dendrite The pattern of growth adopted by a *Crystal* during *Solidification*. It involves the progressive outward growth, in three dimensions, of solidifying branches and multiple sub-branches until the solid crystal is complete.

De-nickelification Corrosion in which nickel is selectively removed from copper nickel alloys leaving a weakened porous material. See Dezincification, a similar but more extensively researched mechanism.

Densification Various techniques for increasing the density of porous materials including compression and *Sintering*.

Density The mass of unit volume of material.

Dental alloys Various alloys used for the repair or replacement of teeth, including mercury amalgams, gold alloys and stainless steel.

Denudation Local *Depletion* of some element particularly when precipitates are formed.

Deoxidation Any process for removing oxygen from a material, particularly the removal of undesirable oxygen during melting and casting.

Deoxidizer A material which when added to molten metal combines preferentially with dissolved or combined oxygen and, usually, forms a compound which becomes part of the *Slag*.

Depletion Reduction, particularly locally, of the quantity of alloying element. The term is used in various contexts and may indicate the loss of an element by diffusion through the matrix over significant distances (on a microscopical scale), or it may refer to localized diffusion leading to precipitation of alloy-rich precipitates. The surrounding material is then left depleted of the alloy element.

Deposit attack Corrosion occurring beneath localized deposits and debris in an immersed environment. The deposit itself is inert but the area beneath it is low in oxygen relative to the surrounding surface and hence is relatively anodic. The metal surface beneath the deposit suffers *Electrolytic* corrosion and, since the *Anode* is small and the *Cathode* large, penetration can be rapid. A form of *Differential aeration corrosion*.

Deposited metal (of weld) *Filler* metal as opposed to fused parent metal.

Depth of fusion/fusion penetration (of weld) The depth to which the *Parent* material has fused, measured perpendicularly from the original surface. In the case of *Resistance welds*, it is the maximum perpendicular distance from the original interface to the boundary of the nugget. See Figures 50 and 51 located at entry on Welding terminology.

Descaling Removal of *Scale*.

Desiccant A material able to extract water or moisture from its environment.

De-stannification Corrosion in which tin is selectively removed from tin *Bronzes* leaving a weak porous material. See Dezincification, a similar but more extensively researched mechanism.

Deuterium An isotope of *Hydrogen*. Also see Atomic structure.

Developer See Dye penetrant examination.

Dezincification A form of corrosion in warm, moist environments affecting brass (copper–zinc alloys). The attack selectively removes the zinc, leaving weak, porous, redeposited copper of similar dimensions to the original. Localized, deeply penetrating attack is termed **Plug dezincification**; more general surface attack is termed **Layer dezincification**. Brass with less than about 15% zinc is not significantly affected and arsenic additions (0.02–0.05%) inhibit attack in the higher-zinc, single-phase, alpha brasses (up to 37% zinc). The alpha–beta duplex brasses can be made more resistant, but not immune, by the addition of tin, typically 1%, and arsenic up to 0.2%. The resistance of hot-worked, duplex alloys is improved by annealing at 525 °C followed by water quenching. *Cathodic protection* can be effective.

Diamagnetic See Magnetic.

Diamond A naturally occurring crystalline form of the element *Carbon*. It is the hardest known material, number 10 on the *Mohs* scale. Apart from its aesthetic value it is used for cutting, grinding and polishing.

Diamond (crystal) structure The crystal lattice structure of diamond and other materials. Each atom is in covalent bond with its four nearest neighbours in tetrahedral formation around it. It is characterized by high strength and rigidity and low electrical conductivity.

Diamond (pyramid) hardness testing Hardness testing using a pyramid-shaped diamond indentor. See Vickers hardness testing.

Die A master component which imposes its shape on other material. Manufacturing processes include *Drawing*, in which bar, tube or wire is pulled through the die orifice, *Extrusion*, in which material is pushed through a die, **Die stamping**, **Forming** or **Forging**, in which material is pressed between shaped die faces and **Die casting**, in which molten metal is poured or injected into re-usable shaped metal dies rather than into expendable sand moulds.

Die casting alloys Any alloy that can be *Die* cast but particularly various low melting point alloys based on aluminium or zinc and having metallurgical characteristics, such as fluidity and low shrinkage, that make them particularly suitable for this process.

Die lines Surface markings on components resulting from imperfections in the *Die*.

Die sinking The machining and engraving of detailed dies for coins.

Die steels Any steel capable of achieving the high strength required for a *Die* without cracking or unacceptable distortion. Such steels are usually high-carbon low-alloy grades with corresponding good *Hardenability*. For hot-working application higher levels of chromium (5% or more) or tungsten (10% or more) are common.

Dielectric A material that does not conduct electricity.

Differential aeration corrosion A form of corrosion resulting from variations in air, more specifically oxygen, content at different points on an immersed component. The *Electrolytic* corrosion occurs at the anodic, low-oxygen, areas. Such attack in crevices may be termed *Crevice corrosion* or, if beneath debris, *Deposit attack*.

Differential thermal analysis A technique for determining the temperature of *Phase* changes. A sample of the material in question together with a sample of another material having the same heat capacity are heated together and their temperatures monitored. At phase changes a major step in the temperature differential between the two is observed.

Diffraction The deflection of a beam of radiation by a grating to form interference patterns.

Diffusion The movement of individual atoms, ions and molecules relative to the remainder of the material. The movement of individual atoms is random but where there is a *Concentration gradient* and a large number of movements occur, there will be a net effect for atoms to move down the gradient. Ultimately,

over a sufficient period of time, a uniform composition will develop. The rate of diffusion is fastest in gases, slower in fluids but it is still significant in solid materials. Consequently, diffusion is a critical factor in many solid-state processes and phenomena. The rate of diffusion is accelerated by increasing the temperature.

Diffusion bonding Diffusion across the interface between solid contacting surfaces producing a bond of useful strength. Clean, close-fitting interfaces are vital. Heating is normally required as is a small amount of pressure. If the pressure is high, for example sufficient to disrupt surface films, the joint is more correctly termed a *Pressure weld*. See also Diffusion welding, which is very similar.

Diffusion coating The formation of a coating by exposing a component to an appropriate environment—solid, liquid or gaseous—from which another element diffuses into the surface to provide beneficial characteristics such as corrosion resistance. The term may include sprayed or electroplated deposits which are subsequently heated to allow diffusion.

Diffusion welding Any welding resulting primarily from solid-state diffusion across the interface of the intended joint. Some interfacial force may be applied but no significant deformation, apart from crushing of minor asperities, occurs. Heating to a temperature below the melting point of the parent materials may be applied. If diffusion produces a low melting point phase which becomes molten the process may be more precisely termed **Liquid phase diffusion welding**. If no molten phase is produced the process may be termed **Solid phase diffusion welding**.

Dilatometer An instrument for accurately measuring the dimensions of a metal particularly during temperature changes.

Dilution (of weld metal) The change in composition of the *Filler* material resulting from mixing with fused parent metal.

Dimethylglyoxime test A spot test to detect the presence of nickel in steel. A spot applied to the surface in question turns red in response to nickel.

Dimpled fracture A fracture surface texture characteristic of ductile fracture. It comprises multitudinous fine-cupped dimples each formed by a ductile shear fracture initiating at small discontinuities such as *Inclusions*. Examination in the *Scanning electron microscope* is usually required to identify such detail.

Dimples/ing Shallow indentations in a surface. They may be a manufacturing defect or they may be deliberate, for example to locate the heads of fasteners. See also Dimpled fracture.

DIN Deutsch Institut für Normen. The organization responsible for standards and specifications in Germany.

Dip brazing/soldering *Brazing/soldering* by immersing the components in a molten bath of *Brazing* metal/*Solder* or, alternatively, where a piece of braze metal has been pre-placed in the intended joint, in a salt bath.

Dip transfer (in welding) See Metal transfer.

Dipole A pair of equal and opposite but separated charges, for example a bar magnet.

Dipping Temporary immersion.

Direct extrusion See Extrusion.

Direct fired Furnaces in which the stream of fuel is injected into, and burnt in, the chamber and combustion products surround the charge.

Direct quenching The practice of quenching carburized components immediately upon removal at full temperature from the carburizing environment. See Case hardening.

Direct stress Simple tensile or compressive stresses developed by axial loads, as opposed to the more complex stresses developed by torsion or bending.

Directional properties Same as Directionality.

Directional solidification *Solidification*

in a manner such that the solidification front is continuously supplied with molten metal thereby avoiding shrinkage cavities.

Directionality The variation in properties observed when some materials are tested in differing directions relative to the axis of the material. The material may be cast, in which case any variation will merit this term. However, virtually all plate and sheet materials have inferior properties in the thickness direction, so in such cases the term usually refers to differences between properties along and across the material. See also Preferred orientation.

Disappearing filament pyrometer See Pyrometry.

Disc piercer See Tube making.

Discard Material scrapped from the tops of ingots, or the ends of extrusions and bars, etc., because of the inevitable deficiencies in quality at such positions.

Discontinuity Any defect in a material. Examples include, on a macro scale, a crack, porosity or inclusion or, on a microscopic scale, a foreign atom or vacancy in the *Crystal structure*.

Discontinuous yielding The variation in plastic deformation from point to point on a test specimen or a component in tension. It occurs because grains are randomly orientated so some will deform before the others. The effect is most obvious with a coarse grain size and in material with a pronounced *Yield point*.

Discrete In its metallurgical context this term usually refers to features, such as particles in a microstructure, that are individually distinct and separate from their fellows.

Dislocation A major, long-range, irregularity in the alignment of planes of *Atoms* on the *Crystal lattice*. Two forms of dislocation, the **Edge dislocation** and the **Screw dislocation**, are illustrated in Figure 8. A perfect crystal lattice would be extremely strong and resistant to *Plastic deformation*. However, the presence of dislocations reduces the

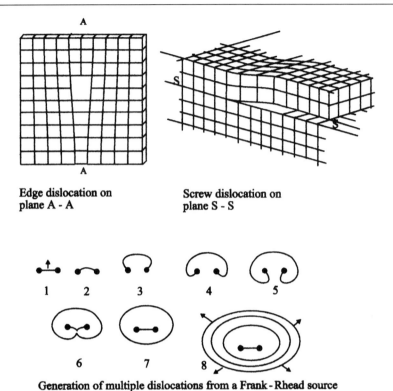

Edge dislocation on
plane A - A

Screw dislocation on
plane S - S

Generation of multiple dislocations from a Frank-Rhead source

Figure 8   Two types of dislocation and the repeated generation of additional dislocations

strength to about a thousandth of the theoretical level and facilitates plastic deformation. The principle can be visualized by considering two planes of atoms which are going to slide across each other. If the structures of the two planes match perfectly then movement will not occur until the load is sufficient to overcome the bonds between every atom and its neighbour on the next plane. However, if a line of atoms is missing from one plane than the load need only be sufficient to overcome the bonds on the single line of atoms next to the missing line. When this happens the missing line, i.e. the dislocation, will move one atomic space. The next line of atoms will then break and the dislocation will move a further atomic space. In

this way the dislocation will sweep across a crystal to leave a step with a height of one atomic space. See Figure 9. An analogy often quoted to illustrate the low stress needed is the requirement to move a long strip of carpet for a short distance. Simply pulling one end requires a very large load to achieve any movement. However, a rod, representing a dislocation, can be put under the width of the carpet at one end and then pulled beneath the carpet to the other end. The load required is small and the carpet will be moved forward by a small increment. The movement of an individual dislocation is measured in terms of its **Burger's vector** which identifies the total lattice translation resulting from the passage of the dislocation. Clearly, large numbers

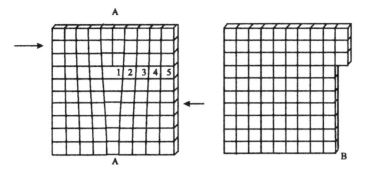

A dislocation at A-A, subject to the forces arrowed, allows slip in five
steps along the plane S-S leading to the deformation observed at B.

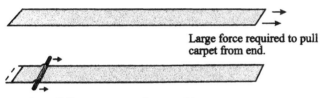

Large force required to pull
carpet from end.

A small force can move the carpet in
small incremental steps.

The top sketch is a grid representing a lattice with atoms at the intersections.
A dislocation in the lattice allows incremental movement at low loads as only
one line of atomic bonds is broken at a time. The effect is similar to that
observed when a long strip of carpet needs to be moved. As shown in the
bottom sketch, the localized hump in the carpet moves along as the rod is
pulled. The carpet is thereby moved forward a small increment by a load
much smaller than that required to pull the carpet as a whole.

Figure 9   Deformation by the movement of dislocations

of dislocation movements are necessary
to achieve any discernible deformation
but in practice metals contain, or can
develop, immense numbers of dislocations. The movement of a series of
dislocations along an individual plane is
termed **Glide**. However, as the dislocations move they become mutually en-
tangled or locked by grain boundaries
and other microstructural features. This
necessitates higher stresses to cause
further deformation. This is termed
'strain hardening'. Ultimately, further
deformation is impossible and an in-

creasing stress will initiate failure at the
most severe 'log jam' of dislocations.
As mentioned, although some dislocations are present in cast or annealed
material many others can be created
by the imposed stress and associated
deformation. The creation mechanism,
termed a **Frank–Rhead source**, comprises an initial dislocation anchored at
its ends. A shear load acting on the
dislocation causes it to bulge outwards
and then pivot round the anchored ends
until it forms a complete **Dislocation
loop**. In the process the original disloca-

tion re-establishes to initiate further loops (see Figure 8).

Dislocation climb A thermally activated, time-dependent mechanism in which the creation and diffusion of *Vacancies* allows a dislocation to move or 'climb' onto a neighbouring plane.

Dislocation pinning/locking The restraint on movement of a *Dislocation* resulting from its interaction with some microstructural feature such as a precipitate, *Interstitial* atom or another dislocation.

Disorder See Order.

Dispersion hardening/strengthening Hardening induced by numerous fine inert particles introduced into the material and evenly distributed throughout the structure including within the grains. The particles are not soluble in the matrix even if the material is melted and hence are not strictly an alloying addition in the usual sense of the term. However, they serve the same purpose as precipitates in that they can lock *Dislocations*, impede slip and increase strength, including *Creep* strength. In addition, as they remain insoluble, they can impede grain growth and even inhibit recrystallization, hence further enhancing high-temperature performance. The particles may be introduced either by addition to the molten metal in the form of an inert powder or by a *Powder metal* technique so that after processing the material comprises the metal matrix plus oxide particles derived from the oxide film which originally covered the powder.

Displacement-controlled loading A term applied to forms of loading during testing (or service) where the load is applied by a rigid structure of, effectively, fixed dimensions so that if the test piece extends after initial loading the load and *Stress* fall. An example is a high-temperature flange joint with bolts passing through it. Because of the difference in cross-sections the tightened bolts experience considerable tensile *Strain* but the flanges are hardly com-

pressed. As the bolts *Creep*, converting elastic strain to permanent creep strain, the tensile stress in the bolts falls so the closure forces on the flanges reduce, eventually leading to leakage. The alternative is **Load-controlled loading** in which the load does not vary as strain occurs. An example is a weight hanging freely on a rod. If the rod extends, the load due to the weight obviously does not change but if the rod cross-section reduces by creep, the stress increases.

Dissimilar metal joint In its general sense, any joint between different metals. Generalizing, such joints can be difficult to make by processes such as welding and often present problems in service because of differences in the chemical, *Electrochemical* and physical properties. The term is often used in a more specific sense regarding welded joints between austenitic and ferritic *Steel* pipework on power plant. The primary motivation to use two materials is cost; the low-alloy ferritic steels are relatively cheap compared with the austenitic steels but the latter have the superior *Creep* properties required for the higher-temperature zones. Sound welds are fairly easy to achieve but high-temperature operation in the 500–600° region causes considerable problems. Carbon tends to diffuse from the ferritic to the austenitic so the ferritic material is weakened; the expansion coefficients differ considerably so differential expansion and contraction causes large stresses which are superimposed on the stresses arising from internal pressure and, finally, the two materials have distinctly different oxidation characteristics. Collectively, these effects lead to premature creep damage, usually in the ferritic steel close to the weld and to a wedge-shaped oxide band penetrating from the exterior, again in the ferritic material close to the weld.

Dissociate Separate. Often, a reversible process or brief duration of separation is implied.

Dissolution The process of one material dissolving another.

Distribution ratio See Partition law.

Distortion An undesirable change in shape, often the result of *Residual stress*.

Divorced pearlite Pearlite whose original lamellar structure has broken down by diffusion, at an elevated temperature but below the lower critical, to form spheroidized carbides in a ferrite matrix. See Steel.

Dolly A massive support held behind material being struck. Its inertia results in the material deforming rather than just bouncing away. The dolly is not normally shaped to the intended profile.

Dolomite Calcium magnesium carbonate, $CaMg(CO_3)_2$, a mineral used for furnace linings and similar refractory applications.

Domain An area in a *Magnetic* material in which all the elementary magnets (electron spins) are in alignment.

Donor (element in semiconductor) See semiconductor.

Dopant (in semiconductor) See semiconductor.

Double anneal A double heat treatment cycle in which *Steel* is first subject to a form of high-temperature normalizing and then to a conventional anneal. The initial heating to well above the upper critical promotes *Homogenization* and coalesces or blunts manganese sulphide inclusions. Rapid cooling to just below the lower critical then inhibits the formation of a coarse grain size. The steel is then immediately reheated to just above the upper critical and finally slow cooled. This produces a fully softened fine-grained material.

Double skin A layer of metal on the surface of a casting separated from the underlying material by an oxide film. It is produced during casting when, after solidification commences at the mould surface, the molten metal recedes temporarily exposing the newly solidified material to the atmosphere. Although the molten metal returns to fill the mould the oxidized layer remains substantially intact as a major weakness.

Dowel A cylindrical pin set into corresponding holes in a pair of mating surfaces to maintain position and alignment.

Down time The period or total time that equipment or production facilities are not available. Planned maintenance activities may be included or excluded depending on site practice.

Downhand welding See Flat welding position, the usually preferred term.

Draft (angle) The slight taper on a mould or die to facilitate separation and removal of the product.

Drag See Cope.

Drag angle The angle between an arc welding *Electrode* and the perpendicular to the plane of deposit.

Drag lines/markings The characteristic linear steps or serrations on a *Thermally cut* face.

Dragout Material inadvertently dragged from a treatment solution or molten bath by the component or handling equipment.

Draw (1) See Drawing.

Draw (2) A large void within a casting.

Draw bench The machinery and associated equipment for *Drawing* tube, wire, rod, etc. It will typically include the table or bench, which carries the die block and replaceable die, the rollers or slides, the device for gripping the product (sometimes termed the **Head** or the **Dog**), the draw chain, the drive motor and associated equipment.

Draw head A set of non-driven rolls through which strip is drawn. The position of the rolls can usually be adjusted to form various simple angles or troughs. See also Draw bench.

Drawability The capacity of a material to be *Deep drawn*. Contrast with *Ductility* which is the measure of the capacity of a material to undergo *Drawing*.

Drawing (1) Processes in which a bar, tube, etc. is pulled through an orifice, termed a *Die*, reducing its cross-section and increasing its length. In practice the material may be pulled through a sequence of dies of progressively reducing diameter.

Drawing (2) *Forging* bar to a reduced diameter. In this context **Drawing-down** is more common.

Drawing (3) (USA mainly) *Tempering.*

Dresser (1) Various tools for hammering and forging operations. In most cases a final shaping operation is involved.

Dresser (2) A tool with a head formed from a number of freely rotating ridged discs. It is held against the working face of a rotating grindstone to dress, i.e. restore, the profile.

Dressing (of weld) Machining or grinding of the surface of a weld to modify its profile. It may be undertaken for cosmetic or technical purposes, for example to reduce the *Reinforcement* height, to flush it with the parent surfaces or to remove defects at *Toes*. See also Flushing.

Drift (1) A tapered bar pushed through a pair of holes to draw them into alignment.

Drift (2) A small plain-faced punch.

Drift test A test in which a hard tapered *Mandrel* is forced into a tube or hole to measure the material's capacity to distend without unacceptable damage.

Drill (1) A **Drill bit**, i.e. a rotating tool with cutting edges at its tip and flutes along the shank to lead away the *Swarf.* It is plunged axially into the workpiece to form, or **Drill**, a cylindrical hole.

Drill (2) A drilling machine, i.e. one in which a *Drill bit* is mounted to drill holes in a workpiece.

Drip test Any test in which the material under test is subjected to a liquid drip of prescribed composition, size and frequency to determine the response of the surface and any coating. The specimen is usually at a defined angle to the horizontal to ensure that the droplets run down the same track. It is usually accepted that the standard **Water drip** test uses a drip of 0.05% sodium chloride in water.

Drive fit/driving fit Same as *Force fit*

except that more force, possibly in a press, may be required.

Drop forging/hammering/stamping Originally processes in which the **Tup** or hammer head was allowed to fall under its own weight. Now the term may be applied to fast-acting powered vertical forges. The tup and anvil faces may be plain but usually carry *Dies* to shape the component.

Dross Oxides and other non-metallic impurities floating on molten metal.

Dry copper *Tough pitch copper* containing excessive quantities of copper oxide particles and hence of low *Ductility.*

Dry ice *Carbon dioxide* in solid form.

Dry joint (soldered or brazed) A joint with inadequate bonding due to the failure of the *Solder* or *Braze* filler to wet the parent metal surface. It results from contamination or lack of *Flux*. See also Cold joint.

Ductile crack/fracture Cracking accompanied by significant deformation and requiring continued input of energy to maintain its propagation.

Ductile irons Cast irons that have greater *Ductility* than normal grey cast iron, particularly spheroidal graphite irons. See Cast iron.

Ductile erosion Metal loss by mechanical abrasion leaving a rippled surface having, on a microscopical scale, discernible surface deformation.

Ductility The capability of a metal to undergo plastic deformation during tensile loading. It is measured in the *Tensile test* as the percentage reduction in cross-sectional area at the point of failure. Material that can be drawn to a point before failing in tension has 100% ductility.

Ductility transition, ductile to brittle transition (1) The temperature at which steel changes between *Ductile* and *Brittle* fracture modes. See Brittle fracture.

Ductility transition (2) Some arbitrary level of energy absorption measured by *Impact testing* and used as an acceptance criterion.

Dump test A test of *Malleability* in which a columnar specimen is compressed longitudinally. It is typically applied to bar for rivets or bolts.

Dummying An initial rough shaping in open dies prior to *Die forging*.

Duplex Having two components, particularly microstructures containing significant proportions of two *Phases*.

Duplex stainless steel Grades of *Stainless steel* with a microstructure of austenite and ferrite in a specific ratio and distribution. They contain high levels of chromium with nickel and, usually, small additions of molybdenum, for example 22% chromium, 6% nickel, 3% molybdenum. With further additions, e.g. copper and tungsten, the steel may be termed **Super-duplex**. Tight control of composition and manufacturing variables produces the required structure. These steels offer excellent corrosion resistance in certain circumstances. See also Steel and Stainless steel.

Duralumin, Dural A proprietary range of *Aluminium* alloys. Sometimes used as a generic term for any aluminium alloy which can be strengthened by heat treatment. See Precipitation hardening and Table 7.

Durville Process A casting process in which the mould is set above and is attached to the vessel containing the molten metal. The mould and vessel assembly is slowly rotated to allow molten metal to flow gently into the mould. This minimizes contact with air and reduces the risk of *Dross* entrapment.

Dwell A period of time spent at constant load during a test or operation which involves a series of load changes. For example, the term is used in *Thermal fatigue* testing where a load is applied, sustained for the dwell period which may be hours or days, and then removed. The cycle is normally repeated a number of times. The specimen is maintained at a high temperature while under load so it sees not only a *Fatigue* cycle but also suffers considerable *Creep*.

Dye Penetrant A liquid, containing a dye, capable of entering narrow cracks. In **Dye penetrant testing/examination** the components suspected of containing cracks are immersed in, swabbed or sprayed with the penetrant. After the component has been wiped clean any penetrant retained in a crack will seep out indicating the defect. Some dyes are **Fluorescent** and the component is viewed, after application of the dye, under ultraviolet light. Sensitivity is usually improved by a **Developer**, a powder or quick-drying coating, applied immediately after wiping off excess penetrant, which absorbs and acts as a contrasting background for the dye.

Dynamic hardness See Hardness testing.

Dynamic loading/testing (and similar terms) Testing, etc. under non-static conditions, for example vibration, fluctuating load or impact.

Dysprosium A metallic element, one of the *Rare earth* group.

E   The elastic modulus, the ratio of *Stress* to *Strain* in the elastic range. See Tensile test.

e   Symbol for *Strain*, the ratio of extension to original length. See Tensile test.

Ears/Earing   The undesirable deeply scalloped edge of *Deep drawn* components due to *Preferred orientation*. A typical cup-shaped component suffering from the defect will have four equally spaced deep scallops separating the ears which are aligned along and transverse to the rolling direction of the original sheet.

Earth   The non-powered return path of an electrical circuit. It may not be via the soil.

EB Welding   *Electron beam* welding.

Eccentric (1)   Not concentric. Most tubing produced by piercing and similar processes is significantly eccentric.

Eccentric (2)   A disc, the centre of which is offset from its axis of rotation.

Eddy current testing   A form of *Non-destructive testing* in which eddy currents are induced in the component under test. The basic principle is that the probe which induces the eddy current moves relative to the component, or vice versa. Any changes in the current induced in the component are detected by the probe and transmitted to the monitoring equipment to indicate the location and size of defects.

Edge dislocation   A line defect with its *Burger's vector* perpendicular to the line of dislocation. See Dislocation.

Edge joint (of weld)   A joint made between the edges of sheet or plate components particularly when they meet at an acute angle, usually less than 30°.

Edge/edging rolls   The rolls that control the quality, dimensions and shape of the edges of the plate being rolled. See Rolling mill.

Edge tool   A tool with one or more sharpened cutting edges.

Effective modulus   The *Elastic modulus* in circumstances where it is time dependent or non-linear. See also Secant modulus and Tangent modulus.

Elastic after-effect   The elastic *Strain* that develops with time in a component subject to a steady *Stress* below its *Elastic limit*. Compared with the immediately observed strain it is very small. After removal of the stress, a similar small amount of strain remains but disappears with time. See also Elastic hysterysis.

Elastic constants   These are the Tensile elastic modulus (Young's modulus) (symbol E), Compression modulus (K), Shear modulus (G) and Poisson's ratio (σ). See respective entries.

Elastic constraint   See constraint.

Elastic deformation/deflection   Dimensional changes that disappear upon removal of the stress. See Tensile test.

Elastic hysteresis   The phenomenon whereby the stress–strain curves on loading and unloading do not coincide other than at maximum and minimum stress. It reflects the energy consumed in the elastic strain process and is a consequence of the *Elastic after-effect*. See Tensile test.

Elastic limit   The maximum stress from

which a material will revert to its original dimensions upon removal of the stress. See Tensile test.

Elastic modulus The ratio of stress to strain in the elastic range. See Tensile test.

Elasticity The property of any material whereby it is able to return to its original shape or volume after being deformed.

Electric arc welding Any welding operation in which heat is developed by a electric arc. It is implicit that the heat generated is sufficient to melt the components being joined in contrast to *Arc brazing*. The arc may be struck between two non-consumable electrodes or between the component and one electrode which can be either consumable or non-consumable. Where the electrode is consumable it will be metal in the form of a short-length rod or a continuous wire (or tube). This metal forms part of the weld and is termed *Filler*. The rod or tube may have a coating (or tube filling) which, if present, contributes to the welding operation in various ways depending on the specific process. A primary role is to release a protective gaseous envelope around the arc and molten metal; it provides *Flux* to form a slag which removes contaminants from the molten weld metal; it may also contribute metal or alloying elements to the weld metal; it can modify the arc conditions and improve weld metal deposition characteristics. Non-consumable electrodes are of refractory materials, particularly tungsten, although carbon is occasionally still used in some processes. Metal non-consumable electrodes are normally carried in a water-cooled holder which also delivers an inert gas such as argon, helium or carbon dioxide (effectively inert with steel). The gas is injected into the arc area primarily to form a protective envelope around the arc and molten weld metal but also in some cases to modify arc and metal transfer characteristics. Although the electrode is not consumed, metal may be added to the joint by a separate

filler rod. See also entries on particular processes including Manual metallic arc, Tungsten inert gas, Metal inert gas and Submerged arc.

Electric furnace Any furnace with an electric power system as its primary source of heat, including electrical resistance, induction and arc.

Electric(al) resistance welding See Resistance welding.

Electric steel (1) Steel made by one of the electric melting processes. There is often an implication of good quality, particularly freedom from inclusions.

Electric steel (2) Steel having low magnetic hysteresis and hence used as sheet for the laminations of transformers and generator stators. Such steels are typically very low carbon with up to 5% silicon and are carefully cold rolled to maximize their magnetic characteristics.

Electric(al) strength The maximum voltage that an insulating material can sustain without arcing or breakdown allowing the passage of a current.

Electrical discharge machining The removal of metal by a high-frequency electric spark struck between the workpiece and an electrode. The workpiece is often immersed in paraffin or a similar material. Complex shapes, deep narrow holes and high precision can be achieved but surface damage due to the high local temperatures may occur.

Electrical resistance pyrometer See Pyrometry.

Electrical resistance strain gauge See Strain gauge.

Electrical resistance welding See Resistance welding.

Electrical resistivity The resistance to an electrical current offered by material of unit length and unit cross-section. It is measured in ohms but also see International Annealed Copper Standard.

Electricians' solder Soft solders used for electrical joints. Various alloys have been used for this purpose, ranging from the eutectic composition, about 63% tin and 37% lead, to much higher lead

grades sometimes hardened with antimony. See Soft solder.

Electrochemical equivalent   The mass
liberated by 1 ampere of current flowing
for 1 second. The electrochemical
equivalent of any element is a function
of its chemical equivalent which is its
atomic weight divided by the valence.
Silver, with an electrochemical equivalent of 0.001 183 grams, is the reference
against which the electrochemical
equivalents of other elements are calculated from the formula:

$$Ey = \frac{0.001\ 183 \times Ay}{Vy \times As}$$

where:

Ey = Electrochemical equivalent of ele-
     ment Y
Ay = Atomic weight of element Y
Vy = Valence of element Y
As = Atomic weight of silver

Electrochemical machining   Processes
utilizing *Electrochemistry* to selectively
remove material to shape a component.
The process may utilize a cathode of
simple shape which traverses the component in a pre-planned pattern to produce the required shape. Alternatively, a
cathode with a reverse image of the
required form may be progressively
moved towards the component. The concentration of current at the points of
closest approach then produces the required image on the anode. See also
next entry.

Electrochemical shaping   This term and
Electrochemical machining are often
used interchangeably but where they are
differentiated 'Machining' is usually recognized as the general term and 'Shaping' refers to the process utilizing a
reverse image cathode.

Electrochemistry   The study of the reactions occurring in a liquid electrolyte, i.e.
an ionic conductor, through which an
electric current passes. This current may
be externally imposed, it may arise between different metals that are in electrical contact directly as well as through the
electrolyte, or it may develop between

areas on a component that differ in some
respect such as composition, hardness or
environment. This electrically induced
activity is termed a **Voltaic cell**, **Galvanic cell** or **Electrolytic cell**. In such cells,
positively charged **Cations** (ions defi-
cient in one or more electrons) travel
through the electrolyte from the **Anode**
to the **Cathode** while a balancing flow of
negatively charged **Anions** (ions carrying one or more extra electrons) travel
through the electrolyte from the cathode
to the anode. The formation of metal
cations at an anode involves loss of
atoms from the metal surface, termed
Electrolytic corrosion, and their arrival
at the cathode can involve deposition, i.e.
Electroplating. The **Electrode potential** or **Electrochemical potential** is the
voltage developed between the pairs of
metals comprising the anode and cathode
of the cell and its value depends on the
individual metals involved. If a standard
electrode is selected, the relative potential of all other metals, in terms of posi-
tive/negative and size, can be measured
and tabulated. Conventionally, the table
is usually presented with the corrosion-
resistant **Noble** metals at the bottom and
the most reactive **Base** metals at the top.
Such a table is termed an **Electrochemical series** or **Galvanic series** although
the two are not identical. A metal may
have two positions in the series, an **Active** state high in the series and a **Passive**
state where it forms an impervious surface corrosion film which resists further
attack. The list below indicates the electropotential relative to the saturated calomel electrode, for various metals in sea
water. Typically, differences of more than
about 0.2 V may give rise to corrosion.

Material	*Voltage*
Magnesium alloys	−1.6
Zinc alloys	−1.1
Aluminium	−0.9
Aluminium alloys	−0.6 to −0.8
Mild and carbon steels	−0.7
12% Chromium steel, active	−0.45

18% Cr 8% Ni austenitic stainless, active	−0.42
Lead	−0.45
Tin and solders	−0.45
Copper and high-copper alloys	−0.25
Nickel and high-nickel alloys	−0.15
Silver	0
Titanium and high-titanium alloys	0
Gold	+0.15
Platinum	+0.15

Electrode (welding) An electrical conductor from which the current leaves the supply circuit to enter the weld zone. In the case of *Pressure welding*, often termed 'spot welding', the electrode is usually of a copper base alloy combing high conductivity with hot strength. An electrode for *Electric arc welding* is defined as **Consumable** or **Non-consumable** depending upon whether it is progressively consumed to provide *Filler* metal to be fused and incorporated into the joint. A **Covered** or **Coated** (consumable) electrode carries an external coating that contributes greatly to the welding process by providing flux, protective gases, filler or alloying elements, or materials that modify the arc characteristics and control the slag characteristics. A **Cored** or **Flux cored electrode** contains similar constituents within the tubular electrode. A **Sheathed electrode** has an external sheath surrounding the coating. There are various **Electrode classification** systems which define the coating, welding characteristics and sometimes the expected properties or other features. Often such classifications refer to the principal constituent of the coating. A **Rutile electrode** has a coating or core with a substantial proportion of titanium oxide, i.e., rutile. These are general-purpose electrodes, relatively easy to handle in manual applications. A **Basic electrode** has a coating or core which is substantially calcium carbonate and fluoride, i.e. chemically basic. Such coatings, suitably handled to keep them dry, release only small quantities of hydrogen compared with rutile, so basic electrodes are sometimes termed **Low-hydrogen electrodes** or, where the quantity of hydrogen is specified, **Hydrogen-controlled electrodes**. Basic electrodes are used in more demanding applications such as highly restrained joints in alloy steels. They tend to require more operative skill than rutile electrodes. **Cellulosic electrodes** have coatings or cores with high cellulose contents providing deep penetration and hence are sometimes termed **Deep penetration electrodes**. **Iron powder** and **Alloy powder** electrodes have coatings or cores contributing iron or alloy elements to the weld metal. Non-consumable electrodes are necessarily refractory conductors such as tungsten, tungsten alloy or carbon.

Electrode negative Direct current arc welding with the electrode connected to the negative pole of the supply.

Electrode positive Direct current arc welding with the electrode connected to the positive pole of the supply.

Electrode potential See electrochemistry.

Electrodeposition The deposition, by electroplating, of a substance onto an electrode for plating, forming, refining or extraction purposes.

Electroforming The deposition, by electroplating, of metal onto a shaped former or die. The die is coated with some conducting release agent so that when a suitable thickness of metal has been deposited it can be lifted off providing an image of the die. Highly complex and detailed shapes can be produced.

Electrogalvanizing Zinc coating of iron or steel by *Electroplating*. See also Cathodic protection.

Electrogas welding Similar to *Electroslag welding* except that the arc and weld pool are shielded by a gas rather than a slag and the arc is maintained throughout the process.

Electrohydraulic forming Processes in which thin components submerged in a liquid are formed by the shock wave produced by an electrical discharge between a pair of submerged electrodes.

Electroless Plating   Any plating in an electrolytic cell where the only electrical current arises from the activity of the cell itself rather than an external source. See Electrochemistry.

Electrolysis   A chemical change induced in an electrolytic cell by the passage of current. See Electrochemistry.

Electrolyte   A fluid capable of being ionized to carry an electrical current. See Electrochemistry.

Electrolytic cell   See Electrochemistry.

Electrolytic cleaning   The removal of scale or other contaminants by making the component an electrode of an electrolytic cell. See Electrochemistry.

Electrolytic copper ˙ One of a number of grades of high-purity copper, typically 99.95% or better, and hence having a high conductivity. They have in common an origin as *Cathode copper*.

Electrolytic corrosion   See Electrochemistry.

Electrolytic dissociation   The process in which a substance splits into positively and negatively charged ions. See Electrochemistry.

Electrolytic lead   A high-purity grade of lead, 99.995% or better, produced by electrolytic refining.

Electrolytic machining   The removal of material in a selective, controlled manner to achieve a predetermined shape in an electrolytic cell. See Electrochemistry and Electrochemical machining.

Electrolytic polishing   The removal, in an electrolytic cell, of small amounts of surface material for decorative or *Metallographic* purposes. See Electrochemistry.

Electrolytic protection   Same as Cathodic protection.

Electromachining   Same as Electrical discharge machining.

Electrometallizing/ization   The deposition, by electrochemical techniques, of a metallic coating onto a non-metallic and, usually, non-conducting substrate.

Electrometallurgy   An imprecise term referring to metallurgical processes which use electricity as a primary power source. In some contexts the term is limited to processes based on *Electrochemistry*, for example electroplating or electrolytic refining. Other contexts include processes where electricity is the critical power source, for example melting in an electric arc furnace. However, secondary uses of electricity, for example for lighting, motor power or instrumentation or even electric arc welding, would not be included.

Electromotive series   A list of the elements arranged in order of their standard electrode potentials. See Electrochemistry.

Electron   The negatively charged subatomic particle that (in a simple model) circles the nucleus of an atom. It has a rest mass of 9.11×10^{-31} kg and a charge of minus 1.6×10^{-19} coulomb. See Atom and Interatomic bonding.

Electron beam analysis   Same as Electon probe analysis.

Electron beam welding/cutting   Processes which utilize a focused electron beam as a heat source. The kinetic energy released by the electrons impacting the target produces very high temperatures giving excellent penetration with minimal distortion. The process normally has to be undertaken in vacuum.

Electron cloud   The valence electrons of a metallic crystal which are not tied to individual atoms but are shared as a general cloud. See Inter-atomic bonding, Band theory and Semiconductor.

Electron compounds   Phases having specific ratios of valence elctrons to atoms, each ratio having similar structures in different materials. **Beta** compounds with a ratio of 3:2 (i.e. three electrons to two atoms), **Epsilon** with 7:4, and **Gamma** with 21:13 are the most common examples. Table 20 gives Greek characters.

Electron density   The number of electrons in unit volume of material. Depending upon the context, the electrons counted may be either the total in the substance or merely the conduction electrons.

Electron diffraction analysis The study of crystalline materials by examining the diffraction patterns produced by electron beams reflected from or transmitted by the crystal.

Electron emission The ejection of electrons from a free surface.

Electron microscope Instrument using an electron beam, in a vacuum, to examine specimens. The image is displayed on a screen similar to a television set but is also readily photographed. In a **Transmission electron microscope** (TEM) the fixed beam travels through the very thin specimen. In a **Scanning electron microscope** (SEM) the beam scans rapidly back and forth across, and is reflected from, the specimen surface. Both instruments offer very high magnifications with resolution of up to about 1 Ångstrom compared with the 0.2 μm of the light microscope. The SEM has the major advantage, compared with the conventional light microscope, of having a very large depth of focus. Consequently, it can examine surfaces that are rough or inclined to the beam, for example fracture faces. Hence much metallurgical SEM work is performed at magnifications below 1000× i.e. within the range of the light microscope. See also Electron probe analysis.

Electron octet The eight electrons forming the filled outer valence shell of an atom. This configuration is very stable; the elements having it are inert and the molecules having it are stable.

Electron pair Two electrons having parallel but opposite spin axes to give a net magnetic moment of zero.

Electron probe (analysis) A facility on an *Electron microscope* for analysing small surface areas.

Electron spin The angular rotational motion of an electron which results in the magnetic moment.

Electronegativity The ability of atoms of an element to retain or gain

electrons in *Inter-atomic bonds* with other elements.

Electronic Phenomena and mechanisms that involve the activities of electrons on a subtle and individual scale and usually at low current. Examples include thermionic valves, transistors and equipment where such components are the primary feature such as television sets, computers and modern cash machines. Phenomena and mechanisms that involve electron flow in bulk, often at high current, are better termed 'Electrical', examples include motors and heaters although these may have control systems that are electronic.

Electrophoresis The movement of colloidal particles or other finely divided particulate matter suspended in a fluid, which is electrically charged by, and then responds to, an external electrical field.

Electrophoretic plating The deposition on an immersed surface of finely divided particulate or colloidal material by application of an electrical field.

Electroplating The application of a metal surface coating to a component in an *Electrolytic* cell carrying an externally applied electrical current. See Electrochemistry. The component forms the cathode; the material being deposited may form the anode or be derived from components of the electrolyte. Usually, the component is immersed in a bath but specialist techniques are available for plating local areas on-site using brush, pad or swabbing techniques.

Electropolishing The development of a smooth reflecting surface on a component by making it the anode of an appropriate electrolytic cell. See Electrochemistry.

Electroslag welding A process in which the two faces to be welded, usually the edges of thick plates, are set vertically and spaced with a wide gap between them. The bottom of the gap is closed by a starter plate and the two sides are closed by moveable water-cooled copper shoes. A granulated *Flux* is places in the enclosure and an arc struck between a *Filler* wire or wires and the starter plate. The arc melts the flux which then conducts the current and the arc is extinguished. The electrical resistance of the flux maintains the heating to keep the flux molten and melt the continuously fed filler. The molten filler beneath the flux layer fuses with the parent plates, the copper shoes commence moving upwards and a weld is formed between the two plates.

Electrostatic precipitator   Equipment in which a gas is passed across a high-voltage field to extract dust. The particles collect an electrical charge and are attracted to catcher troughs which direct them into hoppers.

Electrostriction   The dimensional change experienced by some materials in an electrical field.

Electrovalent bonding   See Inter-atomic bonding.

Electrowinning   Processes for extracting metals from ores by electrolysis. See Electrochemistry.

Element   A substance that can exist as a single atom. See Atomic structure and Table 16. There are 92 naturally occurring elements such as hydrogen, iron, chlorine, uranium, etc., plus a dozen or so man-made.

Elevation   In the context of engineering drawings it is a drawing of the appearance as viewed from the end (end elevation), or the side (side elevation), as opposed to alternative views such as plan (viewed from above) or section.

Elongation   The amount by which the original gauge length of a tensile test specimen extends, usually expressed as a **Percentage elongation**. See Tensile test.

Elutriation   A process for separating particles on the basis of their size by mixing them with a liquid and allowing them to settle. The larger particles settle first.

Embrittlement   Any phenomenon or mechanism which renders a metal liable to fail in a non-*Ductile*, i.e. brittle, manner.

In some contexts it indicates a susceptibility or cracking at some later date, in others it indicates that cracking is already present. See also Brittle fracture, Hydrogen embrittlement, Temper brittleness and Caustic embrittlement.

Embossing   The formation of a pattern on thin material by pressing between *Die* faces.

Emery   Aluminium oxide, particularly in its natural and possibly contaminated state and in finely divided form for use as an abrasive.

Emissivity   The rate of radiation from unit area of surface in unit time. It is normally necessary to specify circumstances such as wavelength, surface condition, temperature, etc.

En steels   The now-obsolete series of steels for engineering applications, detailed in editions of British Standard 970 prior to 1970. The current BS 970 adopts a different system. See Engineering steels.

Enamel   Generally, a paint or other coating that is claimed to be harder or smoother than normal. In a more restricted sense, a coating process in which the coated component is *Baked* to improve coating characteristics such as smoothness, bonding, hardness, etc. See also Vitreous.

End mill   A rotating cylindrical cutting tool with multiple cutting edges on its flat (or nearly so) end face.

End quench test   A test for assessing the *Hardenability* of steels in which a standard bar specimen, usually 1-inch diameter, is heated to the austenitic range and quenched on one end face by a standardized jet of water. This produces a variation in cooling rate along the length of the bar. Measurement of hardness along the bar then indicates the hardenability characteristics of the steel.

Endothermic reaction   A chemical reaction in which heat is absorbed.

Endurance   (Limit)   The maximum stress amplitude that can be tolerated to survive a defined number of fatigue cycles, often 10 million. An endurance limit may also be specified in the case of steels subjected to stresses which exceed their fatigue limit. See Fatigue.

Endurance ratio The ratio of fatigue limit to ultimate tensile strength. For ferritic steels of UTS up to about 1100 MPa (75 Tons/in^2) it is approximately 0.5. See Fatigue.

Energy barrier See Potential barrier.

Energy product See Magnetic.

Energy trough See Potential barrier.

Engineering steels In the UK, engineering *Steels* have been specified in BS 970 for many years. Prior to 1970, this Standard listed in a not entirely logical manner a range of steels, termed the En series, with individual steels numbered after the En prefix, e.g. En 8 or En 25. Despite the limited logic the system was deeply ingrained and has been slow to die, so steels are still occasionally referred to by these old descriptions. The 1970 revision of BS 970 developed, in Parts 1 to 4, a more logical system with steels having a six-character designation comprising three initial digits, a letter and two final digits, for example 080M40, which is explained below. These designations indicate, with some anomalies and to a limited extent, the type of steel, its condition of supply and its composition. The first three digits indicate alloy type as follows:

001 to 199 are carbon or carbon–manganese steels with the figures being 100 × the mean manganese content range, i.e. 0.8% in the example above.

200 to 240 are other carbon–manganese, free-cutting, high-strength low-alloy steels, the second and third figures indicate the mean or minimum sulphur context × 100.

250 is Carbon–manganese silicon spring steel.

280 is microalloyed steels.

300 to 399 are austenitic stainless steels essentially based on the AISI series.

400 to 499 are ferritic and martensitic

stainless steels essentially based on the AISI series.

500 to 999 are alloy steels other than stainless varieties.

The fourth character, a letter, indicates the condition of supply as follows:

A—Close limits on chemical composition

H—Hardenability requirements

M—Mechanical property requirements, as in the example above

S—Stainless grades

The final digits are, for most grades except the austenitics, 100 times the mean of the carbon content range, i.e. 0.4% in the example above. Where the steel has an 'M' in its designation signifying that mechanical properties are a condition of supply there may be an additional letter indicating these properties. For example, the steel above could be Steel 080M40 Condition R. This is a carryover from the earlier obsolete BS 970 and can be summarized as follows. Condition N—normalized condition, mechanical properties vary with individual steel.

Condition	UTS min. MPa
P	550
Q	625
R	700
S	775
T	850
U	925
V	1000
W	1075
X	1150
Y	1225
Z	1550

Enthalpy The total internal and external energy content of a system.

Entropy The thermodynamic measure of the state of disorder in a system.

Environmental stress cracking (alternatively Stress cracking) Cracking occurring as a result of the effect of a corrosive environment and stress acting concurrently. This term is common with reference to polymeric (plastic) materials but is occasionally used in the case of metals as an alternative to *Stress corrosion*. The stress may be externally imposed or it may be residual from manufacturing operations.

Epitaxis/epitaxy The growth of a new *Crystal* at the surface of a pre-existing crystal where the *Lattice* orientation of the new crystal is dictated, across the intervening boundary, by the lattice of the pre-existing crystal.

EPNS Electroplated nickel silver. Silver electroplated onto *Nickel silver* for cutlery and similar applications.

Epoxy resins A range of high-strength polymers (see Plastics). They are typically of high strength, high electrical resistance and have good resistance to chemical attack. Depending on individual formulations thay can be used in bulk, as a coating or as an adhesive.

Epsilon compound See Electron compound.

Equiaxed Having approximately the same dimensions in all directions, particularly *Grains*.

Equicohesive temperature The temperature at which damage development within *Grains* is matched by that at grain boundaries for a given *Strain* rate. Above this temperature a material fails in an intergranular mode, below in a transgranular mode.

Equilibrium The state in which all changes are in balance and no further change will occur unless some additional external factor is applied.

Equilibrium diagram A diagram for an alloy system displaying the relationship, under equilibrium conditions, between composition, temperature and pressure (the latter usually ignored in the case of metals) and the various phases. Also termed 'phase diagram'. See Phase and Lever rule for more information, Eutectic and Steel for examples.

Equilibrium potential The potential of an unpolarized reversible electrode.

Equilibrium segregation The tendency for solute atoms in a solid solution to diffuse to particular sites such as grain boundaries. This may result in depletion of the solute in the near-grain boundary regions. See Solution.

Equivalent/equivalence factor Various factors applied to alloying elements to allow comparison of their relative effect on some aspect such as microstructure or hardenability. See Carbon equivalent as an example.

Equivalent section/round In the context of *Hardenability* of steel, it is the circular section equivalent to the noncircular section under consideration.

Equivalent weight The mass of a substance that will react, directly or indirectly, with unit mass of hydrogen.

Erg The unit of energy in the CGS (centimetre–gram–second) metric system; it is the work done by a force of one dyne moving one centimetre. $1 \, \text{erg} = 10^{-7} \, \text{J}$.

Erichson (cupping) test A test for determining the *Ductility* of sheet metal particularly for cupping or *Deep drawing* applications. The sheet under test is held by a hollow ram against a die face having a central hole. A ball-ended cone plunger in the bore of the ram is then forced into the sheet until a crack appears. The height, in millimetres, of the cup at crack initiation is a measure of ductility.

Erosion Removal of material from a surface by the abrasive action of a fluid with or without entrained particles. If there is a corrosion contribution the mechanism is termed **Erosion–corrosion**. See also Impingement and cavitation.

ERW Electrical resistance welding. See Resistance welding.

Etch figuring In *Metallography*, deliberate *Etching* to an extent that would normally be excessive to produce a pattern of pitting which reveals some feature of interest. Such features may be on a micro scale, for example the facets of the pits can reflect crystallographic planes, or on a macro scale as the distribution of pitting can indicate bulk *Segregation* or flow.

Etch pits Pitting resulting from excessive *Etching* unless *Etch figuring* is intended.

Etching Processes exposing metals to aggressive environments to cause selective attack of the exposed surface. This may be for decorative purposes or, in *Metallographic* examination, to reveal microstructural features. Unless otherwise specified the term usually implies chemical attack in a liquid. Other techniques include electrochemical dissolution, selective oxidation in air at elevated temperature and ion bombardment.

Eutectic, eutectoid A eutectic is an intimate mixture of two *Solid solutions* formed, on a solidifying from a single liquid solution, at the eutectic temperature. Metals mix together in different ways in both the liquid and the solid states. One simple case is where two metals are completely soluble in each other when fully liquid but form two solid solutions when fully solid. This is a **Eutectic system**, an example of which is the lead–tin system illustrated in Figures 36, located at the entry on Soft solder. The pure metals melt and freeze at specific temperatures but most mixtures of the two melt and freeze over a range of temperature. As increasing amounts of the secondary metal are added to the primary constituent there will be a continuous reduction in the *Liquidus* temperature, that is, the temperature at which the alloy ceases to be fully liquid. In most cases the cooling alloy does not solidify at a single temperature but passes through a pasty stage in which solidification progresses until at the *Solidus* temperature the alloy is fully solid. For compositions beyond the limit of solid solubility the solidus temperature is constant so, as more of the secondary metal is added, the falling liquidus temperature will meet the solidus. The composition at this point is termed the **Eutectic composition** and the temperature is the **Eutectic temperature**. All liquid remaining when the eutectic temperature is reached must transform before the temperature can fall further so a eutectic is an example of an *Isothermal* transformation. The eutectic composition is the only one in this alloy system, apart from the pure metals, which solidifies at a single temperature rather than over a range. The **Eutectic structure**, under the microscope, usually appears as an intimate mixture of the two solid solutions deposited in distinct patterns, for example alternate platelets. Similar reactions occur in solid metals when a solid solution, on cooling, forms two other solid solutions. This is termed a **Eutectoid** reaction and an example is shown in the iron–carbon diagram, Figure 37, located at entry on Steel.

Eutectic welding/brazing Joining operations utilizing a *Filler* metal which is a *Eutectic* composition of the two different metals being joined and so has a significantly lower melting point than either.

Eutectoid steel A steel with a *Eutectoid* composition, i.e. about 0.86% carbon, depending upon other alloying elements. See Steel.

Evaporation The formation of a gas from a liquid or solid (although direct evaporation from the solid may be termed 'sublimation'). In a metallurgical context the term often refers to coating processes in which, in a vacuum, a metal is evaporated on a hot filament and deposits on a cold target component. The component may be metal or other material and the thin coating is typically bright and lustrous.

Excess (of weld) Weld metal that stands proud of the straight line joining the weld to parent metal *Toes*. See Figures 50 and 51, located at entry on Welding terminology. Some authorities prefer this to the alternative term, 'reinforcement', but others consider that it has an undesirably pejorative implication bearing in mind that some overfilling is normal, usually fully acceptable and preferable to insufficient filling.

Excitation (1)   The promotion of an electron to a higher than normal energy level resulting from some energy input.

Excitation (2)   The delivery of electric power to a coil or similar system which in turn induces a current in an associated part. For example, an **Exciter** produces power which is delivered to the rotor of an electrical generator which in turn induces the main power output from the generator stator.

Exfoliation, exfoliating corrosion   The repeated loss of corrosion product, as thin layers, from a metal surface. Sometimes termed **Lamellar corrosion**.

Exothermic (reaction)   A chemical reaction which releases heat.

Expanded metal   Sheet metal in the form of narrow strips surrounding approximately diamond-shaped holes. It is manufactured by punching rows of short slots across the width of a sheet, each row being staggered by half a pitch. The sheet is then pulled apart length ways to form the holes. This twists the narrow strips and in some cases the expanded sheet is rolled again to flatten them.

Expanding metals   Alloys, usually based on *Bismuth*, which expand rather than contract on solidification.

Expansion (1)   The characteristic of most metals to increase dimensions as they increase in temperature.

Expansion (2)   Any process of increasing some dimension, for example a tube end required to mate with a fitting may have a tool inserted into the tube bore to force it open.

Expansion joint   A joint between two components which is capable of accommodating the anticipated thermal expansion. It may be merely some elastically compressible material, a bellows or more complex mechanical device.

Expansion rollers   Rollers that support a beam, bridge or similar mechanism and allow it to expand freely relative to its surroundings.

Explosive forming   Processes in which the energy released by detonation of an explosive is utilized to shape a compo-

nent. The process is usually undertaken in submerged expendable vessels. Strain rates are very high but, where the material is suitable, large, complex and accurate shapes can be produced in simple cheap moulds.

Explosive nail   A nail that is driven into the surface by an explosive cartridge carried in the nail holder.

Explosive rivet   A rivet that after being set through the pre-drilled holes is clenched by detonation of a cartridge within its shank.

Explosive welding   A form of *Pressure welding* in which detonation of an explosive causes two components to impact together forming a joint. Typically, one component is substantial and fixed and is impacted by a relatively light plate termed the **Flyer (plate)**. In practice careful selection of the explosive and the set-up are critical. In particular, the angle and rates of approach of the two surfaces and the detonation rate of the explosive have to be such as to ensure that a liquid jet is formed at the interface. This has a scouring effect, removing surface contamination and allowing intimate mixing of the two surfaces. See Figure 10. The process is particularly suitable for *Cladding* but has been used for installing tubes in tube plates.

Extensometer   A device for measuring linear dimensional changes; high precision is implied.

Extra hard/spring   Grade designations for the highest strengths of copper commercially available.

Extraction   The first stage in obtaining a metal from the ore in which it is chemically combined with other elements. The result is typically impure and requires refining.

Extraction replica   A technique for producing samples for *Electron microscopy* in which a coating, often carbon (hence the term **Carbon extraction replica**), is applied to the prepared surface. The specimen is then etched, often electrochemically, in a reagent chosen to selectively attack the matrix leaving behind

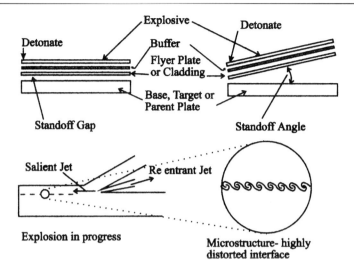

Detonate — Explosive — Buffer — Flyer Plate or Cladding — Detonate — Base, Target or Parent Plate — Standoff Gap — Standoff Angle — Salient Jet — Re entrant Jet — Explosion in progress — Microstructure- highly distorted interface

The two initial set-up options are shown at the top. In either case the process is the same, shown in progress at bottom left with the final structure at bottom right.

Figure 10   Explosive welding

the particles of interest on the carbon film.

Extrados   An architectural term referring to the upper, outer, curve of an arch. It is sometimes used in engineering circumstances to refer to the exterior surface of a bend in a bar or, particularly, a tube. The inner face of the bend is the *Intrados*.

Extrapolation   Various techniques, graphical, mathematical, etc., for predicting the properties of a material beyond the range established by testing. For example, long-term creep behaviour is often predicted from a series of short-term tests.

Extreme pressure lubricants   Lubricants capable of maintaining separation of the two metal surfaces under very high pressure particularly when the flow of liquid lubricant is interrupted. They include solids such as graphite or molybdenum disulphide or liquids which develop some form of solid film on the bearing surface.

Extrinsic semiconductor   See Semiconductor.

Extrusion   (1) Various processes in which a metal *Billet* is formed into lengths of bar or tube etc., having a uniform cross section. Figure 11 shows the various systems described below. In the basic extrusion process the billet is contained within a cylinder and forced, usually by a hydraulic ram, though an open *Die* shaped to the required product profile. In **Direct extrusion** a hydraulic ram enters at one end of the cylinder to push the billet forward through a die in the far end. In **Inverse** or **Indirect extrusion** the die is located in the head of the hollow ram and the billet, located against the closed cylinder end, does not move and so avoids *Extrusion defect*. Tubes can be produced by setting a mandrel in the ram or, particularly for complex hollows, by **Port hole dies** or **Bridge dies** in which the billet divides into a number of streams to pass over the locating system for the bore former. The streams then merge prior to passing out through the die and over the former. In **Impact extrusion** a small blank of metal is set in a die and impacted by a punch. The metal may emerge through an orifice in the die or may it may flow back up the side of the punch, as in the manufacture of toothpaste tubes.

Extrusion (2)   Bar or tube or other sections produced by the extrusion process.

Extrusion (3)   The microscopically small projections or steps of material on the surface of components subject to fatigue. They are produced when repeated reversal of *Slip* causes material to be squeezed from the surface.

Extrusion defect   Defects appearing in the last portion of an *Extrusion* to emerge from the die. It is a characteristic of direct extrusion caused when the contamination on the exterior of the

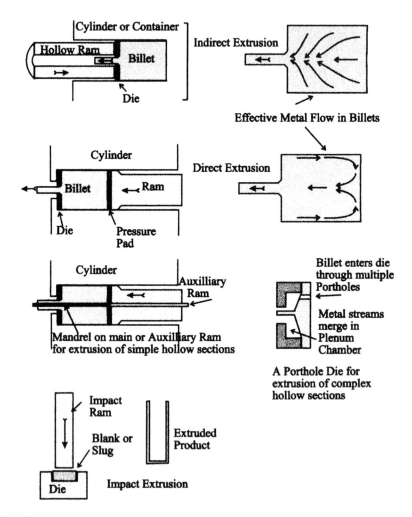

Figure 11   Extrusion processes

billet is effectively held back by friction against the cylinder wall and is then swept across the face of the ram to emerge initially at the centre of the extrusion section. As extrusion continues the defect forms an annulus progressively moving towards the exterior surface. See Figure 11. Normally the last 10–20% extruded from each billet is discarded because of this phenomenon. It is also termed **Back end defect**.

Eyeletted hole   In an engineering context this usually refers to a hole with a raised integral rim. Apart from any decorative aspect, a correctly formed rim can reduce chaffing on material passing through or across the hole, improve the stiffness of sheet and reduce *Fatigue* susceptibility. More generally, the term may refer to a hole which is protected by an added eyelet as in the eyelet applied to the lace holes of shoes.

F

Face centred cubic A crystal structure in which *Atoms* are close packed, i.e. in contact with six others in the same layer and with three others on the layers above and below, and in which the layer alignment repeats every third layer. Close packed hexagonal is similar except that the layer alignment repeats on alternate layers. See Crystal structure.

Facet A flat, usually brightly reflective, portion of surface on a *Crystal* or fracture face.

Facing (1) A machining cut made across the end face of a rotating bar.

Facing (2) Some form of surface coating, for example the fine sand or slurry applied to a mould to improve finish.

Factor of safety See Safety factor.

Faggot A bundle of rods or bars. The term may imply some careful, tidy arrangement. **Faggot(ed) iron** was the result of rolling or forging carefully packed bundles of *Wrought iron.*

Fail safe The concept whereby it is recognized that if a failure occurs it will do so in a manner that does not involve hazard although there could be acceptable economic cost. In some contexts the term implies that the mode of failure is acceptable, as in the case of the braking system on some vehicles in which the brakes are held off against spring pressure so that if the pressure system leaks the brakes are applied. In other cases it implies that in the event of a failure some secondary systems will intervene or provide support before a hazard develops. See also Leak-before-break.

Falling weight test A crude form of *Impact testing* in which a weight is allowed to fall on to the centre of a bar or rail, etc., supported at its ends.

False brinelling Local surface depressions, corresponding in location and shape with point loads, caused not by overload but by *Fretting.* It is sometimes observed on the race surfaces of ball bearings which have been under load and apparently stationary but subject to repeated low-amplitude movement or vibration. Contrast with Brinelling.

Faraday's laws (of electrolysis) No. 1: The amount of material deposited or dissolved is proportional to the quantity of current that passes, quantity being current × time. No. 2: The amounts of different substances deposited or dissolved by a given quantity of electricity are proportional to the *Chemical equivalents* of the substances.

Fast fracture Any fracture occurring virtually instantaneously as opposed to mechanisms such as fatigue and creep which develop over a period of time. The term is also often used loosely as a synonym for *Brittle,* as opposed to *Ductile,* fracture although reference can equally be made to 'fast ductile fracture'.

Fast neutrons Neutrons with energies greater than 0.1 MeV.

Fatigue The development of one or more cracks by the repeated application and removal of a *Stress* significantly less than the tensile strength of the material. Components are normally expected to fail at a stress equivalent to their ultimate tensile strength (see Tensile test). However, if a component is subjected to repeated cycles of application and removal of a lower stress it may ultimately fail by fatigue. Damage occurs by the initiation of a fatigue crack, usually at the component surface, which propagates over a considerable period of time until the remaining intact section is insufficient to carry the load. The damage is often highly localized at major features such as manufacturing or welding defects or section changes. 'Fatigue' or colloquially 'metal fatigue' does not mean that the bulk material away from the crack has deteriorated in any general sense. Many millions of cycles of loading, perhaps over years, may be required to cause failure and, generalizing, in the absence of large initiating defects a considerable proportion of the total component life is occupied in developing a crack to a detectable size. Once initiated, cracks usually propagate at an increasing rate. The number of cycles to failure, in specified circumstances of loading, is termed the **Fatigue life** and, as might be expected, the higher the stress, the less

the number of cycles required to cause failure. Consequently, failure occurring within a few thousand cycles is conventionally termed either **Low cycle fatigue** or **High strain fatigue**. If more than about 10 000 cycles are required the mechanism can be termed either **High cycle fatigue** or **Low strain fatigue**. The fatigue properties of a material are commonly determined by testing a series of test pieces over ranges of stress (more accurately, stress amplitude as defined below). The results, in terms of the number of cycles to failure at each stress amplitude, are then plotted on an S/N **curve** (see Figure 12). For most steels there is a level of stress amplitude below which a component will survive an infinite number of cycles. This stress is usually termed the **Fatigue limit**. The ratio of fatigue limit to the tensile strength is termed the **Fatigue ratio** and is typically about 50% for steels up to about 1100 MPa (75 tons/in^2) UTS. For many other metals a fatigue limit cannot be defined as failure will eventually occur if sufficient cycles are experienced. In these cases it is normal to determine an **Endurance** (or

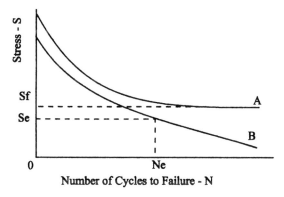

Curve A, characteristic of steels, shows a Fatigue Limit - the Stress at Sf.
Curve B, characteristic of many non-ferrous metals has no Fatigue Limit.
The Endurance for metal B is a stress of Se for Ne cycles.

Figure 12 Typical fatigue curves, often termed S/N curves

Endurance limit). This is the maximum stress amplitude that can be tolerated to survive some specified number of cycles, often 10 million. **Endurance ratio** is then the ratio of endurance to the UTS. The term 'endurance' may also be used of steels, possibly as an alternative to 'fatigue limit' but particularly where the stresses exceed the fatigue limit. Up to this point it has been assumed that the fatigue cycle is very simple, merely the application and removal of a load. However, the cycle is often more complex than this. There is always a cyclic stress but it may range from tensile to compressive and there may also be an additional constant stress. In these complex circumstances the various stresses are normally resolved into the **Mean stress** and the **Stress amplitude** or **Stress range** (merely double the stress amplitude). These and other related terms can be illustrated by the following examples and also Figure 13. A crane cable experiences a **Repeated stress** or **Pulsating stress** ranging from nil when unloaded (ignoring the hook and the cable self-weight) to high tensile when loaded—an example of high-amplitude, moderate mean stress loading. A lift (elevator) cable experiences a **Fluctuat-**

ing stress comprising a constant stress due to the cabin plus an additional stress due to passengers—an example of high mean stress and medium-amplitude loading. A clock hair spring is deflected backwards and forwards either side of its unstressed rest position so a point on its periphery experiences an **Alternating stress** or **Reversed stress** ranging from highly tensile to highly compressive—an example of high-amplitude, zero mean stress loading. The combined effect of the mean stress and the cyclic stress can be determined from various empirical relationships established, with varying degrees of conservatism, by Goodman, Gerber or Soderberg as illustrated in Figure 14. Reference may also be made to the **Stress ratio R** which is the ratio of minimum stress to maximum stress or to stress ratio A which is the ratio of alternating stress amplitude to mean stress.

FATT See Fracture appearance transition temperature.

Faying surface A surface that contacts another surface, particularly when the two are to be welded together.

FCC Face centred cubic. See Crystal structure.

Feed (1) The material entering a process or its rate of introduction.

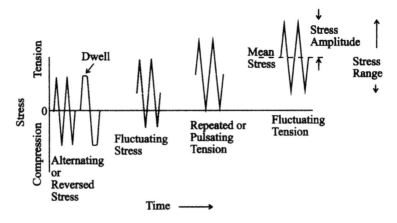

Figure 13   Fatigue loading systems and terms

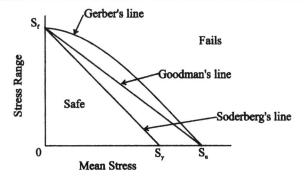

S$_f$ - Fatigue limit at zero mean stress.
S$_u$ - Ultimate tensile strength.
S$_y$ - Yield strength

Figure 14   The various diagrams for determining the combined effect of a static load and a superimposed cyclic load

Feed (2)   The rate at which material is cut. Depending on context it may refer to the depth of cut or the rate at which a tool traverses a surface or a combination of the two.

Feeder   A channel through which molten metal is fed to a *Casting*.

Feeder head   A reservoir of material that remains molten to feed *Shrinkage* in the solidifying casting below.

Feeler gauge   A strip of metal of high precision thickness to measure or set gaps.

Feldspar   A mineral comprising aluminium silicate with other silicates, usually barium, calcium, sodium and potassium.

Felt metal   A substantial felt fabric reinforced with fine metal strands. The strands may be loosely woven or bonded for improved strength. The term is sometimes loosely extended to similar materials where the reinforcement is plastic.

Ferric oxide   Haematite, the reddish brown iron oxide, Fe_2O_3.

Ferrite (1)   Iron or steel having a *Crystal lattice* structure of *Body centred cubic* form. See Steel.

Ferrite (2)   Iron compounds similar to magnetite, Fe_3O_4, in which some of the iron is replaced by other metallic elements to give a material having significant ferromagnetic characteristics.

Ferritic steels   Any steel in which the predominant phase is ferrite. Depending on the context, the term may be used merely in contrast to austenitic steels in which case it will include pearlitic steels and even steels capable of forming martensite on fast cooling. Such steels range from virtually pure iron to complex high alloys. In other contexts the term will be confined to steels which have a microstructure entirely of ferrite, a low carbon content and which cannot be hardened to any significant extent by fast cooling. However, some of these steels may contain alloying elements in considerable quantities, for example the 12% chromium, low-carbon steels are sometimes termed 'ferritic stainless steel'. See Steel.

Ferro-alloys, as in **Ferro-chrome** or **Ferro-manganese**   An iron master alloy containing large quantities of the indicated alloying element for addition to the molten metal during iron and steel alloy production.

Ferrograph   A device for measuring and displaying on a screen the magnetic characteristics of a material, particularly hysteresis curves.

Ferromagnetic See entry on Magnetic.

Ferrous Iron-based materials.

Ferrous oxide Black iron oxide, FeO.

Ferrule A cylindrical reinforcement or sleeve.

Fettling Various processes for improving or restoring a component or equipment. For example trimming *Flash* or other excess material from castings or locally repairing components, furnaces linings, etc.

Fibre (1) The variation in mechanical properties, structure, etc. observed when wrought material is tested or examined in different directions relative to the primary direction of working. For example, plate has a greater strength and ductility in the longitudinal direction than in the thickness direction.

Fibre (2) Filaments. Thus **Fibre metal** comprises metal fibres packed and *Sintered* together possibly with another material filling the interstices.

Fibre stress The stress acting along the main axis of a fibre. The fibre may be an individual filament or it may be a local unit section of a component that is not uniformly stressed.

Fibrous fracture (surface) Silky, dull grey and ductile as opposed to bright and crystalline.

Fibrous surface A rough surface with linear features similar to rope.

Fick's Law The quantity of atoms diffusing across a unit plane in unit time is inversely proportional to the concentration gradient.

File hardness (test) A simple system for determining hardness by abrading the component in question with a series of files of graded hardness. The component is considered to have a hardness equivalent to the hardest file which fails to abrade it.

Filiform corrosion A form of corrosion which develops beneath surface coatings, particularly paint. It progresses as lines or fingers from initiation points which commonly are sites of coating breakdown or of precoating contamination.

Filigree Delicate ornamental wire work.

Filled shell See Atomic structure.

Filler (1) Material added in the molten state to a *Welded, Brazed* or *Soldered* joint by a *Filler rod* or by a consumable *Electrode*. It becomes incorporated into the joint and its main functions are to form the bond and fill the joint gap.

Filler (2) Bulking materials such as powder or fibre added to plastics.

Filler rod *Filler* (1) in the form of short rods. They are used in processes such as oxy acetylene welding or TIG welding and are often held in one hand while the other holds the torch.

Fillet Features relating to the corner between two surfaces meeting at an angle of, very approximately, 90°.

Fillet weld A weld formed between two surfaces at approximately 90° to each other such that *Filler* material is deposited to form a weld that is substantially triangular in section. See Figures 47(b) and 51 located at entry on Welding terminology.

Filtration The separation of a liquid plus solid mixture by passing it through a filter. The liquid produced is the **Filtrate**.

Fin A thin projecting piece of material. It may be deliberate as in a cooling fin or it may be inadvertent, for example the fin or *Flash* formed on a die forging at the interface of badly fitting dies.

Fine gilt A thin gold surface layer applied by coating items with a mercury gold amalgam and then heating them at about 360 °C causing the mercury to evaporate, leaving an adherent gold film.

Fine-grain steel Any *Steel* having in intrinsic tendency to retain a fine austenitic grain size during normalizing and annealing at the temperature appropriate to the composition. The characteristic is common in steels deoxidized with aluminium which leaves a large quantity of very fine aluminium nitride particles which impede grain growth.

Fine gold 24-carat gold, i.e. commercially pure.

Fine silver Silver having a *Fineness* of 999, i.e. commercially pure.

Fineness The purity of gold and silver in parts per 1000.

Finishing temperature The temperature at which *Hot working* is discontinued.

Finite element analysis A technique for determining the levels and distribution of *Stress* in components. Its approach is to develop a mathematical model, in the form of a grid or mesh, which divides the structure into many small, simple segments, usually termed 'elements'. These are analysed individually and collectively, usually by powerful computer programs, to determine the stress induced in each element when a force is applied. The technique can deal with complex and cracked three-dimensional structures and components. See also Fracture mechanics.

Fire bars The grating supporting a solid fuel fire.

Fire box The combustion chamber of a fire tube *Boiler*.

Fire cracking Cracking during heating.

Fire-refined copper Copper that has been refined in a furnace rather than by electrolytic means. Fire refining involves melting the metal first under oxidizing conditions to oxidize impurities and then under reducing conditions to reduce the excess oxygen.

Fire side In its narrower sense it is the surface facing the fire or approaching hot gas stream. This definition as applied to a tube carrying water encompasses only a half of the exterior surface the remainder being termed the **Back face** or **Back side**. The bore surface is then the **Water side** or **Steam side**. In a more general sense the term encompasses the entire exterior surface.

Fire side corrosion Severe corrosion occurring on the *Fire side* of, usually, a tube in a steam-raising boiler or similar plant. It is the result of a reaction between the metal and the gas stream or, more often, aggressive ash deposits on the surface.

Fire tube boiler See Boiler tube.

Fire welding See Forge welding.

Fireclay A siliceous refractory mineral predominantly kaolin ($Al_2O_3.2SiO_2.2H_2O$). It can be moulded while wet but sets hard on firing at high temperature.

Firecracker welding Electric arc welding in which a coated consumable *Electrode* is laid flat on the line of the joint and an arc struck between one end of the electrode and the adjacent metal. The arc then travels along the electrode, depositing filler metal and forming the joint without further operator intervention. See also Gravity welding.

Fisheye fracture Sub-surface *Fatigue* cracks that have initiated failure by some other mechanism, perhaps simple overload but often *Brittle fracture*. Against a contrasting surface texture, the dull concentric fatigue markings have a fisheye appearance. Note the potential confusion with *Fisheyes*.

Fisheyes Bright marks, typically a millimetre or two in diameter, observed on some fracture surfaces and evidence of *Hydrogen embrittlement*. The hydrogen embrittlement cracks may initiate other damage mechanisms, including fatigue, and so increasing the potential confusion with Fisheye fracture.

Fission Splitting into a number of parts. See Nuclear.

Fission fragments Sub-atomic particles resulting from *Nuclear* fission. The term often implies larger particles such as nuclei comprising multiple protons and neutrons.

Fission poisons Materials in nuclear fuel, particularly those produced in the *Nuclear* fission process, which have a high neutron capture cross-section and hence wastefully absorb neutrons which would otherwise contribute to the chain reaction.

Fission products Elements produced by the *Nuclear* fission of the heavy elements such as uranium. They are usually the elements of atomic number 34 to 58.

The term often implies radioactive isotopes of these elements.

Flakes   Internal cracks in steel resulting from cooling and transformation stresses and the presence of *Hydrogen*. They may be termed *Fisheyes* when they are exposed on a fracture surface.

Flame cleaning   The use of a soft, broad, oxy-fuel gas flame to remove deposits of paint, grease, rust, etc. from components or large structures prior to painting.

Flame cutting   Processes utilizing a torch similar to that used for gas welding but in which the primary cutting action is the chemical reaction of the material with oxygen. A fuel gas is also involved but its main contribution is to raise the component to reaction temperature at the commencement of the cut. Also termed Oxy(gen) cutting.

Flame gouging   See Gouging.

Flame hardening   Surface hardening of steel components by rapidly heating the surface with a gas flame, usually followed by quenching. See Case hardening.

Flame plating   A process in which an oxy-acetylene mixture carrying a suspension of powdered refractory material is detonated to project the powder at high temperature and high velocity onto the surface to be coated.

Flame scaling (1)   Descaling by a process similar to Flame cleaning.

Flame scaling (2)   Flame heating freshly galvanized steel wire to promote intermetallic bonding and improve surface finish.

Flame temperature   The maximum temperature within a flame, usually at the tip of the inner cone.

Flammable   Inflammable.

Flange   A raised rim at or around an edge and usually perpendicular to the principal axis. It is often formed on tubes. See Figure 45, located at entry on Tube manipulation.

Flange(d) joint   A joint formed between two components one or both of which is flanged. The joint may be made by bolting, welding or brazing.

Flare test   A test in which a tapered mandrel is forced into the end of a tube to measure its expansion capability. Similar to Drift test.

Flaring   A *Tube manipulation* process in which the tube end is conically expanded. See Figure 45, located at entry on Tube manipulation.

Flash (1)   The projecting fin of material formed at the parting line of casings and forgings.

Flash (2)   Molten material expelled during *Flash welding* and similar processes. Also termed 'spatter' particularly when it forms adherent droplets on the component being welded.

Flash Anneal   *Annealing* of short duration applied to wire or thin sheet.

Flash (butt) welding   An electric welding process in which the joint faces are first brought together to strike an arc and then held in light contact or even drawn slightly apart to allow heating before they are finally forced together again. It is essentially a *Pressure weld* as any molten metal is ejected in the final closure.

Flashover   The passage of electric current across the surface of an insulator.

Flask (1)   Any vessel for containing fluid.

Flask (2)   In the context of nuclear waste transportation, a metal vessel of massive proportions, high integrity and tightly sealed.

Flask (3)   A framework to contain bulk material such as moulding sand.

Flat fracture   A fracture surface approximately perpendicular to the principle stress and substantially level. The term is unambiguous where the fracture mode is *Crystalline*. However, it is also used where the fracture comprises multiple, microscopical, ductile shear *Dimpled fractures* in which case it is used in contrast to *Ductile* or *Shear fractures* whose plane of fracture is at approximately 45° to the principle stress.

Flat position welding   The position in which the components being welded and/or the weld surface are approximately horizontal and the weld is made from above. This is usually taken to mean a *Weld slope* not greater than 5° and a *Weld rotation* not greater than 10°. Downhand position is the same.

Flats An imprecise term usually indicating rolled steel products 3–6 mm thick and up to about 600 mm wide.

Flattening mill A *Rolling mill* producing sheet metal.

Flattening test A test in which a hollow component, welded tube, etc. is flattened to a specified degree which is not always contact of the internal surfaces. Acceptance criteria depend on the application but usually specify limitations on cracking.

Flaw Defect.

Flaw detection Any technique for detecting defects such as cracks and voids. See Non-destructive testing.

Flexible manufacturing system A number of machines and other plant items, often computer controlled, that can be readily arranged and utilized to produce a variety of components.

Flex (1) Flexible insulated cable for conducting electricity, particularly for domestic equipment.

Flex (2) To bend, usually with the implication that the degree of bending is within the elastic range so that the component returns to shape following removal of load.

Flexural strength The *Tensile strength* and related properties deduced from *Bending tests*. Brittle materials likely to break in the grips of a tensile machine are often tested in this manner.

Floatation A technique for separating materials suspended in a liquid, in particular to concentrate the mineral content of ores. The finely particulate ore is mixed into the fluid with oil or similar agent and the mixture agitated. The oil attaches to the mineral and air bubbles assist in floating the particle to the surface where it is retained by surface tension before being skimmed off.

Floating plug A short *Mandrel* used in *Tube* drawing to control the bore diameter. The plug is free to move, being designed to position itself and remain in the die throughout the drawing operation. It is usually inserted into the back end of the tube and blown into position by compressed air.

Flocculation The process of agglomeration of fine suspended matter in a liquid.

Flong Papier-mâché (pulped paper) used as a moulding material for casting low melting point metals, in particular tin-base printers' type metals.

Flow brightening The brief remelting of a surface to produce a bright finish. It is practised on electrodeposited tin coatings on steel or copper base items.

Flow lines The lines on a *Polished* and *Etched* section through a wrought material that reveal the deformation pattern during previous working operations. They are caused by the varying etching characteristics of areas with minor differences in composition or by lines of *Inclusions*. See also Sulphur print and Nature print.

Flow spinning See Spinning.

Flow stress The *True stress* at which plastic strain commences. It is increased by prior plastic strain. Flow stress and *Fracture stress* both fall as temperature rises but the rates of fall may differ. Some materials, for example steel, have a flow stress higher than the fracture stress at low temperature but the reverse at higher temperatures. Thus at low temperatures they fracture before plastic strain can commence, i.e. they fail in a *Brittle* manner. At the higher temperatures they plastically deform before failing, i.e. they behave in a *Ductile* manner. See Tensile test.

Flow welding Process in which bulk molten metal is poured into the joint gap which is usually of substantial width. The weld metal needs sufficient superheat to melt the joint faces to form a fusion weld. Alternatively, the weld metal may be allowed to overflow to waste until the joint faces have fused sufficiently. See also Burning, Burning on and Thermit welding.

Flower(s) The bright *Spangle* on hot dip *Galvanized* components.

Fluctuating stress See Fatigue.

Fluid Any liquid or gas. Its shape is defined by the containment vessel.

Fluid lubrication The condition where closely approaching surfaces in relative motion are held apart by a film of fluid lubricant. See Oil wedge.

Fluidity The ability of a liquid to flow into and fill small cavities.

Fluidized bed A bath of granular material through which gas is blown, usually via a porous bottom plate, to make the material flow like a liquid. For heat treatment applications the component is quenched into the bed to achieve fairly high rates of heat transfer, similar to an oil bath but cleaner. The gas may be air or an inert gas. Fluidized beds are also utilized as a combustion device in which coal particles are fluidized and burned by the injected air.

Fluorescence The emission of light or other electromagnetic radiation by a material while it is being irradiated by incident light or other electromagnetic radiation of shorter wavelength such as ultraviolet and X-rays. If emission continues after radiation has ceased the phenomenon is termed Phosphorescence.

Fluorescent penetrant/dye See Dye penetrant.

Fluorine A gaseous element, one of the halogen group.

Fluoroscopy An X-ray technique in which the image is immediately displayed on a screen rather than being recorded on film or tape.

Fluorspar A mineral, calcium fluoride, CaF_2, used as *Flux* in a number of metallurgical processes. It is number 4 on the *Mohs* scale of hardness.

Flush weld A weld with surfaces following the profile of the parent materials. The profile may be formed directly by welding or, more often, by subsequent grinding, machining, etc.

Flushing (of weld) Machining or grinding the surface of a weld to bring it flush with the parent surfaces, The term 'underflushing' is ambiguous, being used to indicate either insufficient or excessive flushing.

Flux (1) A material or compound introduced in various processes, usually involving melting or heating, to react with undesirable materials such as impurities in the bulk material or contaminants on the surface. The material formed by the combination of the flux with the contaminants forms a *Slag* which is readily separated from the product.

Flux (2) The density of magnetic lines of force.

Flux (3) A measure of nuclear intensity, the number of particles per unit volume times the mean particle velocity.

Flux (4) Rate of heat input. Units include J/s and Btu/h.

Flux cored electrode See Electrode.

Fly ash Finely particulate ash from the combustion of pulverized coal.

Flyer plate See Explosive welding.

Flying shear A machine for cutting continuously moving sheet, usually as it emerges at high speed from a *Rolling mill*. The shear cutters, one above and one beneath the sheet, reciprocate along the sheet and, on their forward movement, are synchronized with the sheet to execute the cut.

Fog quenching *Quenching* in fine mist or spray. It gives a cooling rate intermediate between air cooling and quenching into oil.

Foil Very thin sheet, usually less than 0.1 mm (0.004 inch) thick.

Fold Apart from its normal use, a surface-emergent defect formed during casting or forging when one area of metal is inadvertently folded over a neighbouring area without bonding.

Fool's gold Iron pyrites, FeS_2, which can resemble gold.

Forbidden bands See Allowed bands.

Force fit See Interference fit.

Forehand welding The technique of manual *Gas welding* in which the flame both points and moves forward to an unwelded joint. Any filler rod points back towards the completed weld. Also termed **Forward welding** and **Leftward welding**, although the latter may give rise to some confusion in the case of left-handed operators.

Foreign atom An atom of an element other than that forming the dominant pattern of the *Crystal* structure.

Forensic science The application of scientific techniques to assist in matters of civil and criminal law.

Forge The equipment in which *Forging* takes place. Equipment ranges from a simple hammer and anvil, as in a black-smith's forge, to major installations capable of applying very large forces. There are various types of forge: Steam, Drop, Hydraulic forge, etc., the name indicating the source of energy for the forging operation.

Forge welding A joining processes effected below the melting point with the materials in a plastic condition and in which the interfacial force between two components is a significant factor. A primary function of the forging action is to disrupt surface films allowing intimate contact and hence bonding between clean metal surfaces. Heating may be involved either prior to or during the forging operation. The force may be steady, intermittent or percussive. **Fire welding** and **Blacksmith welding** are forms of forge welding.

Forging Deformation of a component set between two approaching tool faces by, mainly, compressive forces. The tool faces may be plain or shaped and may or may not meet. The process is termed **Closed die forging** where the die faces are of complex shape and a carefully sized billet is set in them to be formed to that shape when the die faces are brought fully together. Other processes are termed **Open die forging**.

Form machining Machining processes which utilize a **Form tool**, i.e. a cutting tool shaped to the final complex profile.

Forming processes Those processes which change the shape of material along its major axis without substan-tially changing its cross-section, for example bending, coiling or twisting.

Forward welding See Forehand welding.

Fouling (1) Unwanted deposits, particularly barnacles and weeds, on the submerged surface of ships or other marine equipment.

Fouling (2) Unwanted mineral or other deposits on, for example, the interior of pipework or boiler gas passages.

Foundry Any place where *Casting* is carried out.

Four-point bending A test, or other circumstances, in which the component, usually a bar, is supported at two points towards its extremities and the load is applied through two points symmetrically located between the supports. A feature of such loading is that the stress is uniform over the length between the two inner points, unlike *Three-point bending*.

Fractography The examination of fracture surfaces under a conventional light microscope or in a *Scanning electron microscope*. Normally no preparation of the surface, apart from cleaning, is involved although *Replication* is common.

Fracture A crack or break or the process of cracking (see Break). Often 'fracture' without qualification has implications of low **Ductility** whereas 'Rupture' may imply significant ductility but there are no fully reliable rules and 'ductile fracture' is quite acceptable.

Fracture analysis Any technique, but particularly numerical ones, based on the data derived from examining and measuring features on a fracture face. For example, detailed measurements of the number and spacing of *Beach marks* on a *Fatigue* fracture can indicate key factors such as the number of cycles, the levels of stress and the date of initiation.

Fracture appearance transition temperature (FATT) The temperature at which a fracture face is 50% *Cleavage* and 50% *Ductile*. See Impact test.

Fracture energy The energy input required to produce unit area increase of crack face. Since any crack has two faces the fracture energy is half the *Fracture toughness*.

Fracture mechanics, also termed **Linear elastic fracture mechanics (LEFM)** The study of the mechanics of crack growth, in particular stress-related matters in the zone of material at the crack tip. Note that although this definition excludes crack initiation in a defect-free material many apparently sound components contain crack-like features. Examples include, on a macro scale, the surface irregularities on welds (see Weld defects) or, on a microscopical scale, the *Inclusions* inevitable in many metals. The principles of fracture mechanics are most readily applicable to low-ductility mechanical damage mechanism such as *Brittle fracture* and *Fatigue* but they are also applied, perhaps with more difficulty, to other mechanisms such as *Stress corrosion* and *Creep*. Clearly, the severity of loading in the presence of a crack is influenced by both the size of the externally imposed

load and the size of the crack and hence a basic requirement is to develop a parameter reflecting these two variables that is measurable and calculable in a manner analogous to *Stress* in an uncracked component. This parameter is termed the **Stress intensity factor, K** and can be defined as the measure of the elastic stress field in the vicinity of a crack tip. It has dimensions of [stress $\times$ square root of crack length] and hence units such as $\mathrm{MN/m^{3/2}}$, $\mathrm{MPa\,m^{1/2}}$ or $\mathrm{ksi}\,\sqrt{\mathrm{in.}}$ Stress intensity should not be confused with *Stress concentration factor* or even **Stress intensification factor** since the latter two terms are not readily measurable and, at best, are quantified only as a multiplication factor. Cracks can extend in three **Modes** depending on the form of loading. These are Mode I—the crack opening mode (simple tensile loading), Mode II—the edge sliding mode (in-plane shear loading) and Mode III—the shear mode (out-of-plane shear loading) (see Figure 15). Of these, the opening mode is by far the most common but,

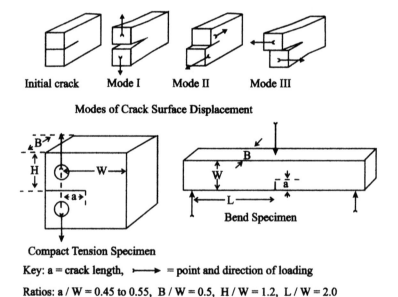

Initial crack   Mode I   Mode II   Mode III

Modes of Crack Surface Displacement

Bend Specimen

Compact Tension Specimen

Key: a = crack length, ▶—— = point and direction of loading

Ratios: a / W = 0.45 to 0.55, B / W = 0.5, H / W = 1.2, L / W = 2.0

Figure 15   Fracture mechanics terms and fracture toughness test pieces for Mode 1 testing

where necessary, the relevant mode is indicated by the appropriate subscript, for example K_I, K_{II} or K_{III}. Basic fracture mechanics deals with idealized circumstances where all stresses in the crack tip zone are elastic and K increases to a **Critical stress intensity**, termed K_C (or, if appropriate, K_{IC}, etc.), at which the crack will extend without any plastic deformation. Such behaviour is, by definition, *Brittle* and for brittle materials K_C is a measure of **Fracture toughness**. In practice some plastic deformation is inevitable but, where it is slight, the basic concepts can be applied with adequate accuracy. However, where the amount of plasticity is large, relative to the crack size, then more complex treatments are required. These are usually based on the J **integral** which can be considered analogous to K with corrections for plasticity. Effectively, elastic conditions are encountered in components which are thick relative to the crack size. In these circumstances the potential local plasticity at the highly stressed crack tip is constrained to a very small volume by the surrounding bulk of lower-stressed material. This causes stresses in the transverse directions and is described as a condition of **Plane strain**, hence the use of the term **Plain strain fracture toughness** when referring to K_{IC}. The term **Plane stress** refers to circumstances where the tensile stress is uniaxial, i.e. there is no stress in the transverse directions and all material in the section carries the **Net section stress**, i.e. the total applied load divided by the full cross-section. Such circumstances are encountered in crack-free material or at a crack that is large relative to the plate thickness. In the latter case bulk yielding ahead of the crack allows stress redistribution to eliminate stresses in the transverse directions. Generalizing, if the net section stress does not exceed about 0.8 of the yield strength, the plastic zone will be sufficiently small for plain strain to apply and hence for calculations based on K to

be reasonably accurate. See also Fracture toughness.

Fracture stress   The true stress at the instant of fracture. See Tensile test.

Fracture toughness   The measure of the resistance offered to the further growth of a crack (see Fracture mechanics). It can be determined in the laboratory to provide a reliable indication of how a component containing a crack will perform in service. The test pieces for determining fracture toughness are usually either three-point *Bend* specimens or thick section, short-height specimens pulled in tension and termed **Compact tension specimens (CTS)**. See Figure 15. Both types of specimens are pre-cracked prior to fracture toughness testing. Typically these pre-cracks are sharp-tipped fatigue cracks developed at machined notches. During testing the applied stress is measured by conventional techniques but the deflection is measured as **Crack opening displacment (COD)** or **Crack tip opening displacement (CTOD)**. Both of these terms refer to the displacement of the crack surfaces at the crack tip although the actual measurement may be made as a **Crack mouth opening displacement** and appropriate corrections made.

Fracture transition   Same as Fracture appearance transition.

Francium   A metallic element, one of the alkali group, with little commercial application. See Table 15.

Frank–Rhead source   A mechanism capable of generating *Dislocations*. The mechanism can be visualized as an edge dislocation anchored at its two ends and subjected to a stress which causes it to bow outwards forming loops around the anchor points. As the loops grow the structure becomes unstable and jumps to form a new continuous loop dislocation round the original source which remains to trigger further loops. See Figure 8, located at entry on Dislocations.

Free carbon   The carbon present in ferrous materials as graphite, including temper carbon. See Cast iron.

Free cutting/machining A metal which has been treated, usually by the deliberate introduction of impurities, to improve its machining characteristics, in particular to ensure that material being cut away is released as small chips rather than continuous strands. The impurities form an even dispersion of particles of a size large enough to serve as **Chip breakers** but not so large that they have an unacceptable affect on mechanical properties. The effect is achieved, for example, in *Steels* by appropriate quantities of manganese sulphide inclusions and in *Brass* by additions of lead. In some cases machining characteristics can be improved by heat treatments that influence the distribution and size of precipitates or that induce some beneficial gain structure. For example some steels can develop a **Blocky structure** with a favourable ferrite/pearlite distribution. Although the effects of such treatments may not be so pronounced as that induced by inclusions they are useful where high-speed machining is required but any inclusions are unacceptable, as in the case of steels for ball and roller bearings.

Free electron theory This suggested that the valence electrons in a metallic bond formed an electron cloud or **Electron gas** in which there was no constraint on their movement around all the atoms in the metal. Subsequently, the *Energy band theory* was developed to better explain phenomena such as *Semiconduction*.

Free ferrite Ferrite in iron or **Steel** that is not intimately associated with carbide forming pearlite.

French chalk *Talc.*

Frettage Fretting.

Fretting, fretting corrosion Damage developing when two surfaces in firm contact experience repeated slight relative movement. The damage is not simple abrasion but more like *Adhesive wear* involving local high temperatures and temporary welding. Material is lost from one or both surfaces as a fine dust,

usually an oxide of the parent material. The reddish-brown dust on fretted steel surfaces is occasionally referred to as **Cocoa** or **Bleeding**. The fretted surface is very susceptible to *Fatigue* crack initiation and the combined mechanisms may be referred to as **Fretting fatigue**.

Friction The force acting between two contacting bodies in motion relative to each other. The laws of friction are (1) the static frictional force is proportional to the load acting normal to the interface, (2) the frictional force is independent of the interfacial area, (3) sliding friction is less than the limiting static friction.

Friction coating/surfacing The deposition of a coating by a rod of material which rotates as it bears heavily against the component to be coated. The rotation first causes friction heating to a temperature near the melting point of the rod material and, second, scours the receiving surface allowing a good bond with the deposited metal. A wide range of materials including hard-facing alloys can be deposited.

Friction cutting/sawing Operations in which frictional heating plays a significant role in melting or softening material being cut. In the case of friction sawing the blade is usually toothed to assist metal removal but in friction cutting the blade may be plain edged.

Friction grip bolting A bolted joint which relies on having sufficient *Preload* in the bolts to develop a high level of friction between the mating faces of the joint. Loads across the joint are then carried by friction in the parent materials rather than by the bolts in shear.

Friction welding Any welding process in which heat is produced by the friction between two contacting components moving relatively to each other. The term usually refers to a production process in which two components are forced together while one is rapidly rotated or otherwise moved relative to the other. The relative movement causes heating, disrupts surface films allowing contact between clean surfaces and produces local deformation. When these effects reach an optimum stage the movement is sharply halted and the interfacial force is considerably increased leading, in suitable cases, to strong joints even between materials not readily welded by other techniques. The friction welding effect may occur unintentionally as a result of *Adhesive wear*, in which case it is usually termed *Seizure*.

Fringes Bands of light of varying intensity or colour produced by *Interference* effects.

Frit/fitting (1) A coating applied to a component surface as granules and then heated to bond the substrate and form a glaze.

Fritting (2) *Sintering.*

Frozen equilibrium A condition in which a material is unable to progress to the theoretical equilibrium state because there is insufficient thermal energy to allow diffusion.

Frozen stress (technique) A stress analysis technique. See Photo-elastic modelling.

Fuel cell A form of *Electrolytic* cell in which electricity is produced by the oxidation of a fuel, usually a hydrocarbon fluid.

Fuel element A component for a *Nuclear* reactor comprising fissionable material in a metal sheath that provides protection and support and retains any fission products.

Fuel gas Any gas which burns with oxygen, to release useful amounts of heat. Common gases include acetylene, propane, butane, hydrogen, methane, coal gas and natural gas.

Full anneal *Annealing* steel by heating into the austenitic region and cooling slowly, usually in the furnace, as opposed to *Normalizing* or *Sub-critical annealing*. See Steel.

Full(y) hard An imprecise term indicating that a material is in the hardest, highest tensile condition that can be economically and reliably achieved by heat treatment and/or work hardening. In this condition the material cannot be deformed to any significant extent; it will behave in an elastic manner virtually until it fails with negligible *Ductility*.

Full mould casting A process in which the sand mould is formed round a polystyrene pattern which remains in position as the molten metal is fed, usually from the bottom. The polystyrene vaporizes, suitable venting being provided, and the metal rises to fill the mould which can be of highly complex form.

Fuller A grooved *Forging* die, particularly when used for forming a waisted section on a length of bar.

Fullerines See Buckminsterfullerine.

Fully killed steel A steel that is fully deoxidized, mainly by silicon but also other elements such as manganese, aluminium or titanium as appropriate.

Fully penetrating weld A weld in which there is complete fusion through the thickness including the root. See Figure 50, located at entry on Welding terminology.

Fulminates Highly unstable compounds which are readily detonated by a slight shock.

Furnace A vessel or chamber in which a reaction or heat treatment occurs at a considerably elevated temperature, for example *Blast furnace* and *Annealing* furnace. For lower temperatures 'oven' tends to be favoured although there are obvious exceptions such as *Coke oven*. There is a large variety of furnaces. Heating may be any fuel, including gas, oil and electricity. The atmosphere may be air, combustion products, controlled (with respect to, for example, its oxidation or carburisation characteristics) or vacuum. The *Charge* may be inserted from above, the end or the side, by hand, by a charging machine or on a rail car hearth. The charge may remain stationary or move forward progressively. Also see Rotary hearth and Walking beam.

Furnace brazing The process of *Brazing*

by heating in a furnace. The braze *Filler* in the form of a shim, sleeve or similar preform, will have been positioned in the assembled joint and a *Flux* applied unless the furnace atmosphere is controlled to avoid contamination and produce a reducing environment.

Fuse (1) A length of wire, or similar material, in a suitable insulating and heat-resistant carrier that is installed in an electrical circuit to protect the circuit if an excessive current is drawn. The cross-section area of the wire is selected to match the maximum continuous current limit. A **Surge fuse** has a similar function but is intended to protect against short-term excursions to high current.

Fuse (2) Generally, to melt.

Fusible alloys See Low melting temperature alloys.

Fusible core A core for *Injection moulding* manufactured from a *Low melting temperature alloy.*

Fusible insert (of weld) A pre-placed *Filler* having specific dimensions to locate snugly in the *Root* of a joint to be made from one side. It is intended to be fully fused to become an integral part of the joint and is not normally machined following welding. Same as Consumable insert. See Figure 47(a), located at entry on Welding terminology.

Fusible plug A plug in the wall of a pressure vessel that is intended to melt and release the pressure if an excessive temperature is reached.

Fusion (1) Amalgamation by some process involving melting. In some contexts it may imply merely that all or part of a previously integral material had melted and resolidified.

Fusion (2) In a *Nuclear* reaction, the joining of light elements to form heavy elements with associated energy release.

Fusion cutting/sawing Same as Friction cutting/sawing.

Fusion face (of weld) A surface of the parent material intended to be fused during welding.

Fusion penetration (of weld) Same as Depth of fusion. See Figure 50, located at entry on Welding terminology.

Fusion welding Any welding process in which the surfaces of the components to be joined are melted and fused together. Additional *Filler* metal may be introduced. It is usually implicit that some or all of the fused metal will be retained in the joint and the term usually excludes processes where the joint surfaces are brought together under pressure.

Fusion zone (of weld) Parent material that has fused during the welding process. See Figure 50, located at entry on Welding terminology.

G

Gadolinium A metallic element, one of the *Rare earth* group. See Table 15 for physical properties.

Gage Gauge (USA).

Galena Lead sulphide, PbS, particularly as the naturally occurring ore.

Galling A severe stage of *Adhesive wear*.

Gallium A metallic element. See Table 15 for physical properties.

Galvanic cell/corrosion/series etc. See Electrochemistry.

Galvanizing The application of a zinc coating to iron and steel components. The zinc provides *Cathodic protection* rather than merely excluding the environment. It can be achieved by a *Hot dip* process, i.e. immersion in molten zinc at about 450 °C beneath a molten zinc chloride flux. The resulting relatively thick coating comprises a zinc outer layer and a zinc iron diffusion layer at the interface. It is strongly bonded to the steel and is particularly suitable for substantial sections for structural applications. *Electroplating* or *Electrogalvanizing*, provides a thinner more accurately controlled coating suitable for smaller precision items. *Metal spraying* can offer a range of thickness and it can be applied on-site if necessary but the adhesion to the steel is critically dependent on the quality of preparation and application. *Sherardizing* involves baking components in zinc powder at just below the 419 °C melting point of zinc to promote the formation of a thin diffusion coating. The term *Cold galvanizing* is sometimes used for the application of paints containing a high proportion of metallic zinc particles. It is claimed that the quantity of zinc and the particular binding agents employed are such that the zinc particles and the substrate are in sufficiently good electrical contact to provide some cathodic protection.

Galvanizing embrittlement The *Embrittlement* of steel resulting from the heating involved during hot dip galvanizing at about 450 °C. A form of *Strain age hardening*.

Gamma The Greek alphabetical reference to various *Phases* in alloy systems, particularly gamma iron, i.e. austenite. See Steel and also Table 20 for other Greek characters.

Gamma loop The closed loop of the austenitic (i.e. gamma) region of iron–chromium alloys. See Figure 16.

Gamma radiation The electromagnetic radiation emitted by some radioactive materials as a result of their *Nuclear* activity.

Gangue The earthy material in an ore.

Ganister A siliceous sandstone used as an acidic furnace lining.

Gap (in welding) (1) The minimum distance between surfaces to be joined as in **Root gap**, the distance between the surfaces at the root (see Figure 48 located at entry on Welding terminology).

Gap (in welding) (2) The distance between the *Electrode* and the workpiece over which an electric arc is struck, as in **Air gap**.

Gas The state of matter in which a material is in a low-density, highly fluid and elastically compressible form.

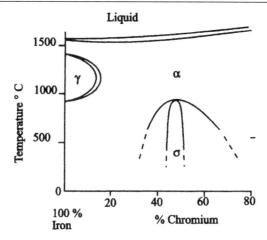

Figure 16   Gamma (γ) loop and sigma (σ) in iron–chromium alloys

Gas carburizing   See Case hardening.

Gas constant   The constant, R, for one mole of gas given by:

$$pV = RT$$

where p = pressure, V = volume and T = temperature, absolute. In SI units, $R = 8.31 \text{ J K}^{-1} \text{ mole}^{-1}$.

Gas cutting   The use of gas torches in which a fuel gas and oxygen are burnt to provide a heat source for metal cutting. By far the most common process is the **Oxyacetylene** flame using acetylene and commercially pure oxygen from cylinders for cutting *Steel*. The general principle is that, initially, an intense oxyacetylene combustion flame heats the steel locally to its **Ignition temperature**, 750–825 °C, and then an additional large quantity of high-pressure oxygen is introduced to cause combustion of the steel. The temperature is sufficient to produce a molten oxide *Slag* which is vigorously ejected by the gas stream giving, under suitable conditions, a clean square-edged cut. The equipment for manual use is similar to that used for *Gas Welding* except that the burner head has a pair of concentric nozzles, the outer providing the initial mixed fuel and the inner delivering the

main oxygen blast. This oxygen blast is usually controlled by a lever valve beneath the hand grip. A similar burner head, or multiple heads, may also be machine mounted and programmed to cut components of complex form with, if necessary, bevelled edges. These machines are often termed **Profile cutters**.

Gas (shielded) metal arc welding   Any welding process in which the heat source is an electric arc struck between the component being welded and an electrode in the form of continuously fed *Filler* wire. A suitably inert gas, supplied from an external source, provides a protective shroud around the welding zone. Also know as metal inert gas (MIG) welding.

Gas pocket/hole/porosity   Voids of various size caused by gas entrapped during *Solidification* of castings.

Gas pore (of weld)   Relatively small internal voids formed by the entrapment of gas during solidification of the weld metal. Voids larger than about 1.5 mm are usually termed *Blow holes*.

Gas shielded arc welding   Electric arc welding processes in which a gas, continuously supplied from an external source, shields the *Weld zone* from contamination, etc. Note that, strictly, this excludes processes in which shielding gases are produced by materials on, or in, the *Electrodes*. Also see Arc welding and Gas metal arc welding.

Gas tungsten arc welding/cutting Processes in which the heat source is an electric arc struck between a nonconsumable tungsten electrode and the component being welded or cut while a suitably inert gas, supplied from an external source, provides a protective shroud for the welding/cutting zone. In the welding process a separate *Filler* wire may be used. Also termed Tungsten inert gas (TIG) welding.

Gas welding Any fusion welding process in which the heat source is provided by the combustion of a fuel gas with oxygen including air. **Oxyacetylene**, i.e. the combustion of acetylene with oxygen, is the most common fuel gas flame as it is the only one capable of melting steel. Alternatives for lower melting point metals include propane, butane, coal gas and hydrogen. In a typical hand-held *Gas torch* or *Blowpipe* the two gases are mixed in the burner head and exit through a nozzle. Two screw valves to control gas supply are located close to the grip which is separated from the burner by metal tubing of length sufficient to avoid overheating the grip area and the operator. The torch is connected by substantial flexible rubber tubes to the gas supplies which may be cylinders or some more permanent installation. Additional *Filler* may be introduced to the joint by a plain metal rod manipulated by the second hand of the operator. Compared with *Electric arc welding* the equipment for gas welding is relatively cheap and portable. Both demand operator skill but the relatively low temperature of the gas flame limits, in practice, the size of weld that can be made. Also, because the rate of heat input is lower, more heat has to be introduced to melt a given amount of material so distortion is greater with a gas flame.

Gassing Internal fissuring produced in *Tough pitch copper* by exposure at high temperature to hydrogen-rich environments. See Hydrogen damage.

Gate (of mould) A channel through which molten metal enters. In this context it does not imply a means of closure.

Gauge Gage in USA (1) A measuring device. It may be an instrument or other device with some system for reading off dimensions, pressure, etc., or it may be a simple precision shape, for example feeler gauge.

Gauge Gage in USA (2) A dimension such as thickness of sheet, wall thickness of tube, diameter of wire and bar or spacing between tracks. There are numerous national and industry systems relating **Gauge numbers** or letters to dimensions.

Gauge length The parallel length of a *Tensile* or *Creep* specimen on which *Elongation* is measured.

Gauss The unit in the non-*SI* metric system of magnetic flux density defined as unit magnetic pole subjected to a force of one degree. 1 Gausse $= 10^{-4}$ T (tesla).

Geiger counter A device for detecting ionizing radiation. Various systems are employed but the common feature is that each ionizing event that is detected produces an electrical signal that may be recorded or emitted as noise.

Gel A material having some characteristics of a solid and some of a liquid. It may be considered as a liquid reinforced by a near-continuous solid network.

General corrosion The usually observed form of corrosion acting over most or all of a specified area of a component and producing a fairly even attack with no significant pitting and no associated cracking.

Gerber diagram See Fatigue.

German silver Various alpha, single phase alloys typified by 52% copper, 26% zinc and 22% nickel. No silver.

Germanium A brittle metallic element. It is a semiconductor much used in electronic devices. See Table 15 for physical properties.

Getter A reactive material enclosed in sealed vessels such as thermionic valves to scavenge the last trace of oxygen or other undesirable gas. The getter material forms, or is located, on a filament that is heated to incandescence after the vessel has been sealed. Various metals are used including magnesium, cerium, zirconium, and titanium.

Ghost bands Bands of ferrite within the ferrite/pearlite matrix of some wrought pearlitic *Steels*. They are a residual effect of casting *Segregation* and usually contain high phosphorus.

Gibbs' phase rule See Phase rule.

Gilding The application of *Gold leaf* or sheet to a surface for decorative effect.

Gilding metal *Brass* with 90% copper and 10% zinc, no gold. It is similar in colour to gold and is usually used in wrought bulk form rather than as a surface coating. It can also be treated to produce a brownish tint similar in appearance to *Bronze*. See Table 8.

Gilt Silver or other metal that has been gilded, i.e. surfaced with a thin layer of gold.

Girder A beam for structural purposes, typically of 'I' section to give maximum resistance to bending for a given weight, usually steel.

Glass An inorganic material which, on cooling from a fluid state, becomes progressively more viscous forming an amorphous, i.e. non-*Crystalline* solid.

Glass paper An abrasive material produced by bonding a layer of graded, powdered glass to a strong paper backing. More useful for wood and similarly soft materials rather than metal.

Glaze (1) The production of a hard, smooth surface or the surface so formed.

Glaze (2) The installation of glass sheet into a framework or glazing bars.

Glide (1) Referring to deformation within a grain, the term usually means plastic deformation by slip along the main planes, usually of closest packing. See Dislocation.

Glide (2) Referring to deformation on a scale larger than an individual grain, the term usually means non-crystallographic movement of one grain over another.

Globular cementite Cementite in discrete globules rather than combined in pearlite. See Steel.

Globular transfer (in welding) See Metal transfer.

Glow discharge spectroscopy A surface analysis technique in which a glow discharge lamp provides a low-pressure argon environment for the specimen which forms the cathode of the high-voltage (800–1200 volt) circuit. The plasma, or glow discharge, of Ar^+ ions bombards the specimen releasing material in a high-energy state. The wavelength and intensity of the resultant radiation are measured to provide a quantitative analysis of the surface material.

Glut Metal bar or section providing filler material in *Forge welding*. The glut becomes fully incorporated as a load-bearing member rather than merely occupying space. See Figure 17.

Gold A metallic element. One of the noble metals, it is one of the only two coloured metals (copper being the other). It is soft and, being the most *Malleable* of all metals, can be beaten into ultra thin *Gold leaf*. The purity of gold is measured as a percentage or, for jewellery, coinage and similar applications, in carats (parts per 24), or fineness

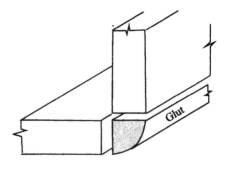

Figure 17 A glut to be incorporated into a forged corner joint

(parts per 1000). Apart from decorative and coinage uses and its eternal 'hoarding' value is has considered industrial application where its corrosion resistance justifies its cost. See Table 15 for physical properties.

Gold bronze Alloy of copper with about 4% aluminium (no gold) having a gold colour.

Gold filled A surface coating of gold of significant thickness and purity applied to the base metal. In the UK the quality is normally indicated by a stamped fraction, the upper indicating the weight percentage of the gold layer and the lower the carat of the layer. Assay regulations for other countries vary.

Gold leaf Thin sheet gold of high purity typically $0.075-0.125$ μm $(3-5$ μ in) thick.

Goldschmidt process The *Thermit reaction* when used for extraction of metal from its ore.

Goniometer An instrument for measuring the orientation of the surfaces and planes of *Crystalline* materials.

Goodman diagram See Fatigue.

Gouging A severe form of localized *Wear* in which material is removed in a single pass or cut leaving a deep groove. *Flame gouging* is the use of some oxyfuel gas flame to melt defects or other unwanted feature on a component surface. The molten materials is vigorously ejected by the main gas stream or an additional stream, leaving a deep groove. **Thermal gouging** is similar but includes processes in which heating is by an electric arc.

GP zones See Guinier–Preston zones.

Grain (1) A portion of a metal or alloy having all of its atoms in alignment forming a *Crystal structure*. Grain is largely synonymous with *Crystal*. Metals are composed of grains, much as a lump of sugar is, except that metals have no gaps at the interfaces and bonding between grains is strong. The interface between grains is termed the **Grain boundary** which may be visualized as a band of material in which the atoms

have a virtually random arrangement as they blend from the lattice of one grain to that of the neighbouring grain. In most cases the grain boundaries are not a weak zone. The exceptions include high temperatures (see Creep) and certain cases of **Embrittlement**.

Grain (2) A unit of weight, the smallest on the apothecaries' scale, 1 grain $=$ 0.002 08 oz troy $=$ 0.002 29 oz av. $=$ 64.7989 mg.

Grain (3) The elongated fibrous cell structure of wood.

Grain boundary See Grain (1).

Grain contrast Variations in appearance of the grains on a surface prepared for *Metallographic* examination. The variation may be produced by techniques including *Etching* or *Heat tinting* and examination may be by normal light or polarized light.

Grain growth The process by which large grains grow, by diffusion, at the expense of smaller grains. In practice the effect is usually confined to materials exposed to high temperature during heat treatment or service. See also Critical strain grain growth.

Grain refinement Any process that reduces the size of the *Grains* in a component. A fine grain size improves most mechanical properties except, usually, *Creep* strength. The grain size of castings, including welds, may be manipulated by techniques such as alloying additions or control of cooling rate. In the solid state, grains can grow by diffusion at elevated temperatures but a reduction in size requires the development of a new grain structure. This can be accomplished if the material has been work hardened and is subsequently annealed to cause recrystallization. However, a metal that has not been work hardened cannot form a new grain structure unless it undergoes a *Phase* change on heating, as in the case of iron and *Steel*.

Grain size Any measure of the dimensions of *Grains* by volume or more usually by area. Data are presented in various terms including average diameter in millimetres or by a numerical designation derived from comparison with standard charts.

Grain size analysis   The measurement of *Grain size* in metals or the measurement of the particle size of powders.

Gram-atom   The mole, the relative atomic mass in grams.

Granular (1) (with reference to volume) Comprising separate, approximately equiaxed grains.

Granular (2) (with reference to surfaces) Having a rough texture, rather like coarse-grained lump sugar.

Granulation   Any process for producing metals as granules including pouring through a mesh or onto a spinning disc from which droplets are flung.

Graphite   A form of *Carbon* with a hexagonal crystal structure having low-friction characteristics. It has good electrical and thermal conductivity, a high melting point and, in the absence of oxygen, is stable with useful strength at high temperatures. See Table 15.

Graphitic carbon   Carbon, in iron base materials, that is in the free form as graphite rather than combined with iron or another element. See Cast iron.

Graphitic corrosion   A form of corrosion affecting graphitic *Cast irons* in aggressive aqueous environments, particularly soils containing sulphate-reducing bacteria. An *Electrolytic* cell is formed in which the iron matrix is progressively corroded leaving a weak mass of graphite and iron oxide. The corrosion product is the same volume as the original material and does not readily detach, so even extensive deep graphitic corrosion may not be obvious on a visual examination.

Graphitization   The formation of free carbon, i.e. graphite, in iron and steel. When it occurs during solidification it is termed **Primary graphitization**. When it occurs in the sold state it is termed **Secondary graphitization**. The process may be deliberately induced in the production of *Cast irons*. In steels,

graphitization may occur as a result of very prolonged exposure to high temperatures, for example about 100 000 hours at about 500 °C for mild steel, perhaps 50 °C higher for 0.5% molybdenum steel. Aluminium deoxidized steel are relatively more susceptible and imposed stress accelerates the rate of graphitization. The term has also been used, perhaps not very accurately, with reference to *Graphitic corrosion*.

Graticule   A grating, network or scale on a lens or transparent disc inserted in the *Microscope*. Its image is superimposed on the material being examined, allowing features of interest to be measured or their position plotted.

Gravity die casting   The process of *Die casting* in which the molten metal is fed into the die by gravity rather than being pressure injected.

Gravity welding   Processes in which a *Metal arc* electrode is carried by a device which locates the electrode tip at the start of the joint and, once the arc has been struck, directs the falling electrode along the line of the joint without further assistance from the operative. The device is usually very simple, for example no more than a bipod set across the joint.

Gray (US spelling)   See Cast iron.

Greek characters   See Table 20.

Green   This term is often used to indicate a product in a preliminary state of preparation, for example **Green compact** (the pressed powder compact prior to *Sintering*) or **Green strength** (the strength of such a compact).

Green rot   The green corrosion products occurring on some nickel–chromium alloys at high temperature in carburizing environments. The formation of chromium carbide depletes the surface of chromium and the nickel then forms a green oxide on the surface and penetrating *Grain boundaries*.

Grey (cast) iron   See Cast iron.

Grey tin   The *Allotrope* of tin stable below minus 13 °C which is hard and friable. In practice it forms when tin is cooled below about minus 20 °C and is inhibited by various alloy additions including lead above 5% or antimony above 0.1%.

Griffith critical crack The sharp-tipped crack, assumed to be present in **Brittle** materials and having, at its tip, a stress concentration sufficient to explain the difference between the high theoretical strength of the material and the much lower strength observed in practice.

Grinding *Abrasion* to remove surface material, particularly when the abrasive is in the form of a wheel which is power driven.

Grinding burn Surface damage resulting from frictional heating during grinding and similar operations. Damage may include cracking, grain boundary oxidation, local softening, local rehardening with brittle martensite (in the case of steel) and the introduction of *Residual stresses.*

Grinding cracks Surface cracking resulting from severe *Grinding*. The frictional heating causes local surface deformation which, on cooling, induces high levels of tensile *Residual stress*. In addition, steel components may be locally hardened. These various consequences render the material liable to low *Ductility* cracking on cooling.

Grinding in The process of rubbing together, with an abrasive, two surfaces that need to mate closely to achieve a close fit. Probably the classic example is the (amateur) hand grinding of the valve ports of car engines.

Grinding stress The *Residual stress* remaining from the surface heating and deformation caused by severe grinding.

Grit Particulate hard material.

Grit blasting Processes in which grit is projected at a surface to remove scale, paint or other contamination. The metal or mineral grit may be entrained in a high-velocity air stream or flung from a high-speed wheel (*Wheelabrator* is one trade name). If the grit is entrained in water various terms (some of them trade names) such as Hydra Blasting, are used.

Grog A hard, granular material added to a refractory to reduce shrinkage and improve resistance to thermal shock.

Grommet/grummett Various devices to prevent chafing or similar local problems. Examples include twisted rope rowlocks on a rowing boat or the circumferentially slit ring of rubber, etc., inserted to sit astride the edge of a cable hole in sheet metal.

Gross energy requirement The total energy expended in producing a material or component from the ore-mining stage to its installation in service.

Group (of elements) See Atomic structure.

Growth (of cast iron) The increase in dimensions of some *Cast iron* heated for long periods above 500 °C. At such temperatures the cementite breaks down, forming ferrite plus graphite and oxygen permeates along the graphite causing internal oxidation. Both effects cause a volume increase which may lead to cracking.

Guillet diagram A diagram predicting a steel's microstructure on the basis of the alloy contents which are defined in terms of their *Equivalent* effects relative to nickel and carbon. See Figure 18.

Guillet equivalent Generally, the factor by which the percentage quantity of an element has to be multiplied to indicate its effect on some characteristic such as microstructure. This term is used particularly regarding the structure of *Brass* evaluated in terms of the equivalent percent of zinc. The Guillet equivalents for zinc in brass are: $5 \times$ aluminium, $2 \times$ magnesium, $10 \times$ silicon, $1 \times$ lead, $0.9 \times$ iron, $0.5 \times$ manganese and minus $1.2 \times$ nickel. The products of each element percentage times its factor are added to the actual zinc content to give the zinc equivalent.

Guinier–Preston (GP) zones The local concentrations of *Solute* atoms that develop at an early stage in the formation

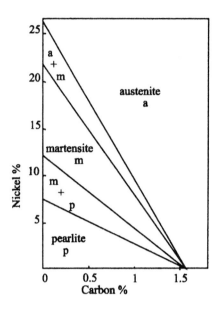

Figure 18   Guillet diagram

of a precipitate from a supersaturated *Solid solution*.

Gunmetals   *Bronzes* containing a small additional quantity of zinc and possibly other elements such as lead to improve machinability or nickle to further improve corrosion resistance; a simple example is 88% copper, 10% tin, 2% zinc. They combine good strength with resistance to atmospheric and marine corrosion and they have good casting and plain bearing characteristics. Once used for large guns they are now used for cast bearings and gears.

Gusset   A reinforcement, usually a plate of triangular form set between two surfaces meeting at an angle.

Gypsum   Hydrated calcium sulphate, $CaSO_4$, $2H_2O$. It occurs naturally as a soft fibrous mineral or in crystalline form as **Alabaster**. It is also a by-product of some systems for removing sulphur dioxide from coal-combustion flue gases.

Habit plane The crystallographic plane on which the process in question occurs. See Crystal structure.

Hackle surface A term usually applied to areas of rough texture on the fracture face of glass but it is sometimes used for similar features on *Brittle* fractures in metals.

Hadfield's (manganese) steel An austenitic *Steel* with about 12–14% manganese and 1.2% carbon. It is tough and wear resistant particularly to gouging *Abrasion*. Before entering service it is usually quenched from about 1000 °C to retain the austenitic phase, in which condition it is relatively soft, about 200 HV. However, severe cold deformation such as abrasion induces an increase in surface hardness to about 600 HV by a process variously attributed to transformation to martensite or, alternatively, to severe strain hardening of the austenite.

Haematite Red iron oxide, Fe_2O_3.

Hafnium A metallic element with little commercial application. See Table 15 for physical properties.

Hairline cracks Fine cracking particularly that due to *Hydrogen damage.*

Half-hard A grade of *Copper* or other non-ferrous material. Many such materials are obtainable in various strength levels achieved by varying the severity of the final cold-work operation. As the amount of work increases so does the hardness and strength, leading to terms such as quarter-hard, half-hard and full-hard.

Half-life The time taken for the level of radioactivity to fall by one half. A further similar period will reduce the level to three quarters of the original and so on; zero is never reached. The same principle may be applied to other physical and chemical reactions.

Half-thickness The thickness of material that reduces the level of radioactivity in the area being shielded by one half. It may be expressed as thickness, i.e. mm, or weight per unit area, i.e., gm per square cm.

Halides Compounds of the *Halogen* elements: fluorides, chlorides, bromides, iodides and astatides.

Hall effect The phenomenon whereby a conductor carrying a current in a magnetic field develops a charge perpendicular to the plane containing the conductor and the field.

Hall–Petch relationship The relationship between grain size and *Yield strength.*

$$\text{Yield strength} = \sigma_0 + kd^{-\frac{1}{2}}$$

where d is the average grain size and k and σ_0 are constants for the material.

Halogens The elements comprising Group VII of the Periodic Table, fluorine, chlorine, bromine, iodine and astatine. See Table 16.

Hammer scale The iron oxide scale that forms on steel being forged in a normal atmosphere.

Hard chromium (plate) A chromium *Electroplating* applied for functional applications, such as wear resistance or dimension restoration, rather than for decorative purposes.

Hard drawn/rolled Tubing, sheet or bar, etc. that has been *Drawn* or *Rolled* to increase its tensile strength to the commercially practicable limit.

Hard facing/surfacing Any technique for depositing hard material onto a surface requiring such protection. Techniques include *Welding, Brazing, Spraying*, etc.

Hard lead Alloys of lead, particularly those with antimony up to about 25% used for batteries and corrosion-resisting structural applications.

Hard metals A term sometimes applied to tungsten carbide and similar materials used for tool cutting tips.

Hard plating See Hard chromium.

Hard soldering An alternative term for silver soldering and brazing in contrast to *Soft soldering* with tin–lead alloys. See Brazing.

Hard zone cracking Cracking in the *Heat affected zone* of a *Steel* weld cooled sufficiently rapidly to form martensite. See Weldability.

Harden Any means for increasing the *Hardness* and, by implication, the tensile strength of a metal. The hardness of a pure metal can often be increased by adding, at the casting stage, a suitable second, alloying, element. Metals and alloys can be hardened by cold working, i.e. *Rolling, Drawing, Forging,* etc. Alloys can be hardened by heat treatment if their composition has been selected to correspond with suitable *Phase* changes. See Steel and Precipitation hardening for examples.

Hardenability This term refers to the ability of a steel to fully harden in thick sections, i.e. at slow cooling rates. Its usual measure is the maximum section size, or *Ruling section*, of a steel that can be fully hardened through to the centre when quenched from the austenitic condition. Occasionally, some alternative combination of properties and position within the section may be specified. It is not a measure of the

maximum hardness achieved. See Steel.

Hardener A master alloy containing a large proportion of the required alloying element. It is usually added to the molten metal at a late stage as a final precision adjustment of composition.

Hardening See Harden.

Hardness (1) Resistance to indentation or abrasion.

Hardness (2) The ability to retain *Magnetism* when the magnetizing flux is removed.

Hardness (3) Colloquially, the measure of vacuum.

Hardness testing Any technique for measuring *Hardness (1)*. Hardness measurement offers a quick, cheap, usually non-damaging check of mechanical properties, particularly tensile strength. It is commonly measured by applying, to the surface in question, an indentor carrying a known load. The size of indentation or the depth of penetration is measured and a hardness figure obtained by direct dial reading or by comparison with standard tables. The use of very small loads requiring measurement at magnifications of $100\times$ or more is termed **Micro-hardness** testing. Common measuring systems are Brinell, Vickers (Diamond Pyramid) and Rockwell; see entries on respective tests for more details. Brinell result are normally reported as, for example, 212 HB indicating a hardness on the Brinell scale of 212. Vickers results are commonly reported as, for example, 212 HV_{30} (or HD_{30} or VPN_{30}), i.e. a hardness on the Vickers scale of 212. The subscript, $_{30}$, indicates the load in kilograms applied during the test but, in most cases, it can be ignored as the tables used for the calculation make due allowance for the load applied. Rockwell results are normally reported as, for example, HRC 30, i.e. a hardness of 30 on Rockwell Scale 'C'. In this case the final letter and the numeral are both important as the numeral indicates the hardness level and the letter indicates which of scales A, B,

C or D, etc. is being used. These well-established tests are accurate and hence referred to in specifications, etc. However, the standard test equipment is relatively unwieldy and alternative portable devices have been developed. Many are essentially variations of established systems but others adopt an alternative approach. One simple test is to abrade a surface with a series of files heat treated to different levels of hardness; the component is then considered to correspond in hardness to the hardest file that failed to abrade it. Another, the Shore Sclerescope, bounces a ball, contained in a glass tube, on the surface and measures the rebound height. Variations on this theme, termed **Dynamic hardness** testers, project a hard-tipped indentor at the component. The flight characteristics of the indentor, inbound and on rebound, are analysed by electronic equipment and presented on a display. The dynamic systems have advantages in terms of portability, speed and accessibility but are prone to pitfalls which introduce inaccuracy. The *Mohs* scale determines the hardness on the basis of whether the material can be scratched by a series of minerals ranging from talc, the softest, to diamond. There are many others tests, for examples see also Poldi and Knoop.

Hartman lines   Same as Lüders lines.

Hastelloy   A proprietary range of high-strength, corrosion-resistant alloys based on metals such as nickel, chromium, molybdenum, cobalt, etc.

HAZ   See Heat affected zone.

Hazard   An imprecise term usually implying a threat to the health of humans.

HCP   Hexagonal close packed. See Crystal structure.

Head   The top, in particular the separate top of an ingot mould.

Header   A vessel, typically a substantial pipe, into which a large number of lesser pipes feed. It is not necessarily set above the remainder of the system and a wall of tubes may be set into both bottom and top headers.

Header tank   A reservoir, open or closed

top, set above the remainder of the system to pressurize and feed it.

Hearth   The floor of a *Furnace* particularly when it is lined with refractory. It may carry molten metal or a solid *Charge* as in a rotary hearth.

Heat affected zone (HAZ)   The zones of unfused parent material at the edge of a weld that have been affected metallurgically by the heat of the welding process (see Weldability). See Figure 50, located at entry on Weld deposit terms. Metallurgical effects include softening, hardening, precipitation, solution, and diffusion. The term does not normally include the much more extensive volumes of material subject to *Residual stresses* despite the fact that these are induced by the weld thermal cycle.

Heat checking/chequering   Multiple surface cracking due to thermal cycling. See Thermal fatigue and Thermal shock.

Heat resisting   An imprecise term indicating a capability to operate at the temperature in question without undue corrosion or oxidation and with a useful load-bearing capacity. The term 'heat resisting steel' is often used with reference to *Stainless steels*, i.e. those with 12% or more chromium particularly when containing other elements to enhance some aspect of high-temperature performance.

Heat spotting   Same as Hot spotting.

Heat tinting   In *Metallography*, heating of polished specimens to cause selective oxidation to reveal microstructures.

Heat treatment   Any treatment deliberately applied to a metal with beneficial effect. The term is usually used in contexts where the heating and cooling cycle alone induces the desired effect; it would not normally be applied where the heating was part of a larger process, for example heating prior to forging.

Heat treatable alloy   An alloy that can be *Hardened* by heat treatment. Virtually all metals can be hardened by *Working* and then softened by heat treatment but this would not merit the term heat treatable.

Heating depth The radial depth heated by induction heating.

Heavy hydrogen See Atomic structure.

Heavy metals Metals and alloys with a high density, although the criterion for inclusion in the category is ill defined. Certainly, a density equal to or greater than that of iron or copper is always indicated and usually a density equal to or greater than that of lead is implied, including gold, molybdenum, tungsten. Alloys developed primarily for their high density for use in instruments, etc. are sometimes referred to as 'heavy metal'. For example, an alloy of 90% tungsten, 7% nickel and 3% copper has a relative density (specific gravity) of 16.5.

Heavy water Water comprising a substantial proportion of deuterium or tritium isotopes of hydrogen. See Atomic structure.

Helium An inert gaseous element, of low relative density, which boils at −268 °C. See Table 15.

Hematite *Haematite*.

Herringbone fracture Chevron marks on a *Brittle* fracture surface. The V points to the fracture origin.

Hertz The *SI* unit of frequency equivalent to cycles per second. Symbol Hz.

Hertzian stresses The compressive stresses developed immediately beneath the surface at positions of point, or line, contact loading, e.g. in bearing races at the ball contact point.

Heterogeneity Variation in composition or structure from point to point in a material.

Heteropolar bonding See Interatomic bonding.

Heusler alloys A series of non-ferrous alloys having strong ferromagnetic characteristics. Typical alloys are based on the ternary series 18−25% manganese and 10−25% aluminium, remainder copper, although some of the aluminium may be substituted by antimony, arsenic, bismuth, boron or tin.

Hexagonal close packed Same as Close packed hexagonal. See Crystal structure.

Heyn stress See Tessellated stress.

High As in terms such as Two-high describing a *Rolling mill*. The total number of rolls in a mill stand.

High angle boundary A *Grain* boundary having the *Crystal* lattices of the two grains at a large angle. Such boundaries are highly disordered and of high energy so they form preferential sites for diffusion and precipitation. However, they strongly impede *Dislocation* movement.

High brass Common *Brass*, i.e. that with the highest zinc content, and hence cheapest, that remains single alpha phase, about 35% zinc.

High-carbon steel See Steel.

High energy rate forming Shaping processes that employ very high energy release rates for a short duration, including explosive, high-velocity impact and magnetic techniques.

High-frequency induction heating See Induction heating.

High-speed (tool) steel (HSS) *Steel* capable of cutting other materials at high speed without unacceptable deterioration. Cutting at high speed generates high temperatures at the cutting tip, causing softening and hence rapid wear of the cutting edge. High-speed tool steels have high carbon contents with large quantities of elements such as tungsten, molybdenum, chromium and vanadium to stabilize the *Martensite* and *Carbide* structures and hence retain their hardness at high temperatures. Such steels need careful and relatively complicated heat treatment to develop optimum properties. In a typical case the tool is hardened as follows: (1) slowly heat to about 800 °C to minimize any distortion, (2) heat rapidly to close to the melting point, i.e. about 1250 °C to dissolve the maximum carbon, (3) after a very brief hold, perhaps a minute or two, quench into oil to achieve **Primary hardening**, (4) temper at 500 °C to precipitate further carbides and, on cooling, form further martensite from the retained austenite which is now of lower carbon content—termed **Secondary hardening**.

High-strength low-alloy steel (HSLA) *Steels* with small, but carefully selected and controlled, alloying additions, that are capable, under closely controlled manufacturing conditions, of giving high strength with good ductility. Typical additions include aluminium, vanadium, titanium and niobium of up to about 0.1% each. Some of these steels are also able to retain the high strength following welding, unlike more conventional steels which suffer softening of the zone adjacent to the weld. Also termed **Micro alloyed** steels.

High-tensile Brass Various brasses based on 60% copper and 40% zinc with minor additions, for example, 1% aluminium, 1.5% manganese and 0.7% iron. The tensile strength is significantly higher (by some 30% in the as-extruded condition) than that of the plain 60/40 brass. See Table 8.

High-tensile steel An unreliable, vague description that has been applied to any steel with *Tensile* properties better than normalized mild steel (see Steel). In the case of bolts the term is usually taken to mean a *UTS* of about 675 MPa (45 tons/in^2), or greater. See Steel.

Hipping Hot isostatic pressing, see entry on this topic.

Hobbing A precision machining process for cutting gear teeth using a rotary cutter, termed a **Hob**, which has multiple teeth arranged helically around its periphery.

Hogging Bending along the longitudinal axis. The term usually implies an unintentional or undesirable deformation of large components such as rolls, turbine rotors, casings or even ships. Depending on the component, such deflection may be caused by *Residual stresses*, external loads or by temperature gradients and it may remain even after the cause has been removed.

Hold temperature The temperature at which a component is maintained after an operation or between operations. See Weldability.

Hold time The duration of the *Hold temperature*. A maximum or minimum may be specified depending on the context.

Holding bath/furnace A bath or furnace in which material that has been melted or has received some primary treatment is held prior to casting or the next treatment.

Hole (1) A vacancy in an *Energy band* that is normally filled.

Hole (2) A vacancy in a crystal lattice.

Hollow (section) A tube or pipe. The various terms are largely interchangeable but 'hollow section' has implications of dimensional accuracy, perhaps a relatively thick wall or a complex cross-section. The term is also used for the product of the initial stage in tube making in which a billet has been pierced for subsequent processing. See Tube making.

Homogeneity The absence of variation in composition or structure.

Homogenization Heat treatment producing *Homogeneity* by diffusion.

Homologous temperature The relationship, as a ratio or percentage, of a temperature to the melting temperature of the metal under consideration, both temperatures being on the *Absolute* scale. For example, for copper with a melting temperature 1083 °C (i.e. 1356 °K), a temperature of 405 °C (i.e. 678 °K) is a homologous temperature of 0.5 or 50%.

Homopolar bonding See Interatomic bonding.

Honing Fine grinding particularly when applied to the bores of cylinders and intended to achieve a smooth high-precision surface.

Hooke's law Stress is proportional to strain. See Tensile test.

Hoop stress The stress acting around the circumference of a cylinder. It can be visualized as the tensile stress induced in an elastic band stretched round a bar. Generalizing, the hoop stress is the highest stress in an internally pressurized simple cylinder; it is approximately double the longitudinal stress. See Barlow's formula.

Horizontal–vertical welding position The position in which a weld run is approximately horizontal on components which (1) for a *Butt* weld, both are near vertical or, (2) for a *Fillet* weld, one is approximately vertical, the other approximately horizontal and the joint is made from above. This is usually taken to mean a *Weld slope* not greater than 5° and a *Weld rotation* between 70° and 90° for a butt, or between 30° and 50° for a fillet.

Horsepower A unit of the rate of doing work. 1 HP $= 550$ ft lb/s $= 745.70$ watts (J/s).

Hot cracking Any cracking at elevated temperature, but particularly cracks which initiate close to the *Solidus*. The term usually implies cracking during a reheating and cooling cycle, including welding, or during some working operation whereas *Hot tearing* more commonly implies cracking during cooling from the molten state. However, the distinction is not rigid.

Hot dip galvanizing The application of a zinc coating to steel components by immersing them in a bath of molten zinc covered by a molten *Flux*. See Galvanizing and Cathodic protection. The zinc deposit often has a bright crystalline appearance, termed **Spangle** which may be deliberately induced by additions to the bath such as antimony.

Hot dipping Generally, any process where a component is treated in a hot liquid.

Hot forming/working Deformation process above the *Recrystallization* temperature. At such temperatures metals continuously recrystallize rather than work harden so large amounts of deformation can be accomplished. Additionally, less energy and less massive equipment is required than for cold working.

Hot isostatic pressing The combined action of heat and pressure acting in all directions. The term may be applied to the *Sintering* of powders but it is commonly used of the process applied to exservice *Creep* damaged components to close creep voids and offer a measure of rejuvenation. The term is often shortened to **Hipping**.

Hot junction See Thermocouple.

Hot pass (of a weld) A weld run made over the root or top surface of a weld at high speed and, in the case of electrical arc welds, at a high current. The usual intention is to improving the surface contour for technical or *Cosmetic* reasons. Additional weld *Filler* metal may or may not be deposited.

Hot pressing The press compaction of powder at temperatures sufficiently high for *Sintering* to occur.

Hot shortness Low *Ductility*, and hence susceptibility to cracking, during high-temperature working and manipulation.

Hot spotting (1) Any localized heating using a flame or other heat source.

Hot spotting (2) A technique for deforming, but particularly straightening, bulky components such as shafts, rotors or beams by local heating. The basic technique involves applying an intense local heat source, usually an oxy-fuel gas flame, to the surface for a limited period. Local thermal expansion in the hot spot zone induces compressive stresses. If these are sufficiently high, bearing in mind the reduced yield stress at high temperature, the material will locally *Yield* in compression. When, subsequently, the temperatures equalize the yielded zone will revert to a tensile *Residual stress* which cause deflection of the shaft. The technique is fairly crude and not without potential hazard during the treatment and subsequent service.

Hot stage A facility on a microscope allowing the specimen to be heated during examination. The equipment may be simple or complex with a capability to enclose the specimen in vacuum or in a protective atmosphere.

Hot tearing Cracking at high temperature. The term is usually used with respect to *Shrinkage* cracking during cooling from the molten state. Cracking during hot working is usually termed *Hot cracking* or *Hot shortness* although the distinctions are not rigid.

Hot top A collar placed around the top of a mould to retain heat so that the metal within remains molten to provide a feed for *Solidification* shrinkage. The hot top will be a low-conductivity refractory material and may incorporate heating devices or *Exothermic* compounds.

Hot working See Hot forming.

Hot working steels *Steels* suitable for tooling for hot forming operations and hence having good strength at temperature with resistance to softening and *Thermal fatigue*. They are typically high in elements such as tungsten, chromium, molybdenum and vanadium in addition to carbon.

Hounsfield Tensometer A proprietary small-scale machine for measuring *Tensile* and other properties.

HSLA See High-strength low-alloy steel.

HSS See High-speed steel.

Hubbing Forming a *Die* by a *Forging* operation using a master die.

Huggenberger gauge/extensometer A proprietary mechanical device clamped to a *Tensile test* specimen to measure the extension. It gives a magnification of over 1000 times.

Huntsman process Same as Crucible steel process.

HV Hardness on the *Vickers* scale.

Hydrazine N_2H_4 A liquid reducing agent used in the treatment of boiler water to scavenge oxygen.

Hydrodynamic lubrication Lubrication in circumstances where the components are separated by a film of oil of significant thickness that is maintained by the hydrodynamic conditions resulting from the relative movement between the bearing and journal. See also Oil wedge.

Hydroforming A forming process in which one side of a sheet of metal is clamped against the open face of a

pressure vessel, usually with an intervening gasket or rubber sheet, and the other side is engaged by a die face. As the die moves forward hydraulic pressure within the vessel supports the sheet forming it to the die profile. The process has the benefit that only one face of the die is necessary whereas in normal die pressing operations two mating faces are required.

Hydrogen A gaseous element usually damaging in metals if present in significant quantities. See entry on Atomic structure for heavy hydrogen isotopes.

Hydrogen damage Any damage resulting from the presence of hydrogen gas. Hydrogen is a very small atom or molecule and hence can easily enter molten metals, including during welding, from moisture in the environment or in consumable materials. It can also enter and diffuse through solid metals in various circumstances including plating, chemical cleaning and corrosion. It enters most easily and is more likely to be damaging if it enters in the atomic form H, rather than the molecular form H_2. In the presence of imposed tensile stresses, either from an external source or *Residual*, it can cause cracking, see Weldability. Even in the absence of imposed stress the internal pressure of the hydrogen itself can cause surface emergent or sub-surface fissures that can subsequently act as initiators for cracking mechanisms such as *Fatigue* or *Brittle fracture*. It can also produce near-surface planar fissures leading to *Blistering*. Even when hydrogen produces no readily observable crack it may reduce the *Ductility* of certain steels, termed **Hydrogen embrittlement** (not the only use of the term), and it can lead to **Delayed fracture**, also termed **Sustained load failure**, under an otherwise acceptable steady load. Steel boiler tubes charged with hydrogen as a result of severe waterside corrosion suffer a reaction between the carbon of the steel and the hydrogen to form methane, CH_4. This causes local *Decarburization* and internal cracking leading to significant weakening. The resultant failure is of low ductility leading again to the term **Hydrogen embrittlement** although this particular usage is not usually recognized outside the power industry. *Tough pitch copper*, that is, copper containing oxygen as finely distributed copper oxide particles, is damaged if exposed to hydrogen-rich atmospheres during heat treatment, welding, brazing or service. The hydrogen reacts with the oxide to form water which causes internal fissuring with serious effects on the strength and toughness of the material. The term **Gassing** is specific to this form of attack although, again, the terms **Hydrogen embrittlement** and **Hydrogen damage** are occasionally used. Some titanium alloys suffer a considerable reduction of impact strength when charged with hydrogen due to the formation of hydride platelets which impede *Slip*. See also Fisheyes.

Hydrogen electrode A reference electrode against which other materials are compared to determine their electrode potential. See Electrochemistry.

Hydrogen embrittlement See Hydrogen damage.

Hydrolysis The dissociation of a compound dissolved in water to form free acid and alkali *Ions*.

Hydrometer Instrument for measuring the density of liquids.

Hydrophilic Attracting or having an affinity for water.

Hydrophobic Repelling water.

Hydrostatic Acting equally in all directions. In **Hydrostatic tension** the three *Principal stresses* are equal.

Hydrostatic compaction/pressing Processes in which powders intended for sintering are enclosed in a flexible bag or similar container and subjected to a high-pressure environment, usually hydraulic. This allows the use of a higher but more even pressure than conventional pressing and reduces wear of the equipment.

Hydrostatic extrusion Extrusion process in which the material to be extruded is in the same chamber as the hydraulic fluid. The material is initially sealed against the die and may be pre-shaped to fit in the die orifice.

Hydroxyl The OH portion of a compound occurring in solution as in the OH' ion.

Hygroscopic Absorbing water.

Hypereutectic, hypereutectoid Having an alloy content greater than that of the eutectic/eutectoid composition. See entry on Eutectic but note that the term is only appropriate where the solute content is small compared with the solvent. Otherwise the alloy can be considered from the viewpoint to the solute and the hypereutectic will be regarded as the hypoeutectic and vice versa.

Hypoeutectic, hypoeutectoid Having an alloy content less than that of the eutectic/eutectoid compositon. See entry on Hypereutectic.

Hysteresis The lag or delay in the change of one variable as another related variable changes. For example, magnetic induction lags behind the change in the magnetizing force. The **Hysteresis loss** is the loss arising from a single cycle of the **Hysteresis loop**.

Hz *Hertz*.

I

IACS International Annealed Copper Standard. See entry on this topic.

Iceland spar A transparent form of *Calcite*.

ID Internal diameter of a tube or similar hollow material and in contrast to OD, the outside diameter.

Idiomorphic Crystals having a shape corresponding to their *Crystal lattice*. They occur when there is no external constraint to their growth.

Ignition temperature The temperature at which combustion can commence. See also Gas cutting.

Ihrigizing A process for forming a hard, corrosion-resistant silicon coating on *Steel* by heating at high temperature in a silicon tetrachloride vapour.

Immersion oil An oil that is interposed between the specimen and the objective lens of a *Microscope*. It has a high refractive index compared with air and hence improves the resolving power.

Impact extrusion See Extrusion.

Impact(er) forging A *Forging* process in which the component is forged between a pair of moving die faces so both sides of the component are worked.

Impact test Generally, any test in which the load is dynamic rather than static. More specifically, a test in which a specially machined test piece is dynamically loaded in a purpose-made machine which registers the energy absorbed in causing failure as a measure of the material *Toughness*. The test pieces usually have a simple V notch to initiate the fracture, hence the term **V notch test** but less common variations include plain bar tests and **Keyhole**-shaped notch tests. The two most popular testing machines are the **Charpy** and the **Izod**. Both of these utilize a weighted pendulum carrying a striker which impacts the test piece located at the low point of the swing (see Figure 19). The Izod striking energy is 167 J (120 ft lbf) and that of the Charpy 300 J (220 ft lbf). The test is commonly used for *Steels* as many exhibit a dramatic reduction in toughness as the temperature falls. This reduction, reflecting a change in fracture mode from *Ductile* to *Brittle*, is measured by recording the energy absorbed by a series of test pieces broken over a range of temperature (see Figure 20). In addition, the characteristics of the fracture faces may be examined to determine the relative proportions of the ductile and brittle (or crystalline) zones. The observations provides another measure of the change from brittle to ductile behaviour by identifying the temperature at which a fracture is 50% ductile and 50% brittle. This temperature is termed either the **Fracture appearance transition temperature (FATT)** or the **Ductile/brittle transition temperature (DBTT)**. Reference may also be made to the **Nil ductility transition temperature (NDTT or even NDT)** which is the highest temperature at which no ductile fracture is observed. This latter term is sometimes also applied to the temperature at which some specified low level of impact strength is recorded, for example 10 ft lb in the Charpy test. See Brittle fracture and associated figures.

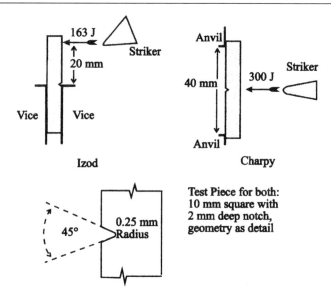

Figure 19   Charpy and Izod impact tests, test piece and testing layout

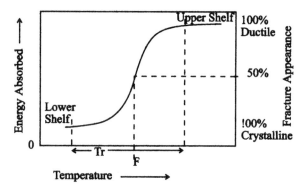

F -Fracture Appearance Transition Temperature (FATT)

Tr- The temperature range variously termed:
Brittle to Ductile Transition Range
Fracture Transition Range, etc.

Figure 20   Graph plotting the results of a series of impact tests in terms of the energy absorbed at the test temperature

Imperfection (in *Crystal Lattice*) Any deviation from perfection in lattice alignment such as a *Dislocation* or any atomic imperfection such as a substitu-tional or interstitial atom in a solid *Solution* or a *Vacancy.*

Impingement corrosion/attack A form of surface damage occurring in flowing liquids resulting from entrained particles or turbulent flow and associated bubbles abrading or disrupting the normally protective surface oxide film. The exposed metal is *Anodic* to the neighbouring oxide and hence is preferentially corroded (see Electrochemistry). Generalizing, the likelihood and severity of attack are increased by increases in fluid velocity, fluid contamination (suspended, dissolved or deposited), surface roughness and irregularities in the system such as bends and inlets to tubes. See also Corrosion–erosion and Cavitation.

Implant   A manufactured device inserted into the human body to replace or reinforce some structure or undertake some function.

Impregnation (1)   Any process for filling the pores of porous casting to prevent leakage.

Impregnation (2)   Any process for inserting grease or other materials into *Sintered* components.

Impressed current/voltage   See Cathodic protection.

Impurities   Any undesirable material in a metal, component, solution, etc.

In situ   In its normal place. In a metallurgical context this Latin term indicates that the item of interest is in its normal location rather than being removed to a laboratory.

Incandescence   The emission of light radiation by a substance at elevated temperature, above about 540 °C in dim background light.

Inclined welding position   A loose term for welding positions not lying within the limits of the four basic positions, *Flat, Vertical, Horizontal vertical* and *Overhead*.

Inclusion   Insoluble particles in the microstructure of a metal that are usually non-metallic and usually undesirable although there are exceptions such as in *Free cutting* materials.

Incoloy and **Inconel**   Ranges of proprietary alloys based on nickel with major additions of other elements, in particular

chromium and iron but also, variously, molybdenum, cobalt, titanium, aluminium, etc. Generalizing, they have good corrosion resistance including resistance to oxidation at high temperatures.

Incomplete fill (of weld)   An insufficient quantity of *Filler* material in a welded joint, particularly a butt joint. Same as Underfill, (see Figure 52, located at entry on Welding terminology).

Incomplete fusion (of weld)   Same as Lack of fusion. See Figure 52, located at entry on Welding terminology.

Indices   See Miller indices.

Indirect extrusion   See Extrusion.

Indium   A metallic element. It is lustrous, soft, corrosion resistant and has semiconducting characteristics. Small quantities added to lead, white metal and copper–lead plain bearing materials improve their corrosion resistance. See Table 15 for physical properties.

Induction hardening/heating/melting   Processes in which heating is by the electrical induction of eddy currents in the material being treated. The general principle is that the material is surrounded by a water cooled conducting coil carrying AC current. Eddy currents are induced in the material causing heating. Low- and medium-frequency currents, up to about 100 kHz, cause general heating through the thickness or volume of material and hence are used for melting and billet heating applications. Higher frequencies induce currents at the surface, the so-called **Skin effect**, and hence are useful for *Case hardening*.

Induction welding   Processes in which the heat for welding is induced by an alternating electric current. A forging action is also normally applied to effect the joint. If the AC frequency is 10 Hz or higher the process can be termed **HF (high frequency)** induction welding.

Inelasticity   See Anelasticity.

Inert   Not chemically reactive.

Inert gas shielded cutting/welding   Processes in which the heat source is an electric arc and the cutting/welding zone is protected from the environment by an inert gas such as argon.

Infiltration A process for filling the pores in *Sintered* materials with some other material, in particular a metal introduced in the molten state and entering by capillary action.

Ingot A casting of simple shape that will be subjected to further working operations. The as-cast ingot is normally cropped to remove the top section containing impurities and other defects.

Ingot iron High-purity iron produced by steelmaking processes.

Ingot mould The substantial vessel into which molten metal is poured to solidify as an ingot.

Ingotism Excessively large *Grain size* in castings, usually the result of casting at too high a temperature.

Inherent (austenitic) grain size The austenitic grain size developed in steel when cast.

Inhibitor A substance which, when added to an otherwise corrosive fluid, prevents or limits attack of exposed surfaces.

Injection moulding Casting by injecting into a *Die*, often of complex shape, an amount of molten metal just sufficient to fill it.

Inoculated (cast) iron *Cast iron* with a fine, evenly dispersed graphite flake structure giving improved mechanical properties compared with a normal Grey cast iron. This structure is developed by **Inoculation**, that is, adding a small quantity of finely powdered refractory material, such as calcium silicide, to the molten metal immediately prior to pouring. The powder particles provide many nucleation sites for the graphite and hence ensure the development of large numbers of fine graphite flakes.

Insert A general term referring to small devices or piece of material set into some larger component by casting-in, pressing, screwing, etc. to fulfil some special local function such as a bearing, screw location, etc.

Insulators Materials that are poor con-

ductors of heat, electricity or sound, etc. See also Semiconductor.

Intellectual property Matters such as copyright, patents, drawings and designs, the ownership of which is legally enforceable.

Intelligent materials See smart materials.

Interatomic bonding The manner in which atoms bond together. In **Electrovalent bonding**, also termed **Ionic**, or **Heteropolar bonding** an element, typically a metal, having a small number of valence electrons combines with one having a large number, typically a nonmetal. For example, sodium has a single valence electron in its outer shell and eight electrons in its full second shell. Chlorine has seven valence electrons in its outer shell. When these two elements combine a sodium atom donates the single valence electron in its outer shell to the outer shell of a chlorine atom. Both atoms then have full, and hence stable, outer shells. The atoms, respectively depleted and increased by an electron and its associated charge, are termed **Ions**, represented as Cl^- and Na^+. These opposite electrostatic charges powerfully attract each other forming a strong bond for the solid compound. In a solution, ionically bonded materials can **Ionize** allowing the two ions to move independently within the solution. Such movement is the phenomenon underlying *Electrochemistry*. In **Covalent bonding**, also termed **Homopolar bonding**, the atoms share electrons to fill their outer shells to form fairly stable molecules. A simple example is hydrogen with its single valence electron; molecules of hydrogen are formed by a two atoms each contributing its electron to a shared outer shell filled with two electrons. **Coordinate** or **Dative covalent bonding** is a variation in which the bond is formed by the sharing of a pair of electrons donated from one of the atoms. In **Metallic bonding** the valence electrons of all the atoms are pooled, effectively forming a cloud shared between the mass of atoms, although see Band theory. Apart from these relatively powerful bonds between atoms there are lesser secondary forces capable of holding molecules together. These are termed **van der Waals' forces**.

Intercept method Techniques for measuring microstructural features. A line is projected across the structure and the intercept points at which it crosses the various phases boundaries are noted. The distances between intercepts are totalled for each phase to give a ratio of the volumes of the phases. Alternatively, the number of features, such as grain boundaries or inclusions, that intercept the line may be recorded to allow calculation of grain size or the quantity of inclusions.

Intercrystalline Any feature at the interface between neighbouring *Crystals*.

Interdendritic Occurring or located at the boundaries between *Dendrites*.

Interdendritic corrosion Corrosion which results from composition gradients from centre to edge of dendrites leading to the formation of an *Electrochemical* cell.

Interface The boundary or gap between two or more surfaces which are in contact or close proximity.

Interference fit A joint in which one component fits inside, and is slightly larger than, the receiving orifice in the other. They may be simply pushed together, termed a **Force fit**, or the outer may be heated (or less commonly, the inner cooled) to allow the inner to be inserted easily, termed a **Shrink fit**. A typical shrink fit interference would be a difference in dimensions of one in one thousand. Less common, and less reliable, is the **Wrung fit** in which the interfaces are slightly tapered and the components are rotated relative to each other as they are pushed together. They are then held together by a combination of *Interference* and *Galling*.

Interference In light and other energy forms transmitted as a wave form, the alternate bright and dark bands which result when two transmissions meet with the waves out of phase.

Interference techniques Techniques of microscopical examination which utilize a single beam of light which is split into two, one of which is reflected on to the surface being examined the other onto a high-quality reference surface. The return beams from the two surfaces are then recombined so that any irregularities on the surface being examined are revealed by interference patterns.

Intergranular Occurring or located at the interface between neighbouring grains.

Intermediate heat treatment/annealing, etc. Heat treatment etc. carried out between stages of *Cold Forming*.

Intermediate phase In an alloy system, any *Phase* with a composition range which does not include the pure metal.

Intermetallic compound In an alloy system, a phase having a narrow composition range corresponding with a simple ratio of the atoms of the two or more elements involved, i.e. $CuAl_2$. Such compounds usually have specific crystallographic structures and are hard and brittle.

Intermittent weld A joint formed, deliberately or otherwise, by a series of longitudinal welds with gaps between. Where such a weld is deliberate, a **Chain intermittent weld** has the welds on one side aligned with the welds on the other and a **Staggered intermittent weld** has the gaps on one side aligned with the welds on the other. See Figure 49, located at entry on Welding terminology.

Intermolecular forces The relatively weak van der Waals' and hydrogen bond forces which bind molecules.

Internal Energy The total energy contained in a material or system.

Internal friction (1) The energy loss involved in the elastic deformation of a material. It gives rise to *Damping* and a measure of its value is the *Mechanical hysterysis* loss.

Internal friction (2)   The resistance to movement of a fluid in a system.

Internal oxidation   The oxidation of parent material or other phases within the body of the component due to diffusion of oxygen from the exterior. The effect is usually damaging but it may be deliberately induced to produce large quantities of fine oxide particles having a *Dispersion hardening* effect.

Internal stress   Same as *Residual stress*.

International Annealed Copper Standard (IACS)   A measure of electrical resistance and its reciprocal, conductivity. The Standard defines pure annealed copper as having, at 20 °C, a specific resistance of 1.7241 $\mu\Omega$ cms and thus a conductivity of 0.58001 reciprocal $\mu\Omega$ cms. This conductivity is then defined as 100% IACS to provide a comparison for all other metals. As examples, pure annealed silver would have an IACS of 105% and a cold worked cadmium–copper an IACS of 85%.

International screw thread   An internationally agreed system of screw threads of standard profile with a radiused root, a truncated crest and dimensions in millimetres.

Interpass temperature   The lowest *Weld* zone temperature encountered between two *Passes*. See Weldability.

Interrupted quenching (1)   Quenching in which the component is removed from the cooling medium well before it has cooled to the temperature of the medium.

Interrupted quenching (2)   Quenching a component into a bath at some elevated temperature, holding it in the bath until it reaches the bath temperature and then removing it for further cooling which may be a further quench.

Interstice   Generally, any gap situated between two features. More specifically, in a metallurgical context, the space between the atoms forming the main *Crystal lattice*.

Interstitial   The space, or an atom located in the space, between the atoms forming the main *Crystal lattice*. Interstitial atoms are much smaller than the primary atoms, for example carbon and nitrogen are interstitials in iron. See Solution.

Intrados   An architectural term referring to the lower, inner, curve of an arch, sometimes used in engineering circumstances to refer to the inner surface of a bend, particularly bent tubes. The outer surface is the *Extrados*.

Intragranular   Any effect such as precipitation occurring within individual *Grains*. Not to be confused with *Intergranular* (between grains) or *Transgranular* (across grains).

Intrinsic semiconductor   See Semiconductor.

Intumesence   A major expansion in volume on heating. The expansion is far in excess of normal thermal expansion of a pure substance and is associated with processes such as gas emission, fluids boiling, etc.

Intrusion   Surface notches and steps formed, on a microscopical scale, by local slip during *Fatigue*.

Invar   An alloy of iron with 36% nickel which has a very low thermal expansion over a range of temperature from about ambient to about 140 °C.

Inverse rate cooling curve   A graph plotting temperature on the vertical scale against time taken for incremental falls or rises in temperature (for example, one degree intervals) on the horizontal scale. Such graphs are plotted for metals being heated or cooled and inflections in the curve indicate the temperature of critical points such as *Phase* changes.

Inverse segregation   *Segregation* where, contrary to normal, an excess of the lower melting point metal is located at the outer surface of a casting.

Inverted 'V' segregation   See Segregation.

Investment casting   A technique essentially similar to the *Lost wax process* and *Cire perdu*. All are casting processes in which a disposable pattern is made from wax. A mould of refractory material is formed round the wax pattern and, after the mould has set firm, the wax is melted and drained out leaving a shaped cavity into which molten metal is poured. The casting has no parting line or *Flash* and, at least in the case of modern investment castings of components such as gas turbine blades, a high precision is implied. The initial wax pattern may be produced by casting in a master mould and used as-cast or it can receive further surface detailing. The final refractory mould is used only once as it has to be broken away. Hollow castings can be formed using ceramic or water soluble cores. The basic process is ancient and 'the lost wax process' has implications of antiquity while 'investment casting' is more associated with modern precision. The term 'cire perdu' is often favoured for art work and by the antique trade. A typical modern production sequence for multiple precision casting involves manufacture of a high-precision metal master die to form the expendable patterns, typically a mixture of wax, resin and filler. The pattern may receive minor surface detailing or fettling prior to being mounted with a number of others onto a 'tree' to form a runner system. This assembly is then coated with a refractory, usually by dipping into a thin slurry, termed 'investing'. The initial coating is of high quality and finish and after it has dried further coats of a refractory are applied to form a **Shell**. The wax is removed by melting or solution and the shell assembly is then heated to harden the refractory. Molten metal, often vacuum melted and cast, is introduced. After cooling, the assembly of castings is parted and individual castings finally fettled.

Iodine A gaseous element, one of the halogen group. It is very corrosive but insoluble in water. See Table 15 for properties.

Ion The electrically charged particle formed by an *Atom* or *Molecule* losing or gaining an electron. See Electrochemistry with reference to ions in solutions and also Interatomic bonding.

Ion(ic) bombardment/plating/scrubbing Processes in which a component in a vacuum or suitable gas at low pressure forms the cathode of an electrostatic circuit and is impacted by positive ions as a means of revealing structure, plating or cleaning.

Ion implantation The development of *Semiconductors* by bombarding the base material with an ion beam to implant ions.

Ionic bonding See Interatomic bonding.

Ionic conduction Conduction arising from the movement of ions rather than electrons or holes.

Ionize See Interatomic bonding.

Iridium A noble metallic element. Apart from limited applications that utilize its corrosion resistance it is used as a hardener for platinium. See Table 15 for properties.

Iron A metallic element. An important characteristics is that it is allotropic, i.e. it exists in the solid state in two crystalline forms (see entry on Steels). In its pure form, iron has limited commercial use but in combination with other elements it has a wide range of attractive properties. Carbon is the most common addition and iron carbon alloys are defined, not very logically, as follows:

Steel contains up to about 2% carbon, possibly with other additions. See entry on Steel.

Wrought iron refers usually, in a metallurgical context, to low carbon iron containing elongated inclusions and made, in the final stages, by repeated folding and forging. See entry on Wrought iron.

Cast irons are iron alloys with carbon contents of about 2–4%. Note that the cast iron industry commonly omits the term 'cast' in descriptions such as 'malleable (cast) iron', 'Grey (cast) iron' and even (cast) 'irons'. See entry on Cast Iron.

See also Blast furnace and Table 15 for properties of iron.

Iron carbide   An iron carbon compound, Cementite, FeC_3. See Steel.

Iron loss   The loss of power in an electrical circuit arising from eddy currents in the magnetic iron core of transformers and similar devices.

Iron powder electrode   A welding *Electrode* containing a large proportion of iron powder which becomes incorporated into the weld metal.

Ironstone   A mineral, usually iron carbonate, containing a large proportion of iron.

ISO   International Standards Organization.

ISO metric fasteners   An ISO-approved system of sizes, tolerances and materials for threaded fasteners. For steel fasteners the material strength is identified by two numbers, the second being a decimal, for example '12.9'. The first number is one tenth of the *Ultimate tensile strength* of the material in kgf/mm^2, the second number is the yield to ultimate ratio as a decimal. Thus in the example quoted, 12.9, the material has a UTS of $120 \ kgf/mm^2$ and the *Yield strength* is the UTS times 0.9, i.e. $108 \ kgf/mm^2$.

Iso- (various combinations)   Equal. Examples below.

Isobar   A line joining points of constant pressure.

Isochronous   Occurring at or related to the same time or the same time interval.

Isometric (crystal structure)   Cubic.

Isostatic   At equal pressure. In some usages the term indicates that pressure does not vary with time but in terms such as *Hot isostatic pressing*, it indicates that the pressure acts equally in all directions.

Isotherm   A line joining points of equal temperature.

Isothermal   Occurring at a single temperature. Isothermal reactions occur over a period of time and involve an input or output of heat.

Isothermal annealing   Austenitising steel and then cooling to just below the critical range until it transforms, at a constant temperature, to a soft structure of ferrite/carbide aggregates. See Steel.

Isothermal transformation diagram   A diagram indicating the phase changes that occur in a steel quenched from the austenitic state to various temperatures. See Steel. These diagrams are also termed **Time, temperature, transformation (TTT)** diagrams for obvious reasons or **'C'** or **'S' curves** because of the shape of the transformation boundaries in some of the diagrams. The concept of the diagram is that it indicates the changes undergone by a steel after it has been quenched virtually instantaneously to the temperature in question and held at that level until all transformation is complete or equilibrium is reached. Transformation to ferrite/carbide structures, including bainites, is diffusion controlled and hence time dependent so the diagram indicates the progress of tranformation to these phases as time elapses. Transformation to martensite is, effectively, an instantaneous shear process, not time controlled but temperature dependent. The diagram therefore indicates the percentage transformation to martensite at any temperature. Figure 21 is a typical example for a low-alloy steel. It can be considered as two sets of C curves, the upper set defining transformation to ferrite–pearlite structures, the lower defining transformation to bainite. The projecting edges of these curves, corresponding to the minimum times to commence trasformation may be referred to as the **Ferrite nose, Pearlite nose** and, most common, **Bainite nose**. The long, flat portion of the bainite start line is often referred to as the **Bainite shelf**. In commercial practice, bulk components cannot be instantaneously quenched so the diagrams have to be used with some caution. However, it is possible to superimpose on the diagram a cooling curve such as the one illustrated. This particular curve, which just misses the bainite nose, corresponds to the **Critical cool-**

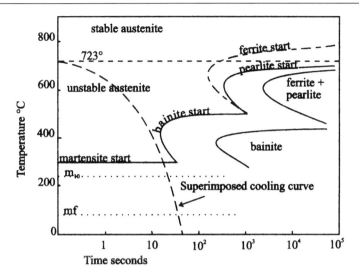

Figure 21 Typical time, temperature, transformation (TTT) diagram for an alloy steel

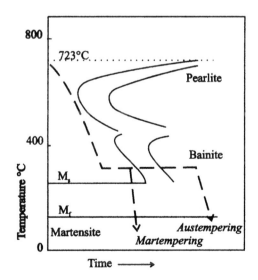

The broken lines represent the cooling of austempered
and martempered steels plotted on a pseudo TTT curve.

See Figure 21 for a more detailed TTT curve.

Figure 22 Austempering and martempering

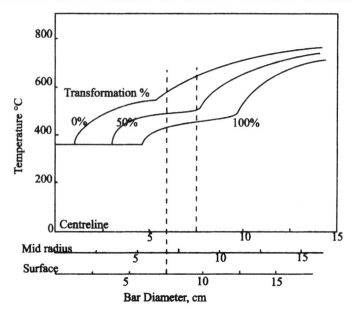

Bar Diameter, cm

Lines drawn upwards from the bar size intersect the transformation
lines to indicate the temperature for the stage of transformation. Thus
the surface of a 7.5 cm diameter bar will transform over the range 570
to 430° C and its centreline over the range from 640 to 450° C.

The composition, quench medium and, usually, the austenitizing
temperature are specified.

Figure 23   Typical continuous cooling transformation curve (CCT) diagram for an alloy
steel

ing rate since any slower rate would not
develop a fully martensitic structure.
The diagrams for alloy steels, which
typically have sluggish transformation,
can also be used to devise treatments
such as *Martempering, Austempering*
and *Ausforming* in which the steel is
rapidly cooled to an intermediate temperature, held for a period insufficient
for transformation to commence and
then cooled to ambient (see Figure 22).
Where it is necessary to determine the
progress of transformation as temperature falls progressively, **Continuous
cooling transformation (CCT)** diagrams are employed. Such diagrams
(Figure 23) depict the progress of transformation occurring at the exterior surface, centreline and midradius of a
range of sizes of round bar subjected to
specified austenitizing conditions and
quenching medium.

Isotopes   See Atomic structure.

Isotropic   Having   similar   properties
along all axes.

Izod test   A form of *Impact test*.

J

J *Joule.*

J integral The property describing the behaviour at the tip of a crack in a material exhibiting significant *Ductility*. It is analogous to '*K*' in a *Brittle* material. See Fracture toughness.

Jacquet polishing Electrolytic polishing particularly for metallographic examination. See Electrochemistry.

Jam nut See Lock nut.

Jet test Various laboratory tests in which a jet of liquid impinges on the surface of the material in question to determine its corrosion characteristics under flowing conditions. See BNF jet impingement test.

Jewellers' rouge Red iron oxide, haematite, Fe_2O_3, in fine powder form used as an abrasive and polish. It is now rarely used for *Metallography* as it tends to smear rather than cutting cleanly like diamond paste.

Jig Equipment that positions and supports components during some operation such as assembly or welding. It may also guide the tool.

Jog A step in a *Dislocation* produced when the dislocation traverses another.

Joggle Misalignment of the surfaces of a pair of butting plates

Joint (bolted/welded) The junction of two or more components which have been, or are to be, held together by some mechanism such as bolting or rivetting or bonded together by some process such as welding or adhesive.

Joint efficiency The strength of a joint made by welding, bolting, etc., relative to the strength of the parent materials, usually expressed as a percentage.

Joint severity (of weld) See Weldability.

Jominy (end quench) test A test for assessing the *Hardenability* of steel in which a standard bar specimen is heated to the austenitic range and then one end is quenched by a jet of water. This results in a progressive variation in cooling rate along the bar. Following cooling the bar is hardness tested along its length to determine the extent of hardening and it may also be metallographically examined to establish the limit of the martensitic and bainitic zones.

Joule The SI unit of energy, symbol J, quantity of heat or work done. One Joule is the work done when the point of application of a force of one Newton is displaced by one metre. 1 J = 0.728 ft lb = 1 watt s = 0.239 calories.

Joule effect (1) The heat resulting from the passage of an electric current. It is expressed as: $J = I^2 Rt$, where I is current in amps, R is the resistance of the conductor in ohms and t is the time in seconds.

Joule effect (2) See Magnetostriction.

Journal (bearing) The surface of a shaft that runs within a *Plain bearing*. The surface is normally a precision dimension of high quality.

K

°K (strictly, the degree symbol is superfluous) Temperature in degrees Kelvin, i.e. the absolute temperature in the *SI* system. 0 °K is −273 °C.

K, K₁c, etc. K is the *Stress intensity factor* defining the relationship, at the crack tip, between *Stress* and crack size. The numeric subscripts identify the mode of cracking, for example '1' indicates the common case of the crack being pulled open. The subscript 'c' indicates the critical level of K at which a crack will extend. See Fracture toughness.

Kaldo process One of the first oxygen steel processes. In the process a stream of oxygen impinges on the surface of molten steel in a rotating converter vessel. The vessel has a *Basic* lining with a lime *Flux* and the process removes both carbon and phosphorus from high-phosphorus pig iron. The use of oxygen avoids the nitrogen pickup associated with air lancing.

Keel block A casting for sample testing made in a Y-shaped mould so that the large head provides ample molten metal to feed shrinkage. The head is discarded and the stem tested.

Keeper A piece of magnetically soft iron set across the poles of a permanent magnet to close to magnetic circuit and hence preserve the magnetism.

Keller's etch 2.5 ml nitric acid, 1.5 ml hydrochloric acid, 1 ml hydrofluoric acid, water to 100 ml. It is used for aluminium alloys.

Kelvin The *SI* unit of temperature. See Absolute temperature.

Kelvin effect (1) The electromotive force (emf) produced by a difference of temperature between two sites of a conductor.

Kelvin effect (2) The release or absorption of heat associated with a electrical current flowing between two sites on the conductor which are at different temperatures. See also Thermoelectric effect.

Kelvin effect (3) The *Skin effect* whereby high-frequency current tends to flow in the conductor surface

Kerf The width of a saw cut or other gap left when material has been cut away. Also the cut face.

Kettle A small vessel with a lose top in which small quantities of low melting temperature metals are melted.

Kevlar A proprietary polymer, an aromatic polyamide fibre having stiff polymer molecules strongly aligned to the longitudinal axis. Its high tensile strength and elastic modulus make it particularly suitable for the reinforcement of *Composites*. The term is also loosely used of the composite material containing the Kevlar fibre. See Table 14.

Key (1) In a metallurgical context this term refers to the ability of a surface to provide good adhesion for paint or other coatings. The key may be improved by mechanical means such as abrading or *shot blasting* or by chemical treatment.

Key (2) In a mechanical engineering context a piece of material set longitudinally, half and half, in the interface between, typically, a shaft and a body mounted on it. It is inserted in the **Keyways** in the two components to maintain location or provide drive.

Keyhole notch A notch, in an *Impact test* specimen, which has a narrow neck at the surface and terminates within the material at a round hole. The fracture is intended to initiate at the hole opposite the neck.

Kill/killing See Killed steel

Killed spirits Saturated solution of zinc chloride used as *Soldering Flux*.

Killed steel (1) Usually, a *Steel* which has been fully deoxidized with a strong deoxidizer such as silicon manganese or aluminium preventing any reaction between carbon and oxygen during *Solidification*.

Killed steel (2) Less commonly than (1), a steel that has been lightly worked to eliminate the pronounced yield point that gives rise to *Lüders lines*.

Kiln Any vessel or enclosed space in which material is exposed to some heating process. 'Kiln' often implies that the charge is not exposed to the heat source but generally a choice between 'kiln', 'oven' and 'furnace' is more a matter of preference in the industry concerned rather than any consistent technical difference.

Kinetic energy The energy possessed by a body by virtue of its velocity relative to some reference point.

Kinetic theory At a temperature of absolute zero atoms and molecules are at rest. As the temperature rises the atoms and molecules absorb energy which they retain as kinetic energy associated with their vibration. As the energy increases, the atoms and molecules vibrate increasingly until at some stage they are able to flow over each other, i.e. the material becomes liquid. A further increase in energy and vibrations allows individual atoms or molecules escape from the bulk, i.e. the material becomes a gas. The impact of the vibrating atoms on the containment is manifested as presure.

Kinks/kink bands Deformation bands associated with *Slip* on multiple parallel *Crystallographic* planes.

Kish (graphite) A coarse graphite produced in the early stage of the solidification of *Cast iron*. It tends to float to the surface but if it is entrapped in the casting it is very weakening.

Knife line/edge attack Local attack in the *Heat affected zone* of welds in some austenitic stainless steels. Similar to *Weld decay* except that it is located in the temperature zone that closely approached the *Solidus*.

Knoop (hardness) test A microhardness test employing a diamond indentor cut to produce an indentation of diamond shape with one diagonal very long and the other very short and hence useful on the edge of thin sheet. See Hardness test.

Knurling A pattern of deep indentations, typically a diamond grid and similar in appearance to a file face, formed on a surface, usually cylindrical, to improve grip. The effect is achieved by forcing hard rollers, machined to the required pattern, against the rotating surface.

Krypton An element, one of the inert gases. See Table 15 for properties.

L

Labile   Unstable.

Lack of fusion (of weld)   The absence of the intended fusion bond between components of a welded joint. It can occur at various positions, see Figure 52, located at entry on Welding terminology.

Ladle   A vessel for transporting molten metal from one stage to the next, particularly from the melting furnace to the mould. It may range from a hand-held 'spoon' to very large mechanized or crane-carried vessels holding many tons of metal.

Ladle addition/treatment/process   Processes for the final refinement and composition control of metal before it is cast, particularly treatment in a ladle or holding bath with no additional heating.

Lagging   Thermal insulation particularly when it is applied on the exterior of the vessel being protected rather than on the interior and in contact with the reaction environment.

Lamellae   Multiple thin plates stacked as laminations.

Lamellar corrosion/attack   The loss of material, as layers, from a metal surface. Corrosion proceeds along a plane just below and parallel with the surface and the released layer is substantially uncorroded. The term *Exfoliating corrosion* is usually applied to cases where the sheets of material released are entirely or predominantly corrosion products.

Lamellar cracking/tearing   Cracks, in rolled products such as plate or bar, that lie beneath and parallel with the surface. They usually follow planes of weakness, particularly areas of non-metallic *Inclusions*, introduced at the original casting stage and deformed to a planar form during rolling. The planes of weakness have little effect on the longitudinal or transverse strength but considerably reduce the strength in the through-thickness direction. Significant loads acting in this direction are not normally encountered except in the case where a weld is applied to the surface of a plate or bar. The initial defect can develop as a result solely of the stresses introduced by the welding operation, in which case cracks will often be confined to the parent metal beneath the weld. However, when a service load is applied through the welded joint the cracks can readily propagate, leading to failure.

Laminar flow   The movement of a fluid stream such that no lateral intermixing occurs, effectively streamline flow.

Laminate   A *Composite* material formed by two or more layers of sheet bonded together.

Laminations (1)   Features, in particular planes of weakness associated with *Inclusions*, aligned parallel to the principal surface.

Laminations (2)   Generally, any sheets intended for assembly as a pile. More particularly, varnished (for insulation) steel sheets piled and clamped together to form the core of transformers, generator stators and some rotors. This construction, together with the use of special steels, minimizes *Iron loss*.

Lamp black   Soot, finely particulate *Carbon*.

Lance   A tube directing some fluid into a process vessel.

Land   A flat surface on a component, particularly one machined to provide a seat or similar function.

Lang's lay   One of the more common systems of laying, i.e. assembling, the individual fibres or wires to form a multistrand rope or cable. In essence all wires and strands follow the same twist direction.

Lanthanides   The rare earth metallic elements with atomic numbers 57 (Lanthanum) to 71. See Table 16.

Lanthanum   A metallic element, the most common of the rare earth group. See Tables 15 and 16.

Lap (1)   A planar manufacturing defect, either a small *Cold shut* from casting or a *Fold*, produced during some working operation, that has been flattened but not welded or bonded.

Lap (2)   A polishing pad carrying embedded abrasives.

Lap joint   The joint formed between two overlapping components such as plate or sheet. The joint may be effected by riveting, *Brazing* or various welding techniques including *Spot welding* through the interface or *Fillet* welding the internal corner between edge and face. See Figure 47(b), located at entry on Welding terminology.

Lap (shear) test   A tensile test in the plane of the parent plates of a *Lap joint* so that the load is carried by the full joint, as opposed to a *Peel test*.

Lapping   Fine grinding and polishing on a *Lap (2)*.

Larson–Miller parameter   An empirical relationship for predicting the complete range of *Creep* rupture properties of a material on the basis of a limited number of tests. Creep life is a function of stress and temperature and, for a particular material at a fixed stress, the parameter P is given by:

$$P = T(c + \log t)$$

where T is the temperature in degrees absolute, t is the time to failure and c is a constant for each material. Normal practice is to undertake a series of tests over a fairly broad range of stress levels and then plot the P value against the logarithm of the stress. This produces a smooth curve which can be correlated with rupture life and temperature as shown in Figure 24. Thus when any two variables are specified the third can be deduced. For example, a designer might need to know the maximum stress allowable to achieve a life of 10 000 hours at 500 °C for the material in question. To determine this, referring again to Figure 24, a horizontal line is drawn across at the 500 °C level to intersect the 10 000-hour life line and from this intersection a vertical line is drawn down to intersect the P curve. A line is then drawn horizontally from the P line intersection to cross the stress scale at the allowable stress.

Laser   Light Amplification by Stimulated Emission of Radiation. A technique for producing a high-intensity beam of monochromatic coherent light.

Laser cutting/welding   Processes utilizing a laser beam as a heat source. The high-energy intensity of the focused beam facilitates deeply penetrating cuts and welds with minimum distortion. In the case of cutting, material can be removed solely by some combination of melting and vaporization. **Gas jet laser cutting** utilizes an additional gas jet to assist ejection of molten material.

Latent heat   The heat emitted or absorbed in a change of state, i.e. from solid to liquid.

Lateral   Acting on or located at a side or acting from one side to the other.

Lateral contraction   The contraction across the width and thickness of a material being loaded in tension. See Poisson's ratio and Anticlastic.

Lathe   A machine tool in which the work piece is gripped in the rotating chuck and possibly supported at its far end by a tail stock. A **Lathe tool** carried in the tool head of the **Saddle** cuts the work piece as the saddle travels along

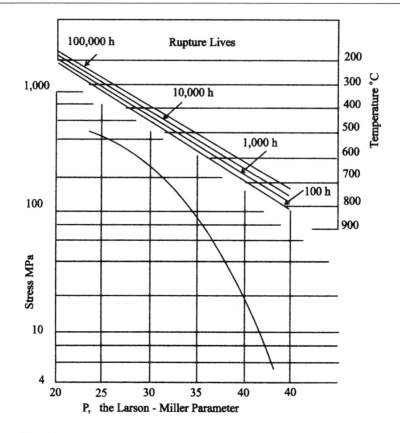

Figure 24   A typical Larson–Miller display for a low-alloy creep-resisting steel

the **Bed** or the tool head traverses the saddle.

Lattice   The three-dimensional grid on which it is visualized atoms are arranged in the *Crystal structure* of solid metal.

Lattice constant   Same as Lattice parameter.

Lattice construction/girder, etc.   An assembly of slender girders, beams, etc. joined in a rigid triangulated manner to provide a lightweight but high-strength construction.

Lattice defects   Defects in the *Crystal structure* such as *Dislocations*, *Vacancies* and *Foreign* atoms.

Lattice parameter   The edge length of a unit cell of the *Crystal structure*.

Laue, von   See Black reflection X-ray technique.

Launder   An open channel through which molten metal runs into a mould.

Laves' Phases   Intermediate *Phases* formed from two metals in the ratio of one to two of the other, such as $MgCu_2$, and where the two atoms differ in size by about 22.5% allowing them to form a *Close packed* structure.

Law of conservation of matter   Matter is not created or destroyed during any chemical or physical process.

Layer (of weld)   A number of *Weld runs* forming a stratum.

LBBT   See Low blow brittle transition test.

LD process See Linz–Donawitz.

Leaching Removal of some constituent of a mixture by dissolution in a solvent.

Lead A dense metallic element which is soft and *Malleable* but not very *Ductile*. In its pure condition it has good resistance to atmospheric corrosion but its slight solubility and its toxic characteristics have virtually eliminated its use for potable water supply. It is widely used in alloys. With tin it produces a range of *Solders* and cheap *White* bearing metals. In other metals such as *Brass, Bronze, Copper* or *Steel* it is virtually insoluble and exists as distinct particles (very fine in the case of steel) providing *Free machining* characteristics. See Table 15.

Lead–bronze An alloy or copper with up to 30% lead used for high load bearings. Contrast with Leaded bronze.

Lead burning A largely obsolete term for lead welding with a gas torch.

Leaded copper/brass/bronze, etc. Metals to which lead has been added, primarily to make it *Free machining*.

Leak-before-break This term recognizes that some forms of failure are acceptable, others are not. In particular, a small leak is often acceptable; a major burst never is. Hence, a pressurized component known to be developing a crack might be considered acceptable to continue operation if it could be established that the crack would announce its progress towards an unacceptable stage by a slowly developing leak. This would allow ample time to shut down the plant before there was a possibility of a major explosion.

Ledeburite The iron carbon *Eutectic* of austenite and cementite formed at about 1130 °C and 4.3% carbon. See Steel and associated Figure 37.

LED Light emitting diode.

LEFM Linear elastic fracture mechanics. See Fracture mechanics.

Leftward welding See Forward welding.

Leg (length) In welding, the size of a *Fillet* weld measured from the intersection of the two components to the *Toe* at

which the fused metal meets the parent metal. See Figure 51, located at entry on Welding terminology.

Lehigh (slow) bend test A test of the *Brittle fracture* characteristics of steel usually after welding. It is basically a three-point *Bend test* using a notched specimen.

Lehigh (weld) cracking test A test of the susceptibility of steel to cracking during welding. The standard test piece is a plate 12 inches by 8 inches, having a series of slots at 1 inch intervals along its two long sides and a central longitudinal slot to accept the weld. The longitudinal slot is of standard profile but its length depends on the plate thickness. The depth of the side slots is selected to produce the required degree of restraint.

Lenticular Shaped like a double convex lens—( ).

Letting down *Tempering* of steel particularly when carried out in a relatively crude manner with temperature estimated by colour.

Levelling Various processes for flattening sheet by light deformation either by uniform tensile loading or by passing the sheet through a series of staggered rollers.

Levelling agents Substances added to *Electrolytic* plating solutions to minimize variations in plate thickness.

Lever rule In a duplex *Phase* field the equilibrium ratios of the two phases α and γ can be determined from the *Equilibrium diagram* as shown in Figure 25. A horizontal line, the **Tie line**, is drawn through the point of interest, X, to meet the boundaries of the α and γ phases. The ratio of $\alpha{:}\gamma$ phases at X is given by the ratio of the lengths of lines BX:AX. As can be seen by considering the vertical line through the composition, the ratio will change progressively as the temperature falls, starting at 100% of α as solidification commences and becoming increasingly rich in γ as the temperature falls.

Levitation melting A form of *Induction*

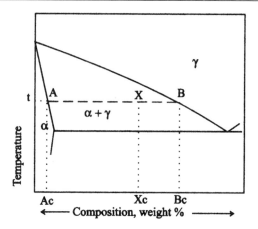

Point X represents material of bulk compositon Xc
at temperature t under equilibrium conditions.
At X the two phases α and γ are present in the
proportions - weights % α : γ = lengths XB : XA
The compositions of α and γ at X are given by Ac
and Bc respectivey.

Figure 25   The Lever rule

melting in which a small quantity of
molten metal can be supported by the
reaction between the magnetic field of
the induction coil and the magnetic field
induced by eddy currents in the molten
metal.

Life assessment   Systems for estimating
the remaining life of components such
as power plant and other high temperature, high duty plant. The term may be
used of a simple calculation of creep life
for comparison with the life experienced
or required. However, it more often refers to a more comprehensive review of
all information that can be obtained
including original materials data and
manufacturers' certificates, site inspection and measurements, removal of
samples for laboratory testing and
examination, review of historical plant
data and reappraisal of design utilizing
improved computer-based stress analysis
techniques.

Life fraction   The portion of life consumed by a particular set of circum-

stances. The concept is sometimes
applied to damage mechanisms where
the variables change. For example, in
Creep both the load and the temperature
can vary, in *Fatigue* the cyclic and steady loads may both vary. The life fraction
approach involves calculating for each
set of circumstances the life consumed
as a fraction of the total life in those
circumstances. Failure is assumed to occur when the fractions total unity.

Ligament   In an engineering context, a
load-bearing part of a structure, for example, the metal remaining between a
number of holes set across the line of
loading.

Light drawn   Tube or wire that has been
slightly cold worked by *Drawing*.

Light metal/alloy   Any alloy based on
the low-density metals aluminium, magnesium and, less commonly, lithium.
Titanium and its alloys have a density
intermediate between these and iron and
copper and, depending on context, may
be included in the category. Alloys of

magnesium, beryllium and lithium are sometimes termed **Ultra-light**.

Lime The material produced when *Limestone* is heated above about 850 °C to drive off water and carbon dioxide. It becomes **Slaked lime** when water is added.

Limestone Naturally occurring calcium carbonate ($CaCO_3$) rock usually with contaminants such as silica, alumina and other carbonates.

Limit of proportionality The stress below which a change in stress produces a linearly proportionate change in strain. See Tensile test.

Limiting creep stress Originally this term referred to the stress below which no measurable *Creep* occurs. Since this stress will depend on the sensitivity of the measuring equipment any remaining use of the term is usually specific to the application and the level of creep that is acceptable.

Limiting stress range Same as Fatigue limit.

Line defect/imperfection Sometimes used as a synonym for *Dislocation*.

Lineage structure A substructure in a grain. Individual areas have differences in lattice orientation that are detectable but not enough to constitute a grain boundary.

Linear damage law A crude estimate of *Fatigue* life under varying loading regimes. Each period of loading is considered in isolation and the *Life fraction* consumed is calculated. Failure occurs when the life fraction total unity.

Linear elastic fracture mechanics See Fracture mechanics.

Linisher A proprietary name for various motorized, bench-mounted, belt grinders.

Linz–Donawitz process The first (probably) commercial process to produce steel on a large scale by injecting oxygen into molten pig (impure) iron.

Liquation (cracking) In its broadest sense this is a synonym for melting but usually it implies transient localized melting of an alloy heated just to its

Solidus temperature. It normally occurs at the grain boundaries and may lead to other forms of damage such as oxide penetration down the boundaries or to cracking immediately or in subsequent service.

Liquid metal attack/embrittlement Generally, any form of damage involving a liquid metal by corrosion, erosion or dissolution. More specifically, especially when called 'embrittlement', the term implies a mechanism in which liquid metal penetrates along the grain boundaries or crystallographic planes of a higher-melting temperature solid metal. Such attack is most likely when the metal is under stress, either applied of *Residual*. The thin planes of penetrating material may initiate cracks immediately or during subsequent service. For example, steel is usually brazed without problems but if, during brazing, it is subject to high levels of stress, either applied or residual, it is susceptible to deep *Brazing penetration* at the grain boundaries.

Liquid phase diffusion welding See Diffusion welding.

Liquidus The line on a *Phase* diagram above which only the liquid phase is stable and below which solidification commences.

Lithium A metallic element with the lowest density of any metal. It has some limited application as an alloy addition in *Light alloys*. See Table 15.

Live load The variable portion of a load system, for example the passengers or cargo in a lift are the live load; the cabin is the fixed, base or static load.

Load The weight or force applied. It is not synonymous with stress. See Tensile test.

Load cell A mechanism for measuring applied loads and, usually, emitting a signal for remote display or recording.

Load-controlled loading See Displacement controlled loading.

Load–extension curve The stress–strain curve, as produced in the *Tensile test*.

Local action *Electrolytic* corrosion

occurring at small, localized variations in composition or hardness.

Local cell   Same as Local Action.

Lock(ed) fit   A severe *Interference fit* not capable of being parted without major damage.

Lock nut   A second solid nut tightened hard against the first nut on a threaded component to prevent loosening in service as a result of vibration or other inadvertent loads. Where the nuts differ in thickness the thicker should be fitted last as it carries the full tightening torque and *Pre-load*. It is incorrect, but not uncommon, for the thin nut to be fitted second. Alternative names are **Jam nut** and **Thin nut**. Another form of lock nut is of thin hard sheet with edges raised to receive the wrench. These are not tightened to full pre-load torque and are fitted second. Also see Self-locking nut.

Long-range order   An *Ordered* structure extending over a large proportion of a crystal.

Long ton   The imperial ton of 2240 lb as opposed to the US 'short ton' of 2000 lb.

Longitudinal   In a metallurgical context, it is the line along which a material is principally extended during a working operation.

Looping mill   See Rolling mill.

Loss angle   The phase angle difference between two interdependent values, for example the difference in angle between voltage and current observed in inductive electrical circuits.

Lost wax process   See Investment casting

Low-alloy steel   A vague term indicating a hardenable steel with up to 5% or so total alloy content other than carbon. See Steel.

Low blow brittle transition test   Similar to the Charpy *Impact test* except that the standard notched test pieces are pre-cracked by an initial impact at a temperature above the transition temperature. A series of test pieces is then tested, as normal, over a range of temperatures to produce the usual graph of

energy absorbed against temperature. The technique is expected to produce a graph with a distinct inflection point at the temperature which, it is claimed, is the lowest safe service temperature for the batch of material. This temperature is termed the **Low blow brittle transition temperature (LBBT)**.

Low-carbon steel   See Steel.

Low-hydrogen electrode (for welding) See Electrode (welding).

Low-hysteresis steel   See Transformer steel.

Low melting temperature alloys   Various alloys with a low melting point, the criterion usually being 232 °C, the melting point of tin. The constituents are usually two or more from tin, lead, bismuth, cadmium, gallium, thallium, indium and zinc. Compositions are often *Eutectics* to ensure a sharp melting point. Some of the alloys are molten at ambient temperature. Applications include fusible plugs, sprinkler activators, tube bending fillers, and jig or die applications. They are also termed Fusible alloys. See Wood's metal as an example.

Lower bainite   Bainite formed at relatively low temperature as a result of cooling at a rate not quite sufficient to form martensite. See Steel and Isothermal transformation diagram.

Lower shelf   See Brittle fracture.

Lower yield   See Tensile test.

LPG   Liquid petroleum gas.

Lubricant   Materials imposed at the interface between surfaces in relative motion. Their primary function is to keep the surfaces apart and hence avoid surface damage and seizure. Clearly they should minimize friction and they may also act as coolants. Lubricants may be solids including graphite and PTFE, semi-solids including grease and animal fats, liquid including oil and water (in appropriate cases), glass (see Ugine process) or gas including air.

Lüders bands/lines   Lines formed on the surface of components due to uneven deformation when only a portion of the component exceeds the *Yield point*. The lines form the boundary of the yielded zones and sweep across the surface as the material continues to yield. *Steels* having a pronounced yield point are particularly susceptible. The lines are aligned at 45° to the tension axis and are observed on components such as lightly drawn tube and pressed sheet. Although not damaging, they may be unsightly but can usually be prevented by prior uniform light working. Also called Stretcher strains and other terms.

Luminous, luminescent Emitting light or other radiation by *Fluorescence, Phosphorescence* and such phenomena.

Luminous flame A flame that emits significant light due to incandescent carbon particles remaining from incomplete combustion.

Luting Sealant such as fireclay applied to fill cracks or joints.

Lye Concentrated solutions of sodium hydroxide or potassium hydroxide or mixtures of the two.

M

M The multiplier prefix, mega, for units of the *SI* and other systems. It is one million.

M$_s$, M$_f$ The temperatures at which the martensite transformation respectively starts and finishes. Numeric subscripts, e.g. M_{50}, indicate the temperature for that percentage transformation. See Steel and Isothermal transformation diagram.

Machinability (1) In a metallurgical context, this term reflects the form of the swarf, i.e. the material removed by the cutting tool. If the swarf is released as many small chips rather than as a long string the metal is regarded as having good machinability and is termed **Free machining**.

Machinability (2) In wide production engineering contexts this term may refer to matters such as the energy required for cutting, the rate of wear of the cutting tool or the texture and quality of the cut surface.

Machining allowance The amount by which a casting, forging or other part-finished component is oversize to allow for machining to the final dimensions.

Machining stresses *Residual stresses* induced during machining operations.

Macro On a large scale as opposed to micro.

Macro(scopic) examination An imprecise term indicating examination with the unaided eye, a hand lens or microscopes at magnifications up to about 30×.

Macro-section A specimen cut and prepared for *Macro-examination*.

MAG welding See Metal active gas welding.

Magnesia Magnesium oxide, MgO. It has a melting point of about 2800 °C and is used as *Refractory* although it has relatively poor resistance to thermal shock.

Magnesite A naturally occurring mineral, mainly amorphous magnesium carbonate and magnesium silicate sometimes used as a *Basic refractory*.

Magnesium A low-density metallic element. It has a strong affinity for oxygen and is significantly corroded in moist environments, particularly if salt laden, but forms a protective oxide film in dry air. In finely divided form as powder or ribbon it burns readily giving an intense white light, hence its use in flares and as photographic flash powder. As a sacrificial anode it offers *Cathodic protection* to steel. It is used in the pure or alloyed state for structural applications and as an alloy addition for other metal, particularly aluminium. See Tables 9 and 15.

Magnesium anode A sacrificial anode that provides *Cathodic protection* to steel.

Magnet A piece of iron, steel or other magnetic material that has been magnetized, either permanently or temporarily, so that it can magnetically attract a non-magnetized but magnetic material. See next entry.

Magnetic Having the capacity to be attracted by, or repelled by, other magnetized materials or magnetic fields. Commonly, metals such as iron are described as magnetic as they are strongly attracted by magnetic fields. More strictly, such metals are **Ferromagnetic** since some other metals, termed **Paramagnetic**, have the same characteristics but to a very weak extent. The remaining metals and all non-metals are repelled by magnetic fields and are termed **Diamagnetic**. In this book, apart from this entry, use of the term 'magnetic' indicates the commonly understood form, i.e. ferromagnetic. A material is termed magnetically **Hard** if it is ferromagnetic and it retains most of its magnetism after the removal of the magnetizing field, a characteristic termed **Permanent magnetism**. If it loses its magnetism it is termed **Soft**. The magnetic properties of elements arise from their *Atomic structures*. An atom can be considered as comprising a central nucleus around which the electrons orbit. Each electron is spinning about its own axis so it develops its individual magnetic field. However, no more than two electrons can exist at a particular energy level. Where two exist they are of opposite spin so their fields cancel out and the atoms having all their electrons paired do not independently align to produce powerful magnetic effects. These are the **Diamagnetic** materials and in a magnetic field their atoms realign to oppose the external field. They therefore carry a slightly smaller magnetic flux for a given field strength than would a vacuum. In those elements having single electrons at particular energy levels the unpaired electrons have magnetic moments that are not cancelled out and hence the individual atoms have magnetic moments. When exposed to an external magnetic field these atoms realign to produce a magnetic flux greater than that in a vacuum. Elements having this characteristic to a small extent are termed 'paramagnetic'; those having it to major extent—iron, cobalt, nickel and gadolinium—are 'ferromagnetic'. The effects of magnetic fields on the various forms of materials are illustrated in Figure 26. The relation-

ship B/H betweeen the **Magnetizing field strength,** H, and the resultant **Magnetic flux density,** B, is termed the **Permeability,** μ. For a vacuum the relationship is linear and μ has a value of $4\pi \times 10^{-7}$ m^{-1} which is regarded as the datum of permeability, sometimes termed the **Free space permeability,** with a value of unity. For non-vacuums a **Relative permeability,** μ_r, is given by the relationship $\mu_r = B/\mu H$, so diamagnetic materials have a value of less than unity and paramagnetic materials a value greater than unity. The **Magnetic susceptibility** is the relative permeability-1. In the case of the paramagnetic and diamagnetic materials the magnetic flux density, B, is directly proportional to the strength of the magnetizing field, H, and when H returns to zero so does B, as shown in (a) of Figure 26. However, as illustrated in (b), when a field is applied to a ferromagnetic material the **Magnetization**, measured as B, rises irregularly to some maximum value shown at S, the **Magnetic saturation**. When the field is removed the magnetization does not return to zero but remains at some intermediate level, termed the **Remanence** or **Residual magnetism**, with a value of B shown at point R in the figure. A reverse, or negative, field termed the **Coercive force,** with a value of H shown at point C, is necessary to return the magnetization to zero. Further increase in the negative field induces negative saturation at S$_2$ and subsequent changes cause the magnetization to follow the complete **Hysteresis loop** shown in the figure. The various ferromagnetic materials have widely differing hysteresis loops. The area within the loop represents energy loss and consequent heating, so steels for *Transformer laminations* have very slender loops. In contrast, a permanent magnet material exhibits a fat loop with high levels of remanence and coercive force. The optimum combination of these two characteristics is the maximum value of $B \times H$ on the curve from

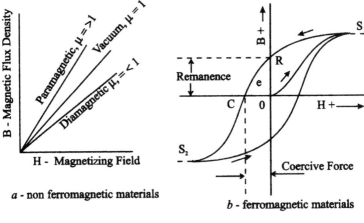

a - non ferromagnetic materials

b - ferromagnetic materials

$$\mu = \frac{B}{H} = \text{Permeability}$$

S - Magnetic Saturation, the maximum value of B + or -
R - The Remanence, B_{rem} , with H zero.
C - The Coercive Force, the negative H applied to return B to zero.
e - The Energy Product Value, the maximum product of B × H on
   the curve R to C.

Figure 26   Magnetic properties of materials

R to C of Figure 26(b) termed the **Energy product value** BH_{max}. This is the measure of the maximum energy that the magnetized material can provide in a mechanism. In some materials such as ferrites having a nearly square loop BH_{max} closely approaches R × C.

Magnetic annealing   The *Annealing* of *Steel* followed by slow cooling in a powerful magnetic field to develop *Magnetic* characteristics such as increased permeability.

Magnetic circuit   The complete path followed by magnetic lines of force from pole to pole.

Magnetic core   The central core of an electromagnet. It is formed from a solid piece, or *Laminations*, of a magnetically soft iron or steel and is magnetic to a significant extent only while there is an electrical current in the coil surrounding it.

Magnetic domain   The zone of material with its atoms all having the same magnetic characteristics. An individual grain will comprise multiple domains.

Magnetic field/force   See magnetic.

Magnetic forming   The use of short-duration, high-intensity magnetic fields to induce deformation. The material to be shaped, which must be a conductor, is subject to a rapidly changing magnetic field which induces, in the body of the material, large eddy currents and associated magnetic fields. The interaction of the induced and external magnetic fields produces large forces capable of deforming the material.

Magnetic particle inspection/crack detection   The use of magnetic particles to detect surface defects in iron and steel. The fine particles, as a dry powder or suspended in fluid, are applied to the surface under test and a magnetic field is imposed by magnets or electric coils. Any defect aligned across the lines of magnetic force effectively forms a pair of magnets which attract the particles hence revealing the location of the defect. Prior to application of the particles it is common to apply a quick-drying white coating, referred to as a *Developer*, to improve contrast.

Magnetic properties See Magnetic.

Magnetism See Magnetic.

Magnetite Black, magnetic iron oxide, Fe_3O_4. In contrast to *Rust*, magnetite can form a protective coating, for example in high-temperature water of good quality. It also forms on the surface of steels being hot worked, in which case it is commonly termed mill scale.

Magnetoelasticity Less common term for magnetostriction.

Magnetostriction The reversible dimensional change of a material due to magnetization. The change may be expansion or contraction, volumetric (Barret effect) or linear (Joule effect).

Magnification In *Metallography*, the ratio of the image diameter to the object diameter. It approximates to the product of the magnifications of the eyepiece and the objective lenses of the *Microscope*.

Major segregation See Segregation.

Magnolia metal A relatively cheap *White metal* bearing alloy suitable only for light duty but retaining its properties to a relatively high temperature. Typically 80% lead, 15% antimony, 5% tin.

Magnox alloys Nuclear fuel *Canning* alloys based primarily on magnesium with small amounts of elements such as aluminium and beryllium added to reduce flammability, limit grain size and enhance resistance to high temperature oxidation in carbon dioxide, the cooling medium in *Magnox reactors*.

Magnox reactors Nuclear reactors of a design cooled by carbon dioxide and having the fuel *Canned* in *Magnox alloy*.

Malleability The capacity of a metal to be deformed by predominantly compressive loads in processes such as rolling, forging or beating.

Malleable (cast) iron Cast Iron capable of accepting some deformation or impact loads without cracking. The term strictly refers to irons in which these desirable properties had been developed by heat treating an initially brittle casting. However, it is now sometimes extended to iron castings which are malleable in the as-cast condition. See Cast iron.

Mandrel/mandril A steel bar inserted into the bore of tubes during drawing and other working operations. In most cases its function is to support and size the bore and it will consequently be a hardened steel; in other cases it merely locates the piercing point or plug. See Tube making.

Manganese A metallic element having little commercial use in its pure state but an important alloying element in *Steel*, aluminium and other metals. See Table 15.

Manganese brass/bronze A term with various interpretations, it usually indicates a *Brass* based on 60% copper and 40% zinc with additions of about 1.5% manganese and often a per-cent or so of iron and aluminium. A better term for this alloy is *High-tensile brass*. Some references confine the term 'manganese bronze' to copper base alloy castings containing about 20% manganese. See Table 8.

Manganese steel Virtually all commercial steels contain manganese up to about 2% but additions above about 1% have significant effects on properties such as strength and *Hardenability* so terms such as carbon manganese, or nickel manganese steel are used. See Steel. However, the term 'manganese steel' often refers to steels with about 12–14% manganese and about 1.2% carbon. They are fully austenitic, and hence non-magnetic, provided they are quenched to retain the carbon in solution. They have good ductility and toughness but if the surface is deformed it develops a very hard layer (variously suggested to be martensite or, alternatively, severely strain hardened austenite) which confers excellent resistance to wear, particularly gouging abrasion but presents difficulty in machining. These steels are also termed Hadfield's (manganese) steel.

Manganese sulphide A very common non-metallic *Inclusion* in *Steel*, often present in appreciable quantities. At the usual hot working temperatures manganese sulphide particles are plastic and, in addition to deforming, they tend to become round ended. Consequently, the stress-raising effect of individual particles is relatively small compared with other forms of inclusion such as alumina silicates which remain sharp tipped. The majority of the inclusions tend to be concentrated towards the centre of the ingot and the resultant raw product and their overall reduction in cross-section is negligible. As a result of these favourable features, moderate amounts of small manganese sulphide inclusions often have a negligible effect on mechanical properties. Furthermore, in some steels, manganese sulphide inclusions are deliberately introduced to provide *Free cutting* characteristics. However, inclusions of any type or quantity are undesirable in steels for some applications; for example any discontinuity in a bearing steel can act as a site for subsurface *Fatigue* crack initiation. High levels of sulphur in any form are undesirable in steels to be welded.

Manifold A pipe or other vessel with multiple inlet and/or outlet connections.

Mannesmann (piercing) mill See Tube making.

Manson–Haferd relationship An empirical method for extrapolating *Creep* data based on a linear relationship between the reciprocal of the absolute temperature and the logarithm (base 10) of the time to rupture.

Manual metal(lic) arc welding (MMA) Metal arc welding in which the consumable *Electrode* rod, installed in a suitable holder, is manipulated by the operative during the welding operation.

Manual welding Any welding process in which the operator holds items of the equipment such as the welding torch or the *Electrode* holder, controls key parameters such as welding current or gas flame and applies skill to produce a quality weld.

Maraging A hardening process applicable to certain highly alloyed *Steels*, the term deriving from 'martensite' and 'age hardening'. Such steels, on cooling from the austenitic region, i.e. about 800 °C, develop a very low-carbon martensite which is softer and tougher than that formed in conventional steels. However, on heating to about 480 °C the highly alloyed martensite undergoes a *Precipitation hardening* process as intermetallic particles precipitate. This produces a steel combining very high strength with good *Fracture toughness*. The term was first used for a range of steels based on about 18% nickel, with typically, about 9% cobalt, 5% molybdenum, 0.6% titanium and 0.1% aluminium but is now used of other alloy ranges offering the same characteristics.

Martempering A heat treatment in which a steel component is rapidly cooled from the austenitic range to a temperature just above the martensite start temperature. It is held at this temperature for a limited time and then further cooling, which can be at a relatively slow rate, causes transformation to martensite. The initial fast cooling suppresses the normal pearlite transformation and the hold at the intermediate temperature allows temperature gradients to even out, thereby reducing *Residual stresses*. The virtual absence of residual stress and the relatively slow final cooling minimize the risk of cracking. *Austempering* is similar except that it allows transformation to occur at the hold temperature thereby forming a bainite structure. See Steel and Isothermal transformation diagram.

Martensite The structure formed in steels by a shear mechanism, rather than a diffusion process, when the steel is cooled, faster than the critical rate, from the austenitic range. The fast cooling retains the carbon in supersaturated solid solution so the face centred cubic ferrite structure cannot form. Instead the austenitic transforms by shear to a body centred tetragonal structure which, under the microscope, has an acicular, i.e. needle-shaped, appearance. See Steel. The term is used for similar crystallographic forms in other alloys.

Mash (seam) weld A *Resistance* seam weld in which the similar thickness sheet components are set up to form an overlap which is mashed, i.e. crushed, during welding so that the final weld thickness approximates to the thickness of a single sheet.

Mass effect In steels, the reduction in cooling rate resulting from an increase in mass. See Hardenability.

Mass–energy relationship The Einstein equation $E = mc^2$, where E is the energy in ergs, m the mass in grams and c the velocity of light in cm/s.

Mass spectrograph An analytical technique in which ions of the material under test are passed across magnetic and electrostatic fields. The path deflection is recorded as a measure of their composition.

Master alloy A concentrated mixture of the alloying elements added, usually at a late stage in the melting and refining process, to adjust the final composition of a casting.

Matrix In a metallurgical context, the predominant *Phase* or material surrounding another phase or feature.

Matt(e) A matt surface is level but dull and diffuses rather than reflects light.

Matte The impure metal sulphide produced at an intermediate stage in the reduction of a sulphur-rich ore to a metal.

Maximum tensile stress Same as ultimate tensile strength. See Tensile test.

Mazac alloy A proprietary range of alloys based on zinc with about 4% aluminium and minor additions of copper and magnesium. They have excellent *Die casting* characteristics.

McQuaid–Ehn test A technique for revealing the inherent austenitic grain size of a *Steel*. The steel under test is pack *Carburized* for 8 hours at about 920 °C and then slow cooled. The prior austenite grain size of the *Hyper-eutectoid* carburized case is then readily observed by metallographic examination.

Mean free path The average distance between collisions for an electron or other particle.

Mean free time The mean free time separating collisions in a gas between molecules or in semiconductors between electrons and impurity atoms.

Mean stress (1) The algebraic mean of the *Principal* stresses.

Mean stress (2) The mid-range level. See Fatigue.

Mechanical alloying The production of a homogeneous mixture by mixing powders. It is implicit that the mixture will be subject to processes such as *Sintering* or even melting.

Mechanical plating Processes for applying a metal plating by mechanical means rather than *Electroplating* or dip processes. Some form of *Peening* is usually involved such as tumbling the component in a rotating barrel with balls or powder of the coating metal.

Mechanical properties Strength-related properties measured by some form of loading including tensile, bend, impact, torsion testing.

Mechanical testing Any testing to determine the mechanical properties of a material particularly *Tensile* testing but including *Bend*, *Impact*, *Fatigue* and *Creep* testing. The term usually including hardness testing since this can give an accurate indication of some mechanical properties.

Mechanical twin See Twin.

Meehanite Proprietary name for a range of cast irons. The term is used casually referring to inoculated irons. See Cast iron.

Mega As a multiplier prefix to units of

the *SI* and other systems of units, it is one million. For example, MegaPascal.

Melt (1) To change from the solid to liquid state.

Melt (2) The molten material in a process.

Melt (3) In *Submerged arc welding*, the molten flux above the weld metal.

Melt run (of weld) A line of parent metal that has been fused without the addition of *Filler*.

Melting point The single temperature at which pure metals and some narrow alloy compositions change from solid to liquid. Other metal alloy compositions change progressively over a temperature range and materials such as glass progressively soften on heating until they are fully fluid.

Melting range The temperature range over which an alloy passes through a pasty condition as it changes from fully solid to fully liquid. It is bounded by the *Solidus* and *Liquidus*.

Memory alloys See Shape memory.

Mendeléev's (Periodic) Table See entry on Periodic Table and Table 16.

Merchant iron Multiple forged and hence high-grade *Wrought iron*.

Mercury The only metallic element that is liquid at ambient temperature. In its pure form its liquid character leads to many applications in instruments such as thermometers, electrical switches and lamps. As an alloy element it forms amalgams such as those commonly used for dental fillings. Its vapour is toxic. See Table 15.

Mercury seal A gas-tight seal provided when a shaft enters equipment via a pool of mercury.

Mercury switch An electrical switch in which a rocking tubular vessel contains a limited amount of mercury which, at one rock position, covers the electrical contacts.

Mesh Apart from its common meanings, in a metallurgical context this term usually refers to the series of numbered sieves of decreasing mesh hole size used for grading powders.

Metal An imprecise term. Generally materials showing most if not all of the following characteristics–crystalline with the atoms in a simple lattice with shared valence electrons, good strength, ductility and malleability, lustrous, high electrical and thermal conductivity.

Metal(lic) active gas welding (MAG) Any welding process in which the electric arc is struck between the component being welded and a continuously fed consumable *Filler* wire and the weld zone is shielded by a gas that is active rather than inert during the welding operation.

Metal(lic) arc cutting Processes utilizing the melting action of an electric arc between a metal electrode and the workpiece.

Metal(lic) arc welding Any welding process in which the electric arc is struck between the component being welded and a metal *Electrode* which is progressively consumed to supply *Filler* to the joint.

Metallic bonding See Interatomic bonding.

Metal fatigue The same as *Fatigue*. The shorter term is usually used by metallurgists.

Metal(lic) inert gas welding (MIG) Welding in which the electric arc is struck between the component being welded and a continuously fed consumable *Filler* wire and the weld zone is shielded by a gas that is inert during the welding operation.

Metal spraying Any process for coating a surface by spraying it with a molten metal as small droplets. The flattened droplets are only mechanically attached as there is insufficient time for diffusion to develop a stronger bond.

Metal stitching A mechanical technique for repairing cracks. As illustrated in Figure 27, pairs of holes are jig drilled on the two sides of the crack. The ligament between the pairs is then slit to produce a wide gap and 'dumbbell'-shaped strengthening pieces inserted. The dumbbells can be a more complex

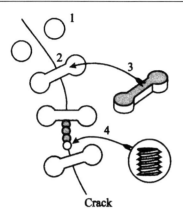

Procedure:

1 Jig drill pairs of holes

2 Slit to join holes

3 Insert 'Dumbbells', multiples if
required by depth

4 If seal required -
Drill along line of crack
Tap holes and insert threaded plugs
Repeat with each hole overlapping
the previous hole and the dumbbells.

Figure 27   The metal stitching process

shape and they may be of a material matching the parent or they may be of a lower-expansion material to provide a crack-closure effect if the component is heated. If a pressure-tight repair is required a further hole is drilled on the line of the crack. This hole is tapped (i.e. threaded) and filled with a threaded plug. This is repeated along the crack with each plug slightly overlapping its predecessor and the dumbbells. The technique can never restore full strength and it is usually confined to low-*Ductility* fractures resulting from a gross overload. However, it can be performed on-site with simple tools and it does not require heating.

Metal transfer (in welding)   The manner in which metal is transferred across an electric arc from a consumable *Electrode* to the weld pool. The term is particularly applicable to *Metal inert gas* welding. High currents cause **Spray transfer** in which individual droplets, not greater in diameter than the electrode diameter, are projected across the arc gap. Lower currents cause **Globular transfer** in which the molten metal accumulates at the electrode tip until eventually a droplet appreciably larger in diameter than the electrode is projected across the gap. The lowest currents re-

sult in **Dip transfer** in which an arc is initiated and the steadily advancing electrode dips into the weld pool causing a short circuit. The electrical surge melts a portion of the tip which detaches to enter the weld pool. The arc re-strikes and the process repeats.

Metallic bonding   The bonding, characteristic of that between atoms of a metal crystal, in which the atoms are arranged on a *Crystal lattice* and the valence electrons form a free cloud rather than being attached to individual atoms. See Interatomic bonding for a less simplistic description of the electron distribution.

Metallizing   A process for applying a metal coating to the surface of a component which may be metal or other material. The term usually implies application by a hot spraying or similar mechanical process rather than by a plating or diffusion processes. However, terms such as *Aluminizing* or *Chromizing* usually refer to diffusion processes.

Metallography   The practice of examining metals by microscopical techniques to establish their structure, reveal defects and provide information on damage mechanisms or other matters of interest. The term **Macroscopic examination** is often used for preliminary examination at low magnifications, say up to 30×. This term is sometimes also implies that the item being examined is in the as-received state, that is, it has not been polished and etched as described below. **Microscopical examination** usually implies the examination, in a light *Microscope*, of **Mounted** and prepared specimens cut from a component. Typical practice will involve cutting a piece of metal, say 10 mm square ×5 mm thick, from the component under investigation. This is then encased, with the face of interest exposed, in a plastic resin to facilitate handling and to protect the edges which are often of particular interest. The face of interest is then ground flat prior to polishing, either mechanically or electrolytically. Mechanical polishing involves careful abrasion on a series of progressively finer grinding papers and polishing pads to provide a flat, mirror finish. Microscopical examination in this state provides some information on microstructure and damage mechanisms. However, some detail may be obscured by the surface deformation and smearing, termed the *Beilby* layer, produced by mechanical polishing. Such detail is revealed by **Etching** which involves exposing the specimen to acid or other reagents that remove the surface film and often react selectively with features of interest. In some cases the specimen may be electrolytically etched. This is essentially similar to the alternative polishing technique, **Electrolyte polishing** which involves making the specimen the anode of a electrolytic cell; see Electrochemistry. With careful selection of reagents and control of the voltage and current a flat, high-quality surface can be achieved. A conventional light microscope has a useful magnification of up to about 1000×. Examination at higher magnification normally requires the use of an *Electron microscope* which uses an electron beam instead of light to examine the specimen. The system for producing the electron beam and projecting the image onto a screen is essentially a cathode ray tube similar to a television set. In the case of a *Transmission electron microscope (TEM)* the metal samples to be examined are very thin and the electron beam travels through the specimen for projection on the screen or photographic film. In the case of a *Scanning electron microscope (SEM)*, the beam scans across, and is reflected from, the specimen surface prior to projection on the screen. Apart from the higher magnification, which is often not utilized, the SEM offers the major advantage that the electron beam has a very large depth of focus so it can deal with a rough, irregular surface lying at an angle to the beam. The light microscope has only a shallow depth of focus so it is effectively confined to flat surfaces aligned normal to the beam. This characteristic makes the SEM a valuable tool for examining fracture surfaces as well as polished samples. The disadvantage of electron microscopy is that the specimen has to be small enough to be inserted into the instrument's vacuum chamber. With suitable ancillary equipment the SEM can undertake analysis either of large areas to give a bulk composition or of small features such as inclusions or precipitates.

Metalloid   A material having some but not all of the characteristics of a *Metal*, also termed a Semi-metal.

Metallurgical notch   See Notch.

Metallurgy   In its widest sense, the study of the extraction, production, use and performance of metals.

Metalock (and similar terms including some proprietary names)   See Metal stitching.

Metastable   A non-*Equilibrium* state in which a metal appears to be stable because it does not transform spontaneously (even over long periods). Essentially similar to Unstable.

Meter (1)   Metre.

Meter (2)   A general term for a measuring instrument.

Methane CH_4, an inflammable gas. It is called **Marsh gas** when it results from the decay of vegetation and *Firedamp* in mine workings.

Metre   The *SI* unit of length.

Metric system   A system of standardized units for all properties, dimensions, etc. with all multiples or divisions to base 10 except time. Now replaced by the *SI system*.

Metrology   The science and practice of measurement.

Meyer index   A measure of the work-hardening characteristics of a material determined by performing a series of *Hardness* tests over a range of loads using a ball indicator. The index is then derived from graph of load against indentation diameter.

MIC   *Mineral insulated cable.*

Mica   A mineral of complex silicates having a strongly lamellar structure. It is has high electrical resistance and is readily expanded to form a low-density thermal insulator.

Micro alloyed steel   *Steels* with alloying additions in small but carefully selected and controlled quantities. Under closely controlled manufacturing conditions these steels are capable of giving a very fine grain size with high strength with good ductility. Typical additions include aluminium, vanadium, titanium and niobium of up to about 0.1% each. Some of these steels are also able to retain the high strength following welding, unlike more conventional steels which suffer softening of the zone adjacent to the weld. They are also termed High strength low alloy steel (HSLA).

Micro analyser   An analytical instrument essentially similar to, and often part of, an *Electron microscope*. It is capable of focusing on and analysing very small features on raw or prepared specimens. Also see Metallography.

Micro inch   One millionth of an inch (symbol μin).

Micro metre   One millionth of a metre, a micron (symbol μm).

Micro shrinkage/porosity/etc.   Internal voids visible only under a microscope. See Pore and Shrinkage.

Microanalysis   Various techniques for analysing small quantities of material.

Microbial attack/corrosion   Any *Corrosion* caused, or assisted, by the presence of microbes, i.e. bacteria etc., in the surrounding environment. Microbes do not directly attack the metal but their presence can modify the environment to make it more aggressive.

Microcrack, microfissure   A small crack not visible to naked eye examination nor to the simpler forms of non-destructive crack detection techniques.

Microhardness   See Hardness.

Micrometer   Any instrument for measuring linear dimensions at high precision.

Micron   One millionth of a metre, symbol μ. This is not a *SI*-approved name but is in common use.

Microprobe analyser   Same as Micro analyser.

Microscope   Instrument for examining a material at a high magnification, in particular the **Light microscope**, or, more strictly, the **Compound microscope** since it comprises two main lenses. The **Objective** close to the specimen produces the initial image which is then magnified by the **Ocular** or **Eye piece** close to the observer. In practice both objective and ocular are of compound form, i.e. formed from multiple individual lenses, some spaced apart, some bonded together. The total magnification is approximately the multiple of the ocular and objective. Biological microscopes project a beam of light through the specimen under examination and then through the lenses into the eye. A metallurgical microscope is similar except that the light beam is injected into the body of the microscope between the objective and eyepiece. It is initially reflected onto the specimen by a clear glass plate and is then reflected back along its original path through the objective, straight through the clear reflector and the ocular into the eye. In any microscope the objective lens is responsible for developing the initial image so its *Numerical aperture*, i.e. its efficiency in gathering light, is critical. The wavelength of light is such that details finer than about 0.2 μm cannot be resolved by the objective lens. This corresponds to an objective lens magnification of about 100× diameters so the **Useful magnification** of the microscope ranges up to about 1000× diameters assuming a 10× ocular. Of course, the image can be magnified further but no greater detail can be resolved. At the highest magnifications an *Immersion oil* may be interposed to fill the very small gap between the specimen and the objective lens to improve the **Resolution**. See also Electron Microscope and Metallography.

Microscopical examination   See Metallography and Microscope.

Microscopy   The science and practice of examining materials under a microscope.

Microsection   A sample of metal cut from an item of interest for *Metallographic* examination.

Microstress/microstrain   *Stresses* and *Strains* which either are of very low magnitude or act over very short distances relative to the context.

Microstructure   The arrangement and appearance of *Grains*, *Phases*, *Inclusions* and other features within a metal. It is normally examined by light or electron microscopes.

Microtome   Equipment for slicing a very thin film or sheet of material from some larger item, usually for some form of examination or test.

MIG   See Metal inert gas welding.

Mil   Either one thousandth of an inch or one millimetre. Both uses are informal and the potential for confusion is obvious.

Mild steel   A vague term for steel with a fairly low carbon content, up to about 0.15%, and no significant quantity of any other alloying element. It has very limited *Hardenability* but is cheap, readily formed and welded and comprises the bulk of steel for general engineering and structural applications. See Steel.

Military transformation   A transformation from one *Phase* to another in which the atoms at the advancing interface act in a relatively restricted and coordinated manner and can be constrained to certain *Crystallographic* orientations or confined by some of the original phase *Grain boundaries*.

Mill (1)   Relating to some manufacturing process or equipment, as in grinding mill or ball mill, or to a building where these activities are performed as in tube mill.

Mill (2)   Machining on a milling machine, i.e. one on which a rotating tool takes a shallow but broad cut as it traverses the workpiece.

Mill finish   Material in the state in which it left the last major manufacturing operation, such as *Rolling*, *Drawing* or *Extrusion*. As examples, mill finished hot-rolled plate will have wide tolerances on dimensions and carry a heavy oxide coating, termed **Mill scale**, while mill finished cold-rolled material will have a smooth surface or even a deliberately textured or embossed surface but will not have been polished and perhaps not thoroughly cleaned. See also Milled finish.

Mill scale   See Magnetite and Mill finish.

Milled finish   Material that has been machined on a milling machine and hence will have a bright surface and, probably, precise dimensions. In some cases the repetitive pattern of circular swirl marks characteristic of some milling operations is seen as aesthetically pleasing. Note, however, that a similar effect is sometimes achieved by abrading a sheet surface with a flat end polishing bob. See also Mill finish.

Miller–Bravais indices   See Miller indices.

Miller indices   A system for identifying planes and orientations in a *Crystal lattice*. It is best considered in three steps and by reference to Figure 28. In the simple case of a cubic system, three axes x, y and z radiate from an origin at the corner of the cube unit cell. At Step

1 the plane of interest is specified by measuring the distance, from the origin, at which it intercepts the three axes. Such distances are measured in terms of the unit cell size, the simple cases in Figure 28 giving 1, ∞, ∞, and 1, 1, 1. Step 2 is to take the reciprocal of the three intercept units from step 1, giving 1, 0, 0 and 1, 1, 1 which are the Miller indices. Step 3, only necessary for planes less simple than those illustrated, requires those reciprocals to be factored to give the smallest possible whole numbers, i.e. $\infty, \frac{1}{2}, 1$ (not illustrated), would be factored to 0, 1, 2, and 2, 4, 6, would be factored to 1, 2, 3. Miller indices are conventionally enclosed in parentheses, the use of simple brackets (111) indicating all planes lying parallel with the reference plane while irregular brackets {111} indicate all planes having similar atomic structures but lying in all directions. Directions, derived by identifying the intercepts of the line on the axes, are indicated by square brackets [111]. If any of the intercepts at Step 1 had been negative it would be identified in the index by a bar above it. The more complex case of hexagonal structures is dealt with by **Miller–Bravais indices** in a similar way by reference to four axes, the w, x and y axes at 120° on the basal

plane and the z axis normal to the basal plane.

Milling (1)   Treating some component in any form of *Mill (1)* such as a *Rolling mill* or a *Ball mill*.

Milling (2)   Shaping in on a milling machine. See Mill (2) and Milled finish.

Mineral dressing   Preliminary actions to prepare an *Ore* for the main process of extracting the metal. The term includes physical processes and, usually, initial chemical concentrating processes.

Mineral insulated cable (MIC)   An electrical cable comprising an outer tubular metal (usually copper) sheath containing one or more bare conductors (again usually copper). The space between the metal items is filled with compacted magnesium oxide powder as insulant.

Miners rule   A means for estimating *Fatigue* life when a component is subjected to a series of different stresses. For each stress level the number of cycles experienced is expressed as a fraction of the number of cycles to failure at that stress. Failure occurs when the sum of the fractions is unity.

Minor segregation   See Segregation.

Mirror surface   A term usually applied to highly reflective areas on the fracture surface of glass but also sometimes used

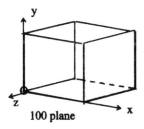

100 plane

The shaded plane intercepts on:
x axis at 1 unit distance from O
y axis at infinity
z axis at infinity

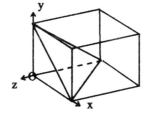

111 plane

The shaded plane intercepts on
all axes at 1 unit distance from O

See text for further explanation.

Figure 28   Miller indices, two examples

for similar features on *Brittle* fractures in metals.

Misch-metal A mixture of rare earth elements particularly cerium, usually about 50%, and lanthanum, usually about 40%. When alloyed with iron or magnesium it is pyrophoric and used for lighter 'flints'.

Miscibility The ability of two or more liquids to mix to form a single homogeneous solution.

Miscibility gap The range of compositions of an alloy system where two *Phases*, existing together in equilibrium, have the same *Lattice structure* although of different *Lattice spacing* and different composition.

Misrun A term applied to various unacceptable casting defects but particularly those due to interruptions in flow leading to local premature solidification with associated gross voids or large *Cold shuts*.

Mist (1) Dispersed fine water droplets in air, sometimes used as a mild quenching medium.

Mist (2) A term usually applied to misty or dull areas on the fracture face of glass but also sometimes used of similar features on *Brittle* fractures in metals.

MMA welding See Manual metal arc welding.

Mock When used as prefix to gold, silver, etc., it indicates a wide range of alloys of base metals claimed to have a superficial resemblance to the metal implied.

Mode 1 etc. See Fracture mechanics.

Moderator In a *Nuclear reactor* context, a material having a low atomic weight and low capture cross-section which slows down fast neutrons allowing them to be captured by the fuel causing *Fission*.

Modification A technique for refining the structure and hence improving the properties of aluminium silicon casting alloys (about 10–13% silicon). A small amount of sodium (about 0.02–0.05%), as sodium chloride or fluoride, is added to the molten metal to refine the nor-

mally coarse *Eutectic* structure that develops on solidification.

Modulus of elasticity The ratio of stress to strain. See Tensile test.

Modulus of resilience The elastic strain energy in unit volume of material at the *Limit of proportionality*.

Modulus of rupture A measure of the strength of a material determined in a three-point *Bend test*. The test is usually applied to *Brittle* materials as it avoids the problems associated with gripping the specimen that arise in the *Tensile test*. For a rectangular specimen the modulus is given by:

$$\text{Modulus} = \frac{3\,ld}{2wh}$$

where l = load at failure, d = distance between supports, w = width of specimen, h = height of specimen.

Mohs hardness A scale of hardness on which a material is rated by comparison with ten minerals ranging from talc, the softest, to diamond as follows:

Mohs number	Mineral
1	Talc
2	Gypsum
3	Calcite
4	Fluorite
5	Apatite
6	Feldspar
7	Quartz
8	Topaz
9	Carborundum
10	Diamond

Molar conductivity The electrical conductivity of a solution containing the *Molecular weight* in grams dissolved in one litre of water.

Molar heat The *Molecular weight* in grams × the *Specific heat* in grams per °C.

Mold Mould.

Mole The *Atomic mass* in grams, also termed the gram-atom. The atomic mass is currently based on carbon so the mole is defined as the amount of substance containing the same number of particles (ions, atoms or molecules) as there are atoms in twelve grams of ^{12}Carbon. The number is 6.023×10^{23}.

Mole fraction The number of atoms of one element, as a percentage of the total number of atoms in a material or *Phase*.

Molecular weight The weight of the smallest quantity of a substance that has all the physical characteristics of the substance, referred to the weight of a molecule of hydrogen defined as 2, or a molecule of oxygen defined as 32, or a molecule of carbon defined as 12.

Molecule Two or more *Atoms* bonded together by sharing electrons to form the smallest quantity of a substance that has all the physical characteristics of the substance. The two or more atoms may be the same element, for example two atoms of hydrogen will normally join together to form a hydrogen molecule, H_2. Alternatively, the atoms may be different, for example two atoms of hydrogen will join with one atom of oxygen to form water, H_2O. See also Interatomic bonding.

Molten metal penetration The penetration, usually at grain boundaries, of molten metal into a solid metal. Usually, the damage only occurs when the material is under stress, either applied or *Residual*. The penetration forms a plane of weakness and cracks may develop immediately or later in service. There are many examples including lead, solder, copper, brass (which is sufficiently common to have its own term *Brazing penetration*) and zinc penetration into steel (it has been suggested that austenitic steel is particularly sensitive to zinc) and of the lower melting point metals into copper and brass. It is a particular form of *Liquid metal attack*.

Molybdenum A dense metallic element. It has good ductility (if pure) and good high-temperature strength but forms a volatile oxide, MoO_3, above about 400 °C. One of its few application in its pure form is as a primer *Metal spray* prior to spraying the main coat of another metal such as nickel or chromium. The hot molybdenum particles lose any

oxide by volatilization and hence they can bond strongly to the substrate as well as forming a good key for the top coat. Molybdenum is widely used as an alloying element particularly for *Steel* in which it stabilizes carbides, inhibits temper brittleness, improves creep strength and improves corrosion resistance of austenitic steels. See Table 15.

Moment The turning action of a force acting about a point. It is the product of the force and the perpendicular distance from the line of action of the force to the point.

Monazite A mineral comprising phosphates and silicates of *Rare earth* metals.

Monel Alloys of nickel containing about 30% copper plus small quantities of iron, manganese and aluminium. The various alloys generally offer good mechanical properties and good corrosion resistance. See Table 10.

Monobloc Manufactured in one piece, for example a large or complex forging, as opposed to comprising a number of sub-members joined by bolting, welding, etc.

Monocoque A form of construction in which all the principal components carry their share of the applied load. For example, the bodywork of a monocoque vehicle is load bearing rather than merely being carried by a substantial chassis.

Monolithic lining/coating A lining or coating, particularly of *Refractory*, in one piece formed by techniques such as pouring, spraying or plastering rather than being built up from bricks or slabs.

Monomer Molecules which may be of some complexity but are not joined into chains to form *Polymers*.

Monotectic A reversible *Isothermal* transformation in which, on cooling, a liquid solution transforms to another solution of different composition plus a solid.

Morphology The study of the small-scale external and internal form or structure of a material.

Mosaic structure   Areas within an individual *Crystal* having slightly different *Lattice* orientations.

Mottled (cast) iron   A *Cast iron* having, usually inadvertently, a microstructure that is grey, i.e. has graphite flakes, in some parts but not in others.

Mould   A vessel or other cavity with its internal form deliberately shaped to receive molten metal for solidification into *Ingots* or *Castings*.

Mould wash   Various fluids, for example a suspension of graphite in water or paraffin, which are applied to the faces of a mould prior to casting to improve surface finish, reduce the damage to reusable moulds and assist stripping. Release agents are similar but this term tends to be applied to more sophisticated products for smaller precision products.

Mounting   In *Metallography*, techniques for handling specimens in preparation for, and during, microscopical examination. The most common practice is to encapsulate the specimen in a resin or similar material, leaving exposed the surface to be prepared and examined.

MPa   MegaPascal. See Pascal.

MPI   See Magnetic particle inspection.

Muck rolled iron   See Pudding process.

Muffle (furnace)   A furnace heated externally so that the material being treated does not come into contact with combustion products.

Mulling   In a metallurgical context, the grinding together of a number of minerals or sands to produce an intimate mixture of fine particles.

Mullite   An aluminosilicate mineral, $3Al_2O_3.2SiO_2$, used as a refractory.

Multi-axial stresses   Any state in which more than one *Principal stress* is applied.

Muntz metal   Brass with 60% copper and 40% zinc. As with all alpha beta brasses it is not readily cold worked so it is used in the cast or hot-rolled condition. Also termed Yellow metal. See Table 8.

Mushet's steel   The first (probably) steel developed for cutting at high temperature, i.e. a *High-speed steel*. It was manganese deoxidized and probably had about 2% carbon, $1-2\%$ manganese and $5-10\%$ tungsten.

Music wire   See Piano wire.

N

n-type semiconductor See semiconductor.

NaK The sodium (Na) and potassium (K) alloy which is liquid at ambient temperature and so is used as circulating coolant in some nuclear reactors.

NAMAS National Measurement Accreditation Service. The organization, in the UK, responsible for calibrating and certificating measuring and test equipment.

Nanotechnology The technology of materials, including their manipulation and assembly into mechanisms, on a very small scale, approaching the size of molecules or atoms.

National Bureau of Standards The body in the USA responsible for establishing standards for scientific and technical data.

Native metal Metals found naturally in bulk and near-pure form, including gold, silver and copper. Sometimes the term is extended to include extraterrestrial material such as meteorites.

Natural ageing (1) Generally, any time-related deterioration.

Natural ageing (2) Age hardening at ambient temperature. See Precipitation hardening.

Natural gas Fuel gas obtained from natural subterranean sources and comprising a wide range of hydrocarbon gases.

Nature print A technique for revealing *Flow lines* in forgings or other wrought material. The forging is sectioned, polished and heavily etched. It is then thinly coated with ink and a print obtained by pressing paper against a surface. See also Sulphur print.

Naval brass Alloy of 60% copper, 39% zinc, 1% tin. See Table 8.

NDT (1) See Non-destructive testing.

NDT (2) Variation on next entry.

NDTT Nil ductility transition temperature. See Impact test.

Necking Local reduction in diameter just prior to failure during *Tensile* loading. It occurs when the rate of work hardening ceases to increase in proportion to the increase in *Strain*. As a consequence, further deformation is confined to the neck which thins until the available *Ductility* is exhausted and the material fails. Also termed **Waisting**.

Negative creep A contraction in dimensions resulting from an alloy undergoing an order/disorder change (see Order). It has been reported in *Nimonic* components in service at about 550 °C but it is not a *Creep* phenomenon although it might be argued that the contraction increases the creep stress.

Negative ion An ion which can be an atom, molecule or radical which has gained one or more electrons and hence is negatively charged.

Neodymium A metallic element, one of the rare earth group. See Table 15.

Neon A gaseous element, one of the rare or noble group of gases. See Table 15.

Net section stress The nominal *Stress* calculated from the total load divided by the total cross-section. It makes no allowance for stress raisers or other source of uneven distribution of stress.

Network structure A structure where one phase forms a continuous, or near so, film around the grains of another phase.

Neuman bands/lines/lamallae Mechanical twins in *Ferrite*, usually indicative of high strain rates. See Twins.

Neutral axis/plane The line or plane, in a component subject to a complex stress pattern, where *Stress* passes through zero as it changes from positive to negative. For example, a symmetrical beam being bent from its natural form has a tensile longitudinal stress at the bend exterior and a compressive longitudinal stress at the inner surface. At approximately mid-thickness the neutral axis has zero longitudinal stress.

Neutral flame (in gas welding) A flame which is neither reducing nor oxidizing in its effect on the workpiece.

Neutral flux A *Flux* which is neither acidic or basic but is added to modify some physical characteristic, particularly fluidity.

Neutron See Atom.

Neutron cross-section A measure of the probability of neutron collision with the atoms in the material in question. The unit is the barn and $1 \text{ b} = 10^{-28} \text{ m}^2$ per nucleus.

Neutron radiography A form of radiography, similar to that involving X-rays except that it utilizes a neutron beam.

Newton The *SI*-derived unit of force, symbol N, equivalent to kg m/s^2.

Newton's Laws of Motion No. 1 A body will remain static or will maintain its motion in a straight line unless an external impressed force acts upon it. No. 2 The change of momentum occurs at a rate proportional to the impressed force and in the same direction as the force. No. 3 Action and reaction are equal and opposite.

Ni-Hard See Nihard.

Ni-Resist See Niresist.

Nibbling Cutting sheet by punching multiple small overlapping holes with a powered, reciprocating, high-speed punch and die set.

Nichrome A range of proprietary alloys the most well known of which is 80%

nickel, 20% chromium. The alloys have good high-temperature strength and oxidation resistance while their electrical resistance is low and their temperature coefficient of electrical resistivity is high. They are used for electrical heating elements and furnace furniture for service up to about 1000 °C.

Nick break test A *Tensile* or *Bend* test in which fracture is initiated at a notch. The term usually refers to tests of welds where fracture is directed through a particular weld zone.

Nickel A metallic element. It has some limited use in a bulk pure form, for example as coinage or electroplate, but has extensive use in alloy systems where it may be the major or minor component. It is the major constituent in *Nimonic*, *Inconel* and other alloys having good high-temperature strength and corrosion resistance. In *Steel* it improves *Hardenability* and *Toughness* and, together with a chromium addition, forms the austenitic *Stainless steels*. With copper it forms the cupronickel alloy series which have good mechanical properties and corrosion resistance. It is also widely used as an electroplate either as a main coating or as an undercoat for chromium. See Table 15 for physical properties of pure the metal, Table 10 for alloys.

Nickel silver Alloys of copper, zinc and nickel; no silver. One group of alloys contains 50–60% copper, 10–30% nickel, possibly with 1% tungsten, balance zinc, also known as *German silver*. Another common alloy contains 45% copper, 45% nickel and 10% zinc. They have decorative and functional applications.

Nickel steel Any steel containing nickel up to about 5% possibly together with other lesser amounts of other metals such as chromium, molybdenum and vanadium. The main contribution of nickel is to improve *Hardenability*. See also Maraging and Austenitic stainless steel.

Nicol prism A compound prism formed

by bonding two pieces of calcite with Canada balsam adhesive. A single beam of light entering a piece of calcite is split into two beams polarized on planes at 90° to each other. In the compound prism the Canada balsam refracts one beam within the prism but transmits the other as polarized light for specialized forms of microscopical examination.

Nicrosilal A proprietary, austenitic, growth-resistant *Cast Iron* with about 18% nickel, 5% chromium, 5% silicon and only 2% carbon.

Niello A technique for forming decorative patterns on metals, particularly silver. The design is engraved or otherwise recessed into the metal and the hollows are then filled with a relatively low melting point alloy of silver, copper and lead together with a *Flux*, traditionally sulphur and borax. After firing to melt the alloy it forms a black area contrasting with the surrounding metal.

Nihard A propriety white *Cast iron* with about 3% carbon, 4% nickel, 2% chromium, 0.5% silicon, 0.5% manganese. It has a high hardness, about 600 Hv, but poor toughness and is used for grinding balls and rings and similar applications demanding resistance to abrasive wear.

Nil ductility transition temperature See Impact test.

Nimocast See Nimonic.

Nimonic A range of proprietary alloys, in cast or wrought form, based on nickel with major additions of other elements, in particular chromium, typically about 20%, with aluminium and titanium plus, in some cases, cobalt, molybdenum, etc. These alloys have good high-temperature characteristics, in particular resistance to oxidation and good *Creep* strength. The **Nimocast** materials are similar materials in the cast form.

Niobium Called columbium in the USA. A high melting point, corrosion-resistant metallic element having limited application in the pure state. It is an occasional alloying element in steels where its strong carbon affinity is useful in the *Stabilized stainless steels*. It is also

added to nickel and other alloys. See Table 15.

Nip The first line of contact between a pair of rolls and the metal entering them.

Niresist A proprietary, austenitic, corrosion-resistant *Cast iron* with about 15% nickel, 2% chromium and 6% copper and 1.5% silicon. See Table 6.

Nital Nitric acid (2% unless otherwise stated) in alcohol for *Etching* steels.

Nitonol A proprietary alloy of 55% nickel and 45% titanium with *Shape memory* characteristics.

Nitralloy A proprietary range of steels designed for *Nitriding* and so containing aluminium, chromium and or molybdenum in the one per cent or so range.

Nitride A compound formed between nitrogen and another material, particularly a metal.

Nitriding See Case hardening.

Nitrogen A gaseous element that comprises about 80% of air. See Table 15.

Nitrogen case hardening Nitriding. See Case hardening.

Noble In *Electrochemistry*, the most cathodic of a pair of metals in the electrochemical series of relative potentials.

Noble gases The six inert gases, helium, neon, argon, krypton, xenon and radon. More usually termed the 'rare gases'.

Noble metals Generally, metals that do not easily react with common environments. Historically, they were gold and silver but will now include the platinum group.

Nodular (cast) iron See Cast iron.

Nomarski interference technique A microscopical technique utilizing polarized light to reveal slight topographical variations.

Nominal This term is used vaguely in technical contexts to indicate concepts such as a token amount or a broad overall view.

Nominal stress The *Stress* as calculated on the basis of the external load acting on the minimum original dimensions of the component or test piece and ignoring matters such as *Stress raising* effects at section changes or deformation under stress.

Non-destructive testing/examination (NDT/NDE) Any test or inspection system which provides information regarding the properties and characteristics of a component without causing unacceptable damage. These terms are most commonly used with respect to flaw detection systems, for example *Dye penetrant* and *Magnetic particle* testing for surface defects and *Ultrasonic testing* or *Radiography* for internal defects. However, they may also include a wide range of techniques meeting the 'no unacceptable damage' criterion including chemical spot tests and other analytical techniques, hardness measurement and dimensional checks, particularly when these are performed on-site.

Non-electrolyte A substance which does not ionize in solution. See Interatomic bonding.

Non-ferrous Metals and alloys not based on *Iron*.

Non-Hookeian Not obeying *Hooke's law*.

Non-magnetic An imprecise term usually indicating that the material is not ferromagnetic. See Magnetic.

Non-metallic inclusions See Inclusions.

Non-propagating crack A crack that has ceased to grow. The term is most common in *Fatigue* where a number of cracks may initiate but often only one continues to extend. However, many forms of crack can become dormant because of some change in the stress field local to the crack tip or because of a change in the environment.

Non-shielded welding *Metal arc welding* processes in which no shielding medium is applied to the *Weld zone* to protect it from contamination, etc.

Non-transferred arc process See Plasma welding.

Normal segregation The *Segregation* of the lower melting point constituents into the areas of a casting that solidify last, usually the centre.

Normal solution A solution having the

Equivalent weight in grams of the substance dissolved in one litre of water.

Normal temperature and pressure (NTP) A reference environment, conventionally, 0 °C and 760 mm mercury. The preferred term is now 'standard temperature and pressure' with the pressure quoted as the equivalent *SI* unit value of 101 325 Pa.

Normalizing For steels, the practice of heating to the austenitic range and cooling in air. This achieves *Recrystallization*, eliminates *Residual stresses* and renders the steel relatively soft, ductile and tough, although not quite as soft as the fully *Annealed* condition. The microstructure will typically comprise fine-grained ferrite and pearlite in proportion to the carbon content. For steel capable of hardening during air cooling the term is inappropriate. See Steel.

Nose As in terms such as 'bainite nose'. See Isothermal transformation diagram.

Notch This term usually has its normal meaning of a geometric discontinuity such as a groove or crack. However, it is sometimes used in reference to a **Metallurgical notch** to describe a significant but localized variation in metallurgical characteristics such as composition, heat treatment, microstructure and hence, mechanical properties. These localized variations can act as a focus for various forms of metallurgical activity including concentrating tensile and creep strain in a manner analogous to a geometric notch. An example is the *Heat affected zone* of a weld.

Notch acuity Same as Notch sharpness.

Notch brittle/ductile These terms refer to the *Notch sensitivity* of a material, i.e. its behaviour and strength in the presence of a notch. The notch may be intentional, as in an *Impact* test, or inadvertent, as with a *Fatigue* crack in service. The basic concept in determining notch sensitivity is the difference in strength obtained from testing two bars of the material in question, one of which is plain and the other is notched. The strength of the notched bar is calculated on the remaining intact cross-section beyond the notch. If the notched bar gives a higher strength the material is termed **Notch ductile** or **Notch strengthening**; if lower it is termed **Notch brittle** or **Notch weakening**. Alternatively, the two strengths may be identified in terms of the **Notch strength ratio** whereby a value less than 100% is, conventionally, notch weak behaviour. It should be noted, however, that the strength of the notched bar may be sensitive to notch geometry, particularly notch tip radius. The term **Notch ductile steel** is sometimes used specifically for steels that meet some *Impact test* specification.

Notch constriction ratio (NCR)   A measure of the severity of a notch in terms of notch depth relative to section thickness.

$$NCR = \frac{P^2 - N^2}{P^2}$$

where P is the diameter of the plain section and N is the diameter of the reduced section at the notch. See also Notch sharpness.

Notch ductility   The *Ductility* of a material for a fracture originating at a notch. The percentage reduction in cross-section area is calculated from the original cross-section area at the tip of the notch. Otherwise the test procedure is the same as for normal *Tensile test*.

Notch rupture ductility   Same as Notch ductility.

Notch sensitivity   See Notch brittle.

Notch sharpness/acuity   The sharpness at the tip of a notch or crack. It may be defined in various ways, for example as a notch tip radius or as a ratio between the radius and the specimen geometry—often specimen diameter: $2 \times$ notch tip radius. See also Notch constriction ratio.

Notch strength/strength ratio   See Notch brittle.

Notch toughness   An imprecise term variously referring to the results of notched

bar *Impact tests*, or to notch ductility or notch sensitivity. See Notch brittle.

Notch weak   See Notch brittle.

Notched bar test   Any test in which the specimen has a notch positioned across the line of principal stress. In some contexts the term is taken to refer specifically to bend tests but more generally it includes Impact, Tensile, Creep and Dropweight tests. See entries on all these topics.

Nozzle   An orifice from which some material emerges. The term usually implies a precision orifice directing a pressurized stream of fluid in a controlled manner.

NTP   See Normal temperature and pressure.

Nuclear   Referring to the nucleus of the *Atom*.

Nuclear power   The controlled use of **Nuclear energy** released by **Nuclear fission**, i.e. splitting of the atomic nucleus, or by **Nuclear fusion**, i.e. joining of nuclei, such processes taking place in a **Nuclear reactor**. See Atom. Fission occurs if an additional neutron is forced into a nucleus of certain elements with a large number of neutrons. This produces an unstable nucleus which splits into two or more nuclei having lesser numbers of neutrons. The process is accompanied by the release of a large quantity of heat energy. For example, if a nucleus of Uranium-235 (the isotope of uranium having a nucleus of 92 protons with 143 neutrons) received an additional neutron the resultant Uranium-236 is unstable. It therefore splits to give Barium-140 (56 protons with 84 neutrons) and Krypton-93 (36 protons with 57 neutrons). At the same time free neutrons are emitted to enter further Uranium-235 nuclei, so continuing the **Chain reaction**. In a commercial nuclear reactor the heat energy is usually used to raise steam to drive a turbine generator producing electrical power. Nuclear fusion occurs when elements with only a few neutrons per nucleus are heated to a very high temperature which provides sufficient energy to initiate the fusion of the two simple nuclei to form an element with a larger nucleus. Again, the process is accompanied by the release of a large quantity of heat. In the sun, hydrogen with only one neutron fuses to form helium with two. On earth the process is the basis of the hydrogen bomb but its development for civil use remains elusive.

Nucleate boiling (of water)   The production, at a heated surface, of steam as discrete bubbles which are swiftly swept away by convection or the flow of water. It is usually the desired process in steam raising plant in contrast to boiling in the form of a broad persistent film, a phenomenon usually termed *Steam blanketing*.

Nucleation   The process by which *Solidification* or *Recrystallization* commences at a point or points in the material termed nuclei.

Nucleus (1)   The central core of an *Atom*.

Nucleus (2)   The group of atoms or foreign particle at which *Solidification* or *Recrystallization* commences.

Nuclide   An atom of an individual element having specific characteristics at the sub-atomic level. See Isotope.

Nugget (1)   Gold or silver in bulk metallic form found naturally.

Nugget (2)   The welded or bonded area of a *Spot, Seam, Projection weld* etc., See Figure 50, located at entry on Welding terminology.

Numerical aperture   A measure of the ability of the *Objective* lens of a *Microscope* to gather light from a wide angle. It is a function of the angle of the cone of light that the lens will accept and the refractive index of the medium between the lens and the specimen, usually air but sometimes oil. For a given magnifying power and focal length the larger the numerical aperture, the better the resolution, i.e. the fineness of detail that can be discerned.

Objective lens The lens in a microscope closest to the specimen being examined. It forms the initial magnified image which in turn is further magnified by the eyepiece or ocular lens. The objective lens therefore controls the resolution than can be achieved. See Microscope and Numerical aperture.

Occluded Closed or shut off or, in a chemical context, absorbed.

Ocular lens The lens in a microscope closest to the eye. It magnifies the image initially produced by the Objective lens. See Microscope.

Octahedral Eight-sided. The **Octahedral planes** of a cubic structure are those having *Miller indices* of {111}.

OD Outside diameter of a tube or similar hollow material and in contrast to ID, the Internal diameter.

OEM Original Equipment Manufacturer.

Offset (yield/stress/strength) Same as Proof stress/Strength. See Tensile test.

Ohm The *SI* unit of electrical resistance.

Ohm's law In a conductor, volts $=$ amps $\times$ ohms, where voltage is the potential drop across the conductor, amps is the current flowing in it and ohms is its resistance. It is applicable to direct current circuits but may require corrections for alternating current circuits. See Power factor.

Oil hardening The process of hardening steels by *Quenching* into oil. See Steel.

Oil hardening steel A *Steel* containing a sufficient quantity of alloying elements to cause it to fully harden when quenched into oil. Unless specified to the contrary it is usually implicit that full hardening should be achieved in a round bar of 50 mm (2 inches) diameter. See Steel and Hardenability.

Oil-less bearing (1) A bearing manufactured by *Sintering* metal powders with a proportion of graphite as a lubricant.

Oil-less bearing (2) *Sintered* metal, produced so as to leave a considerable volume of voids, and subsequently impregnated prior to installation with a lubricant which may be a grease or even thick oil. Note that it indicates not an absence of oil but only that the bearing does not need further oil addition in service. See also Oilite bearing.

Oil wedge When at rest, the shaft journal carried in a *Plain bearing* may penetrate the oil film and contact the bearing material. However, assuming a satisfactory design and adequate lubricant supply, a rotating shaft induces hydrodynamic forces in the lubricant which develop a standing wedge of oil slightly offset from the bottom centre line. The shaft then rides on this wedge. The bore of the bearing may contain a circumferential recess of complex form over much of its length to develop and stabilize the wedge.

Oilite bearing A proprietary sintered powder metal bearing containing a considerable volume of voids, typically 30%. It may be impregnated with oil prior to installation or it may be fed in service, the void contents providing a reservoir during oil flow interruption.

Oliver A blacksmith's forge with two or more dies, the lower halves of which were set on a single anvil, the upper halves being foot-actuated by individual treadles.

Olsen test   A test of deep drawing capability similar to the *Erichson test*.

Open (gap) joint   A joint to be welded in which the components have a significant gap deliberately set between them prior to welding. *Filler* metal is deposited in the gap.

Open-arc welding   Any welding operations during which the electric arc is visible.

Open-hearth process   A largely obsolete process in which steel is produced from impure *Pig iron* and/or scrap in a *Reverberatory furnace*, i.e. one having a large shallow hearth and a low roof constructed to deflect and reflect the flame down onto the charge.

Opening mode   See Fracture mechanics.

Optical microscope   See Microscope.

Optical pyrometer   See Pyrometry.

Orange peel   A coarse-textured surface with dimples that are rounded similar to the exterior of an orange rather than angular like the surface of a cube of sugar. It is observed on badly sprayed paint or on coarse grained sheet metal that has been deformed. In the latter case it results from the slightly differing deformation characteristics of neighbouring grains.

Order   In the context of *Crystal structure*, it is the phenomenon in certain alloys in which the atoms of the elements in a *Solid solution* arrange themselves in a repeated simple pattern on the atomic lattice, often termed a **Superlattice**. For example, two elements present in the appropriate proportion could position themselves at alternate sites on the three-dimensional lattice.

Ore   A natural mineral containing one or more metallic elements in sufficient quantity to be economically extracted. The metal may be chemically combined with unwanted elements as with iron ore or it may be merely mechanically mixed as with gold ore.

Ore dressing   A preliminary stage to the extraction of metal from an ore in which

material with no metal content, such as earth, is removed.

Orientation   The alignment of some feature relative to its surroundings, for example the alignment of the *Crystal lattice* relative to the axis of the material. In an as-cast material the orientation of the lattice in individual grains will, in most cases, be random but working processes such as rolling tend to bring the orientation of all grains into alignment. This is termed **Preferred orientation**. It causes properties such as *Ductility* to vary in different directions. Typically, a rolled sheet will have more ductility in the direction of rolling than at an angle of 45° to the rolling direction. This can causing problems in processes such as *Deep drawing*.

Osmiridium   An alloy found as a native metal in platinum ore. As the name indicates it is predominantly osmium (about 30–40%) and iridium (about 50–60%) with, usually, some lead as well as other elements of the platinum group. It is very hard and wear resistant and is used for small high-value applications such as the tips of pen nibs.

Osmium   A metallic element, one of the noble platinum group. The densest of all elements, it has a very high melting point and is very hard and brittle. It is used in pure or alloy form for high-value applications such as the tips of pen nibs. See Table 15 for physical properties.

Osmosis   The percolation of solvent through a semi-permeable membrane into a more concentrated solution so that the concentrations on the two sides of the membrane tend to equalize.

Out-gassing   Various processes for removing gases dissolved or otherwise entrapped in a solid or liquid. Usually heating and/or vacuum treatments are involved.

Out-of-position welding   Any position other than *Flat*.

Over-age   See Precipitation hardening.

Overhead welding position   The position in which the weld runs approximately horizontal but is made from beneath the components. This is usually taken to mean a *Weld slope* not greater than 15° and a *Weld rotation* between 168° and 180° for a butt and between 115° and 180° for a fillet.

Overheating Heating to an excessive temperature such that the grain size is excessive and properties are impaired. The term usually implies that the damage is reversible by further heat treatment, possibly in conjunction with a working operation. In contrast, *Burning* implies more severe heating following which quality cannot be fully restored.

Overlap (of weld) A weld defect caused by molten metal running onto adjacent parent metal but not fusing to it.

Overstraining The application of a *Strain* beyond *Yield* particularly where this is deliberate to develop beneficial *Residual stresses*.

Overstressing Generally, any excessive *Stress* but particularly the damaging application of one or more cycles of *Fatigue* with a stress in excess of the fatigue limit.

Oxidation In its metallurgical sense, the reaction of a metal with oxygen to form its oxide. In its electrochemical sense, the loss of electrons from a metal during any chemical reaction. The term encompasses a wide range of reaction rates, extending from the slow surface oxidation of iron in air, through burning to explosions.

Oxidative wear A form of *Abrasive wear* in which metal is removed leaving metal oxide particles on the surface. The metal may be oxidized prior to or following removal but the oxide particles may have a beneficial effect in that they prevent metal-to-metal contact and hence reduce the potential wear rate. Also see Fretting.

Oxide A compound of any element with oxygen. The manner in which metals form oxides determines the corrosion characteristics of a metal. The noble metals such as gold do not readily form oxides and hence are considered corrosion resistant. Other metals such as

aluminium are highly reactive with oxygen and hence form an oxide film as soon as they are exposed to air. However, the oxide film is thin, tenacious and impervious to oxygen and so further attack is stifled. Such metals, which include aluminium, titanium and chromium, are also regarded as corrosion resistant in the environments in which the protective oxide is formed and maintained. Finally, some metals such as iron in a moist environment react with oxygen to form a oxide which is bulky, not impervious particularly if it cracks and is readily detached. Such corrosion coatings provide no protection and attack continues until the metal is completely oxidized.

Oxide rooting Oxidation penetrating along grain boundaries at and close to the surface of a component.

Oxide wedging (1) The build-up of a bulky hard scale in a crack or interface that forces the surfaces apart. The simplest case is where two components are clamped together and the interface develops a bulky oxide or other solid corrosion product with sufficient volume and compressive strength to force the components apart and perhaps break the bolts or other fasteners.

Oxide wedging (2) This term is also used to describe the damage observed in welded joints between austenitic and ferritic steels such as are used for some pipework in power plant. Such *Dissimilar metal joints* can develop a wedge-shaped oxide band in the ferritic material immediately adjacent to the weld. However, the damage mechanism in this case is more complicated than a simple jacking action because of (i) carbon diffusion from ferritic to austenitic steels, (ii) the complex stress system resulting from the pipe internal pressure and from the thermal expansion differential between ferritic and austenitic materials and (iii) the difference in oxidation characteristics between the austenitic and ferritic steels.

Oxidizing agent A substance that causes

Oxidation of another material and, in the process, is itself reduced.

Oxidizing flame   A welding, cutting or other flame with more oxygen than is necessary for correct combustion and which therefore tends to oxidize the component under treatment. It produces the highest flame temperature.

Oxyacetylene, oxypropane, oxy-fuel gas/ cutting, welding, brazing (and similar terms)   See Gas cutting, Gas welding and Powder cutting.

Oxygen   An element boiling at minus 193 °C and hence gaseous at normal temperature and pressure. Air is a mixture of gases comprising about 80% nitrogen, about 20% oxygen and much smaller quantities of other gases. Water is a chemical compound of oxygen and hydrogen. Many of the metals in commercial use are extracted from oxide ores and many corrosion processes involve a reaction in which metal reverts to its oxide. The rapid oxidation of a material, i.e. burning, is a primary source of heat. Oxygen is normally diatomic, forming O_2 molecules, but is also encountered as ozone, O_3. See Table 15.

Oxygen cutting   Processes in which the primary cutting action is the chemical burning reaction of the component with oxygen. A fuel gas may be involved but its primary contribution is to initiate the process by heating the component to *Ignition temperature*. See Gas cutting, Powder cutting and also *Thermic lance*.

Oxygen lance/blown processes   Various processes for producing refined *Steel* from *Pig iron* and/or scrap, in which the molten metal is exposed to oxygen injected from beneath, as in the *Bassemer* process or is delivered from above by a water-cooled tube, i.e. a lance.

Ozone   See Oxygen.

P

p **type semiconductor** See Semiconductor.

Pa Pascal, The *SI* unit of pressure and stress. $1 \, Pa = 1 \, N/m^2$. 1 N (Newton, the unit of force) $= 0.224\,809$ lbf.

Pack carburizing See Case hardening.

Pack rolling The process of rolling two or more sheets as a pack. Unlike *Cladding*, there is no intention that the sheets will bond to each other, this being inhibited by oxide or oil films.

Palladium A metallic element, one of the noble, platinum group. In its pure form it is malleable, ductile and corrosion resistant although it oxidizes in air above about 400 °C. It can absorb about 1000 times its own volume of hydrogen at atmospheric pressure and is readily permeated by the gas above about 300 °C so is used as a filter, typically in fine tube form. For jewellery it may be used in its pure form or, particularly for larger items, alloyed with about 3% of either ruthenium or molybdenum. It is used as a catalyst on its own or in combination with platinum. With 50% gold it forms palladium gold used as a dental material. See Table 15.

Paraffin and whitewash test An early form of *Penetrant test* in which the suspect component was immersed in warm paraffin, withdrawn and wiped dry and then painted with whitewash. Any cracks were revealed by paraffin seeping from them to stain the whitewash.

Paramagnetic See Magnetic.

Parent metal In welding, the original unjoined metal and its microstructure.

Parfocal lenses Sets of lenses which have the same focal length so that they can be interchanged without the need to refocus the *Microscope* or other optical equipment.

Parkerizing A proprietary treatment for forming a corrosion-resistant *Phosphate conversion* coating on steel.

Partial pressure The pressure exerted by individual gases in a mixture.

Parting line/face The plane of division between multi-piece moulds or dies.

Parting sand The fine dry sand applied to the *Parting faces* of a mould to allow them to be split apart easily.

Partition coefficient See Partition law.

Partition law This states that 'the concentrations of any individual molecular species in two immiscible phases maintain a constant ratio to each other at constant temperature'. It is, of course, implicit that the two phases are sufficiently mixed or otherwise in contact for a time sufficient for equilibrium to be reached. The ratio of the two concentrations is the **Partition coefficient** or **Distribution ratio**.

Pascal The *SI* unit of pressure, stress. $1 \, Pa = 1 \, N/m^2$. 1 N (Newton, the unit of force) $= 0.224\,809$ lbf.

Pass (1) One movement of the material through a pair of *Rolls* or alternatively through the series of pairs of rolls comprising a *Rolling mill*.

Pass (2) One line of one layer of weld deposit. Some processes such as the *Metal inert gas* process can deposit a complete pass without interruption but in the *Manual metallic arc* process a number of *Electrodes* may be required to make one pass.

Passive   Generally, a metal not reacting significantly with the environment, particularly in *Electrochemistry*. More specifically, a metal which can be reactive with the environment but which has developed some surface characteristics such as a protective oxide film which prevents further attack.

Patenting   An isothermal heat treatment process applied to medium- and high-carbon *Steel* wire prior to its final drawing operation. The steel is heated to the austenitic region, typically 950 °C, and then cooled by quenching into a salt or lead bath at about 480–550 °C to produce a microstructure of upper bainite, i.e. ferrite with a fine distribution of carbide. This has a very high *Ductility* and can be cold drawn with total reductions in diameter of 90%, nearly double that possible with annealed wire. This produces wire of very high tensile strength, in excess of 1600 MPa, with adequate ductility.

Patina   Generally, surface discoloration or a coating arising from atmospheric deposition or corrosion. The term usually implies that the effect it is more decorative than deleterious; in some cases it may even be induced by chemical treatment. More specifically, the term refers to the green corrosion product formed on copper in sulphur-rich atmospheres.

Pattern   The shaped former on which a *Mould* is built. It corresponds in shape to the intended casting but has dimensions proportionately increased by the *Shrinkage allowance*.

Pattern metals   Various *Low melting temperature alloys* used for *patterns*. They are cast in a master mould or die and, after the working mould has been formed around them, they are melted and the metal poured out for re-use.

Patternmaker's allowance   See Shrinkage allowance.

Patternmaker's shrinkage   See Shrinkage.

Patternmaker's rule(r)   A rule(r) with calibrations oversize by the *Shrinkage allowance* appropriate to the intended cast metal.

Pauli exclusion principle   An electron state can contain no more than two electrons and they will have opposite spins.

Pearlite   The iron/iron carbide *Eutectoid*. See Steel.

Pearlitic iron/steel   See Cast iron and Steel.

Pebble surface/pebbling   A rounded surface texture similar to, but usually coarser than, *Orange peel effect*.

Peeling   The progressive detachment of a surface layer.

Peel test   A test of *Spot, Seam* and *Projection* welds in which the two joined components are prised or peeled apart in such a manner that the parting load is carried by individual spot welds or at one point on the seam. Also termed a **Slug test**.

Peen plating   See Tumbling.

Peening   Repeated local impacting to produce plastic deformation of the surface layers. It may be performed by hand-held, round-end hammer, for example on weld deposits and the adjacent areas. Alternatively, the component, or specific areas, may be bombarded with peening shot in automatic machines, termed **Shot peening**. Performed correctly, peening induces beneficial compressive *Residual stresses* in the surface although there are balancing tensile stresses below. Significant deformation, deliberate or otherwise, can occur in thin components. The intention is to deform the surface to a carefully judged extent but without scratches or notches. Consequently, the shot is spherical and of controlled size and broken particles are removed to avoid surface damage. In contrast, shot blasting is a relatively crude process intended to remove surface scale or to roughen the surface and accordingly may use broken, irregular shot or even angular, sharp-edged grit.

Peierls/(Nabarro) force   The force necessary to move a dislocation from one equilibrium position to another.

Pellini drop test A comparative test of the *Brittle fracture* resistance of plate or strip. The test piece comprises a full-thickness strip having, on one face, a brittle weld deposit notched to a depth that just reaches the original plate face. It is supported, notch down, at its two ends and impacted on its centre line by a falling weight. A key feature is that the height of the supports is selected to allow a deflection of the plate of 5° at the stage where the specimen face contacts the rig base. The weld is judged to initiate a crack at 3° deflection and hence the test determines whether the plate can deflect a further 2° without fracturing at the test temperature.

Pellini explosive test This is similar in concept to the *Pellini drop test* except that the load is applied by the shock wave from an explosion and the explosive charge is shaped and positioned to apply load over a large area of the specimen rather than just the centre line.

Peltier effect The release or absorption of heat as an electric current crosses the junction from one conductor to a dissimilar one. See Thermoelectric effect.

Penetrameter A device for assessing sensitivity in *Radiography*. It comprises a stepped plate or a stepped series of holes in a plate and it is placed on the component under test during irradiation. The extent to which the individual steps can be resolved indicates the size of defect that can be detected.

Penetrant See Dye penetrant.

Penetration In fusion welding, it is the depth to which a joint has been fused measured from the outer face. In spot, seam welds, etc. it is the depth of the *Nugget* measured from the interface. See Figure 50, located at entry on Welding terminology.

Penetration bead/run/pass (of weld) The first run of a multiple run weld deposited from one side which penetrates and fully fuses the *Root*. See

Figure 50, located at entry on Welding terminology.

Percentage The proportion relative to one hundred.

Percentile The boundary line cutting off the indicated percentage or the data falling outside the line.

Percussion welding A welding process in which heating is effected by a short-duration high-intensity electric arc between the surfaces to be joined, immediately following which the surfaces are brought together with a percussive action.

Period/periodicity (of elements) See Atomic structure and Periodic Table.

Periodic Table A table of elements presented in periods, i.e. with elements having similar chemical characteristics arranged in columns. Various presentations have been devised, one of which is Table 16. Also see Atomic structure.

Peritectic A reversible *Phase* change in which, on cooling, a liquid phase reacts with a solid phase to form another solid phase. See Figure 29.

Peritectoid A reversible *Phase* change in which, on cooling, two solid solutions react to form a third.

Permanent instability See Thermal instability.

Permanent load The loading due to self-weight and other non-varying sources as opposed to additional temporary loads.

Permanent magnet A material that retains a significant level of magnetism after the magnetizing field has been removed. See Magnetic.

Permanent set Plastic strain. See Tensile test.

Permeability See Magnetic.

Permittivity The ratio of the capacity of a condenser measured with and without the dielectric material in position between the plates. Also termed the Relative dielectric constant.

Pewter Various alloys primarily of tin with some or all of the metals antimony, copper, zinc and, in inferior grades, lead.

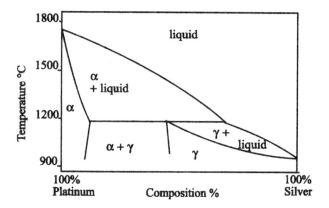

Figure 29   A peritectic reaction, the platinum–silver system

One example is tin with 8% antimony and 4% copper, 1% zinc.

pH (value)   The pH of a solution is the negative logarithm to base 10 of the molar hydrogen ion concentration. It is a measure of the acidity/alkalinity of aqueous solutions with a range from about −1 to 15. A neutral solution has a pH of 7, higher values being alkaline, lower values acid. The 'p' derives from the German 'potenz' for power and the 'H', obviously, from hydrogen.

Phase   A portion of a system which is chemically homogeneous and physically distinct. For example, steam, water (as liquid) and ice are phases of water. If a small amount of salt is dissolved in liquid water a single-phase **Solution** will exist. However, at a certain level the solution will be **Saturated** and no further salt can normally be dissolved. Addition of further salt will result in two phases, the saturated solution and salt particles. In abnormal circumstances it is possible for the amount of salt that is dissolved to exceed the saturation limit. Such solutions are termed **Super-saturated**. Similar effects occur in solid metals where a quantity of one metal can dissolve in another to form a **Solid solution** which is a single phase of completely uniform composition and which appears featureless under the mi-

croscope, apart from grain boundaries. The simplest case is where the two metals are completely soluble in each other at all compositions, an example being the copper-nickel system. However, most systems are not so simple and, as the composition is changed, a series of different phases will be formed. In certain cases, two or more elements will combine with each other in specific ratios, for example one atom of copper with two atoms of aluminium form $CuAl_2$. These are termed **Intermetallic compounds** rather than solid solutions but they are still phases. With many solid solutions, the solubility of the secondary component decreases as the temperature falls. If such an alloy containing the maximum amount of the second element is allowed to cool slowly the second element will come out of solution by a **Precipitation** process. The resultant particles or **Precipitates** form a second phase although, if cooling is rapid, precipitation may be impeded resulting in a super-saturated solid solution. A **Phase diagram** or **Equilibrium diagram** is a graphical presentation of the phases that are stable, under equilibrium conditions, for a range of temperatures and compositions of an alloy. The simplest phase diagram is the **Binary** in which temperature is plotted against composition, usually ranging from 100% of one element to 100% of the second element. Figure 30 shows the diagram for the copper-nickel system where the two metals are soluble in each other at all compositions. Figure 36, located at the entry for Soft solder, is an equilibrium diagram for two elements having limited solubility in each other and forming a *Eutectic* and no other feature. Many diagrams are considerably more complex and some consider only part of such a system, an example being the iron-carbon diagram, Figure 37 located at the entry on Steel. More complex systems are displayed in **Ternary** diagrams (three elements), **Quaternary** diagrams (four elements) and so on. Other forms of phase diagram depict the phases arising under non-equilibrium conditions. See Isothermal transformation diagram.

Phase contrast   A technique of *Microscopical examination* which utilizes the small differences in phase between light reflected from irregular surfaces to provide an image of surface features.

Phase diagram   See Phase.

Phase rule (Gibb's)   This states, for the general case and under equilibrium conditions, that:

$$P + F = C + E$$

where P = the number of phases present

F = the number of degrees of freedom or variance

C = the number of components, i.e. stable chemical substances

E = the number of environmental factors, i.e. two—temperature and pressure.

In the case of metal reactions the consequences of varying pressure are insignificant and can be ignored, leaving temperature as the only environmental factor so the equation becomes

$$P + F = C + 1$$

In this case the components are the pure elements that comprise the alloy.

Phosphating/phosphatizing   The production of a thin *Conversion* coating of metal phosphate on the surface of, usually, steel. The coating offers useful corrosion resistance on finished components and acts as a lubricant on material that is to be cold drawn.

Phosphor   Either a casual term for *Phosphorus* as in *Phosphor Bronze*, or a material that is *Phosphorescent*.

Phosphor bronze   Alloys of copper and tin deoxidized by the addition of phosphorus. The wrought alloys, with up to about 8% tin and 0.4% phosphorus, have good strength with excellent corrosion resistance. Casting alloys with up to about 13% tin and 1% phosphorus also

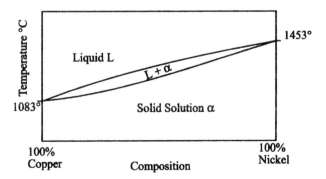

Figure 30   A continuous solid solution between two metals, the copper–nickel system

have good bearing characteristics. See Table 8.

Phosphorescence The same as *Flourescence* except that emission of radiation continues for some period after irradiation from the external source has ceased.

Phosphor(ized) copper A grade of commercially pure copper deoxidized with phosphorous. The small amount of residual phosphorous, typical 0.02%, seriously reduces electrical conductivity but the material is used in many non-electrical applications.

Phosphorus A non-metallic element, undesirable in steel but a deliberate addition to some copper alloys. See Table 15 for physical properties.

Phosphor(us) banding See Ghost banding.

Photo-elastic modelling/stress analysis A technique in which a model of some component is formed in a suitable plastic and deformed while being viewed under polarized light. The bright and dark areas that are produced in the plastic by interference effects indicate the size and direction of *Stresses*. It is possible to stress the model, which can be, for example, a complex pressure vessel, while warm and allow it to cool under stress to 'freeze in' the *Strain*. The model or sections cut from it may then be examined at leisure.

Photomicrograph/macrograph Photographs taken with a *Microscope* at high/low magnifications.

Photon A quantum, i.e. a single indivisible portion, of light energy.

Physical metallurgy The science and technology that is concerned with metals at an atomic and microscopical level.

Piano wire High-tensile (up to about 2000 MPa) high-carbon (about 0.9%) *Steel* wire usually produced by *Patenting*.

Pick-up Undesirable material collected or entrained during some process. Examples include material transferred between tool and component during ma-

chining or forming, and, during welding, elements entering the weld metal from the parent metal.

Pickle embrittlement See Pickling.

Pickling The removal of scale and other contaminants from the surface of a metal component by immersion in appropriate chemicals, commonly acids or alkalis. It is implicit that the metal component is not significantly attacked and often inhibitors are added to achieve this. If carried out *Electrolytically* it is often termed electrolytic cleaning, derusting, etc. In adverse circumstances *Steel*, particularly if of high hardness, can become charged with hydrogen from the surface reaction causing a form of *Hydrogen embrittlement* called **Pickle embrittlement** or similar terms. The hydrogen can be allowed to diffuse out by *Baking* at about 200 °C for a few hours.

Pickup See Pick-up.

Picral Picric acid in alcohol used to *Etch* ferrous materials particularly pearlitic *Steels* and most *Cast irons*. Unless otherwise stated a solution of 4% picric acid is usually implied.

Piezoelectric effect The interaction, in some crystals, between *Strain* and an electrical field whereby application of a strain induces an electric potential and vice versa.

Pig bed/casting machine The bed of sand or the continuous belt of moulds into which molten iron is poured to solidify as pig iron. See next entry.

Pig iron Impure iron produced in a *Blast furnace*. So called bacause the iron was originally run off from the furnace into multiple branched channels formed in a bed of sand. The multiple channels were thought reminiscent of a family of piglets feeding from the sow.

Pigment Fine particulate material added to paint or plastics to provide the colour.

Pilger process See Tube making.

Pin joint A structural joint that is free to hinge so that it does not transmit a moment.

Pin metals Any metals for pins, originally cold-drawn *Common brass* wire, sometimes chromium plated, now more usually hard-drawn medium carbon *Steel* chromium plated or, for higher-quality purposes, hard-drawn *Austenitic stainless* steel.

Pinch pass A *Rolling* pass producing a very small thickness reduction. Usually it is the last pass in the production of plate or sheet and is intended to provide close dimensional control. For steel it can avoid *Lüders lines* during subsequent processing.

Pinchbeck Brass with 5–15% zinc, remainder copper, for cheap decorative applications, hence a derogatory term for cheap jewellery.

Pinholes Fine porosity in castings or coatings.

Pinning The locking into position of a *Dislocation* by some feature such as a interstitial atom.

Piobert lines Same as Lüders lines.

Pipe (of weld) Same as Wormhole. Also see Crater pipe.

Pipe/piping (1) The elongated vertical cavity at the top central zone of a casting or ingot. When a molten metal *Solidifies* and cools it contracts. The metal tends to solidify first at the sides and bottom of the casting leaving insufficient molten metal to fill the top central area and a cavity or pipe is formed.

Pipe/piping (2) A tube or hollow section. The various terms are largely interchangeable but 'pipe' usually implies a fairly large diameter and a simple cross-section.

Pit casting/moulding Casting or moulding by pouring into a refractory lined pit.

Pitting Local loss of metal, usually with the depth of penetration considerably exceeding the surface extent. Corrosion is usually responsible but a form of *Fatigue* produces pitting on gear teeth and similar surfaces subject to sliding/rolling loading.

Pitting factor/ratio The ratio of the depth of the deepest pit to the average

depth of corrosion calculated from the total weight loss.

Plain bearing A bearing in which the shaft runs directly, albeit lubricated, on a non-moving bearing, as opposed to a *Rolling element bearing*. See also Oil wedge.

Plain carbon steel Any steel containing sufficient carbon to allow it to be hardened by rapid cooling but not containing a significant amount of any other alloying element. See Steel.

Plane strain/stress See Fracture mechanics.

Planishing Various techniques for producing a smooth surface but particularly techniques involving repeated hammering by hand or light machinery.

Planishing hammer A hammer having a smooth, usually polished, working face for flattening metal surfaces.

Plasma The condition when, at very high temperature, the atoms in a gas become highly ionized, i.e. partially dissociated releasing some electrons. It is sometimes termed a fourth state of matter (with solid, liquid and gas). A band of plasma carries the current in an electric arc. See also next entry.

Plasma arc welding/cutting/spraying Processes somewhat similar to *Tungsten inert gas* welding in that an arc is struck from a non-consumed tungsten electrode carried in the torch which also supplies inert gas to protect the weld zone. However, in the plasma processes the electrode is surrounded by a water-cooled nozzle which constricts the arc, increasing the arc current density. This increases the temperature of the portion of the gas passing through the arc giving a 'flame' temperatures of about 15 000 °C. This very high temperature compared with the normal TIG 'flame' allows much deeper penetration and hence thicker plate can be welded in a single pass. If the constricted arc is struck between the electrode and the constricting nozzle it is termed a **Non-transferred arc**, if between the electrode and workpiece it is termed a **Transferred arc**.

Plastics (materials)   A non-metallic material which, when subject to heat and possibly pressure, becomes sufficiently plastic to be moulded, extruded or otherwise formed to shape. Many, but not all, are *Polymers*. Table 14 gives some examples and data.

Plastic bronze   Various copper–tin bronzes with high levels of lead to offer some conformity to misalignment. A typical alloy would have about 10% tin, 15% lead, remainder copper.

Plastic constraint   See Constraint.

Plastic deformation/flow/strain   The permanent change in dimensions remaining after all stress has been removed. See Tensile test.

Plastic instability   See Tensile test.

Plastic replicas   See Replica.

Plasticity   The ability to undergo plastic deformation without cracking.

Plastimet   Composite materials in which plastic is reinforced with metal usually in an attempt to combine the formability and corrosion resistance of plastics with the strength of metal. The plastic usually forms the bulk matrix with the metal in various possible forms including fibres, woven strip or powder.

Plate   Flat, rolled metal usually 6 mm or greater thickness. See also Plating.

Platen (1)   A flat working surface of a press or other equipment. It may contact the workpiece or carry tooling and moveable jigs, etc.

Platen (2)   A series of tubes or bars joined by welding to form a plate.

Plating (1)   Metal in plate form.

Plating (2)   The application of a film or coating of some material to another material, for example chromium is applied to brass for mainly decorative purposes and zinc is applied to steel to provide corrosion resistance. See Electroplating and Galvanizing.

Platinum   A metallic element, the principal metal of the noble, platinum group. It has excellent corrosion resistance and finds wide application in a variety of corrosion- and heat-resisting and electri-cal applications and as a catalyst. See Table 15.

Platinum black   Very fine platinum powder, a very effective catalyst of some chemical reactions.

Plowing   Deep grooving by severe *Gouging* and usually involving significant local mechanical deformation.

Plug   A short *Mandrel* over which a tube is drawn in the **Plug drawing** process. It sizes the bore at the same time as the *Die* sizes the exterior. The plug may be retained in the die by a small mandrel in the tube bore or it may be designed to self locate in position—termed a **Floating plug**. See Tube making.

Plug corrosion   See Dezincification.

Plug weld   A weld formed between overlapping components the upper of which has a fully penetrating hole. Weld metal is deposited in the hole to form a joint.

Plumbago   Graphite powder. It may be mixed with a binder to form bulk items.

Plumber's solder   An alloy of about one third tin, two thirds lead. It solidifies over a wide range, 250–183 °C, so it has a long pasty stage during cooling, allowing it to be wiped to form a smoothly curved surface. See Soft solder.

Plunge milling   Machining using a milling cutter that cuts only on its end face rather than as it traverses.

Plutonium   One of the man-made transuranic elements. It is radioactive with a half life of 24 000 years and is toxic. Its applications are confined to nuclear power and weapons.

Point defects   Irregularities in the *Crystal lattice* which affect only a single site or local area, e.g. a *Vacancy* or a foreign atom, either *Substitutional* or *Interstitial*.

Pointing   Any technique for reducing the end diameter of rod, tube or wire for initial insertion through the die to commence *Drawing*.

Poisson's ratio   The ratio of transverse *Strain* to longitudinal strain. When a metal is pulled in tension it extends longitudinally and contracts in the transverse directions. Numerically, it is

between 0.25 and 0.5 for elastic deformation and 0.5 for plastic deformation.

Polar bonding   Same as *Ionic bonding*.
See Interatomic bonding.

Polarization (1)   In an *Electrochemical*
context, the change in value of the electrical potential between the open- and
closed-circuit conditions.

Polarization (2)   In the context of light
and optics, polarized light is composed
of waves in a single plane as opposed to
normal light which has waves in all
planes.

Poldi test   A simple hand-held *Hardness*
testing device. It comprises a 10 mm
diameter, hard steel ball located in a
head piece so as to register on the
component surface to be tested and, at
the same time, on the surface of a square
bar of known hardness. The head is
struck causing simultaneous indentations on the bar and on the test surface.
The two indentations are then measured
and the results read against charts to
provide a measure of hardness of the
component.

Pole figure   A technique for presenting
information on the three-dimensional
orientation of crystallographic planes by
means of a stereographic projection (see
entry on this topic). The basic stereographic projection usually deals with a
single crystal. The term 'pole figure'
usually refers to projections of multiple
crystals to reveal features such as preferred orientation. See Orientation.

Pole piece   The magnetic material, forming the extension of a magnet, that is
intended to concentrate and direct the
magnetic flux.

Poles   The sites on a magnet, usually the
two extremities, from which the lines of
magnetic flux emerge.

Poling   The stirring of impure molten
copper with wood poles which release
reducing gases to combine with and
remove the oxygen in the copper.

Polishing   Generally, any process for producing a bright, scratch-free, reflecting
surface. In *Metallography*, the preparation of a flat reflecting surface for mi-

croscopical examination. The aim is to
achieve a cutting action to reveal microstructure rather than a smearing action
which would obscure it.

Polishing lap/stick/pad, etc.   The device or equipment on which a polishing
medium is carried to polish some component.

Poly-   As a prefix it means many. Often
materials with the prefix are *Polymers*.

Polycrystalline   Composed   of   many
Crystals.

Polymer   A non-metallic material composed of large complex interlinked molecules. They are usually plastics. See
Table 14 for the properties of some
common plastics.

Polymerization   The joining of molecules to form extended chains and cross-
links.

Polymorphic   Having more than one solid form where the change from one to
the other is not reversible.

Pop riveting   A technique for making
riveted lap joints in sheet or other thin
material with access from one side only.
The T-shaped hollow rivet carries a pin-
headed mandrel through its bore, the
mandrel shaft emerging at the head end
of the rivet. A hole, of a diameter just
sufficient to allow passage of the rivet
shank and mandrel head, is drilled
through the pair of sheets. The shank of
the rivet, with the mandrel head projecting beyond it, is inserted through the
hole and the mandrel is then drawn back
so that its head expands the projecting
shaft bore. The rivet firmly nips the
sheets before the mandrel breaks leaving
its head on the far face of the sheets.

Pore/porosity   Voids   in   a   material.
Usually the term implies voids which
individually are of small size relative to
the section and particularly those resulting from gas entrapment during *Solidification*. In the case of highly porous
materials the porosity is often reported
as a percentage by volume.

Porous metal/bearing   Materials made
by *Sintering* so that they contain large
quantities of porosity that is interlinked.

They are used for a variety of applications ranging from filters to bearings. Lubricant may be pre-charged into the bearing or it may be force-fed in service.

Portable hardness testing Techniques and equipment for *Hardness testing* onsite rather than using large unwieldy laboratory equipment. With due care the quality of the results is adequate for most purposes.

Porthole die See Tube making.

Positive ion An ion, which may be an atom, molecule or radical, which has lost one or more electrons and hence carries a positive charge.

Positron Subatomic particles having the same mass as an electron but a positive charge. See Atom.

Post (weld) heat treatment Any process of heating following welding or some other process. The benefits of such heating include *Stress relief*, softening of hardened zones and diffusion of hydrogen from the component. This term usually refers to circumstances where the weldment is allowed to cool fully before being reheated, perhaps in some special facility away from the welding site; compare with the next entry.

Postheating The heating maintained after welding to control cooling rate, allow tempering, hydrogen diffusion and *Stress relief*. If the welded component is allowed to cool before being reheated the term **Post (weld) heat treatment** is more appropriate. Also see Weldability.

Pot annealing Same as Close annealing.

Potassium A metallic element. It is violently reactive with water and has negligible commercial application in its pure state. See Table 15.

Potential barrier The atoms in a *Crystal lattice* interact so that each is normally located at the minimum energy level it can reach and, consequently, is said to lie in an **Energy trough** or **Potential trough**. If an atom has to be moved it needs an injection of energy to initiate its movement out of the trough and over the potential barrier (or **Energy barrier**) into the next trough.

Potential difference The difference in potential, measured in volts, between two points in an electrical circuit when subject to an electromotive force or carrying a current.

Potential drop techniques A series of techniques for measuring the propagation of a crack by measuring the voltage drop in a constant current imposed across the crack zone. Four contacts are made to the crack zone. One pair placed either side of the crack apply the current; the second pair, similarly placed, measure the potential drop. Depending upon the power supply the techniques are termed **Alternating current potential drop** and **Direct current potential drop**, usually abbreviated to **ACPD** and **DCPD** respectively.

Potential energy The energy contained in some body by virtue of its position relative to another body in circumstances such as a gravitational field.

Potential trough See Potential barrier.

Potentiostat An instrument for supplying electrical current of constant voltage but variable amperage.

Poultice A local accumulation of debris which retains moisture and possibly other aggressive contaminants and causes corrosion of the underlying metal surface. *Differential aeration* corrosion is a major contributory factor to the attack. It is a common cause of corrosion on the undersurfaces of road vehicles.

Pour welding A flow welding process. See Burning (2).

Pourbaix diagram A diagram identifying the corrosion characteristics of a metal in aqueous solutions. It plots electrode potential on the vertical axis and pH on the horizontal. Zones on the diagram indicate circumstances in which the metal acts in a passive, immune or a corrosive manner. See Figure 31.

Powder cutting/injection process *Oxygen cutting* in which a powder, usually iron, is injected into the cutting stream.

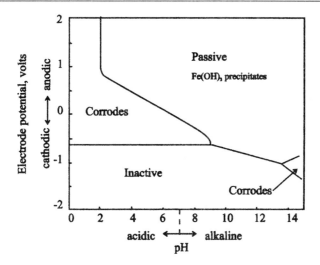

Figure 31 Pourbaix diagram for Fe–H$_2$O system

The iron assists the *Exothermal* reaction as well as providing fluxing and scouring actions. The process can cut high-chromium and stainless steels which, because of their resistance to oxidation, are not readily cut by the normal equipment.

Powder metallurgy The manufacture of metal powders and, from them, bulk components by processes such as *Sintering* rather than melting.

Powder factor The multiplication factor applicable to voltage (volts) × current (amps) to give power (watts) in an alternating current circuit. In an AC circuit with any significant amount of inductance or capacitance the wave forms for voltage and current will be out of phase with each other. The vector of the degree of out of phase is represented by ϕ and the power factor is then $\cos\phi$. Hence power = volts × amps × $\cos\phi$.

Praseodymium A metallic element, one of the *Rare earth* group. See Table 15.

Precious metals Historically silver and gold, more recently also the platinum group—platinum, palladium, rhodium, ruthenium, iridium and osmium.

Precipitate Particles that emerge from a super-saturated solution. See Phase and Solution.

Precipitation hardening The hardening of an alloy as a result of a phase precipitating, or partially precipitating, from a super-saturated solid *Solution*. Some alloys have compositions selected so that all or most of the minor constituents are in solid solution at high temperature but form a second phase at ambient temperature under equilibrium conditions. These characteristics can be manipulated to induce hardening of the alloy by a heat treatment cycle. The effects can be illustrated by reference to aluminium alllloys containing, for example, small amounts of copper, silicon and magnesium. The initial stage of the cycle is **Solution treatment** in which the metal is heated to about 480 °C for about half an hour and then cooled rapidly. This retains the alloying elements in super-saturated solid solution. In this condition the material is relatively soft and ductile allowing it to be formed to final shape. The material subsequently undergoes a precipitation process in which it hardens progressively, with an associated loss of ductility, as the alloying elements commence to come out of solution. This can occur, in certain alloys, over a few days at normal temperature, a process referred to as **Natural ageing** (see Figure 32). Alternatively, with other alloys it is induced by heating for a few hours at about 160 °C, referred to as **Artificial ageing, Precipitation treatment, Warm hardening, Warm ageing** or similar terms. Maximum hardening is achieved when the growing precipitates remain partially *Coherent* with the parent phase lattice. This approximates to the stage just before the emerging precipitates become visible under an optical microscope. See also Guinier–Preston zones. If the temperature or time at temperature is excessive the precipitation process continues towards completion, the material progressively softens and it is termed **Over aged**. See Table 7 for typical properties of aluminium alloys.

Precision casting (1)   A casting having at least one critical dimension of sufficient accuracy as to not require machining.

Precision casting (2)   The lost wax process, see entry on this topic.

Preferred orientation   See Orientation.

Preheating   Any deliberate heating before a process intended to confer some benefit. It is normally implicit that the preheat temperature will be exceeded during the subsequent process. The term is widely used in numerous contexts, for example the initial heating during the hardening of *High-speed tool* steels. However, it is a particularly critical matter in the context of welding *Steels* where adequate preheating is vital to avoid cracking in many circumstances. See Weldability.

Preload/prestress   The tightening load and resultant *Stress* applied to threaded fasteners, such as *Bolts, Screws* and *Studs*, that clamp components together. A high level of preload maintains pressure tightness, largely eliminates the risk of fastener *Fatigue*, reduces the risk of the fastener becoming unscrewed because of vibration and, in structural work, ensures that the load between members is carried by interfacial friction rather than by shear loading of the fastener. Usually, maximum mechanical and economic efficiency are achieved when the specified prestress matches the material's *Yield stress*. However, in practice, tightening equipment and techniques are of limited precision and hence prestress is commonly specified as about 80% of yield stress to avoid the risk of gross yielding leading to *Necking* and failure. An alternative approach, in special cases, is to develop a technique which deliberately extends the fastener beyond yield by a controlled amount without causing necking. Fasteners operating in their *Creep* range are usually tightened to

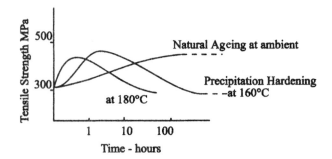

Figure 32   Hardening of a naturally aged aluminium alloy and another alloy precipitation hardened at optimum and excessive temperatures

lower levels of prestress selected for the particular application; a strain of 0.14% is often selected for ferritic steels. On first consideration it might be thought that the preload is directly additive to the service loads applied to the assembly, for example by the internal pressure acting on a pipe flange. However, this is not usually the case and, generalizing, only a small proportion of the service load is additive to the preload. This is a consequence of the small cross-section area of the fastener relative to the much larger load-bearing area of the flange and the associated differences in elastic strain. As a result, until such time as the service stress nearly matches the prestress, the joint will remain pressure-tight. It follows that large fluctuations in service stresses will induce only small fluctuations in the fastener stress so fatigue will be avoided provided the service stress does not closely approach the prestress. Some automatic but crude control of preload results from the wrench size being proportionate to the fastener diameter but more accurate techniques are necessary for high-integrity applications. These include torque wrench, mechanical or ultrasonic measurement of fastener extension, measurement of turn of nut angle after an initial 'nip' has been achieved and the observation of deformable washers or collapsible projections on nut faces. All have their respective advantages and problems.

Premium Material or treatment better than the normal standard, or the additional price for such material or treatment.

Prep Preparation (of a weld).

Preparation (as in joint preparation or weld preparation) The various activities involved in preparing a joint for welding or the geometry and dimensions of the surfaces to be welded.

Press fit See Interference fit.

Press forging A forging operation in which a substantial component is slowly pressed to shape. This achieves a more

uniform deformation throughout the section compared with *Drop forging*.

Pressing A light forging or drawing process in a press producing shallow deformation in a sheet metal or the component so produced.

Pressure (die) casting Any casting processes in which the molten metal is injected under pressure into re-usable moulds. The process offers dimensional accuracy, good surface detail and finish with a fine *Grain* size and minimum *Porosity*. It is suitable for low melting point alloys, particularly those of zinc and aluminium, having relatively low *Shrinkage* during solidification. Pressures can be as high as 670 bar.

Pressure welding Any welding process in which the joint is effected primarily by the interfacial force between components. The effects of the force include disruption of surface films to expose clean material, plastic deformation and mechanical mixing. Heating may be involved but little or no fused metal is incorporated in the joint. Examples include Electrical resistance spot welding and Flash butt welding.

Prestress See Preload.

Prestressed/pretensioned concrete See Reinforced concrete.

Prevailing torque fasteners Threaded fasteners that do not rely solely on the *Preload* to prevent loosening. Systems include local thread deformation, non-matching of mating thread forms, non-metallic inserts, or high-friction coatings. See also Lock nuts.

Primary metal A metal newly extracted from its ore or other natural source and not containing *Scrap*.

Primary solid solution See Solid solution.

Primary stress A directly applied tension or compression stress as opposed to a secondary tensile stresses resulting from bending or torsion.

Primes Material of the finest quality, particularly sheet and plate free from surface blemishes.

Principal stresses The stresses acting in

the three mutually perpendicular planes in the absence of a shear stress.

Prior austenite grain boundary   The locations, in the phase now existing, of the grain boundaries of the austenite that existed previously. See Steel.

Prismatic plane   The plane of the longest axis of a non-cubic crystal.

Probe analyser   An analytical facility on an *Electron microscope.*

Process annealing   The same as *Subcritical annealing* of *Steel.* It is so called because it is often applied part-way through processing when work-hardened steel requires softening with minimum scaling.

Process metallurgy   An imprecise term that usually indicates only the primary metallurgical processes of extraction, refining and casting but may be extended to include secondary manufacturing processes such as working and heat treatment.

Process scrap   See scrap.

Profile cutter   See Gas welding.

Profile examination (1)   The examination of a component in profile, i.e. in silhouette.

Profile examination (2)   The examination of a surface illuminated by a beam of light impinging at a shallow angle so that the shadows magnify surface irregularities.

Profile gauge   A device comprising a large number of needles arranged in-plane like the teeth of a comb but held in a spine through which they can slide when pressed firmly. The edge formed by the needle tips is pressed against the surface of interest so that the needles slide to record the profile.

Progressive fracture   Cracking that extends slowly such as *Fatigue* as opposed to fast *Brittle* fracture.

Projection welding   A *Resistance welding process* in which one or more projections, formed on one or both surfaces to be joined, provide the site for electrical contact and force application. The projections are usually crushed flat during welding.

Proof (piece)   The test product of a die to confirm accuracy and quality.

Proof load   A load, in excess of the maximum designed service load, applied to a component or structure to prove fitness for service. The amount, ratio or percentage by which the proof load exceeds the maximum service load may be referred to as the *Safety factor.*

Proof stress   The stress required to induce a specified percentage permanent strain. See Tensile test and compare with Proof load.

Proof test   The test carried out on a component or assembly to confirm that it is capable of performing to the design intent. The term usually implies mechanical testing (see Proof load) but can include electrical and other tests.

Propagation   In a metallurgical context, usually the growth of a crack.

Proportional limit   The same as Limit of proportionality. See Tensile test.

Prosthesis   A manufactured component for insertion into the human body to replace or support a defective item.

Proton   See Atom.

PS   Proof stress. See Tensile test.

PTFE   Polytetrafluoroethylene. A thermoplastic having a relatively high melting point, good mechanical properties, good resistance to chemical attack and very low coefficient of friction. Above about 300 °C it can produce flourine and if melted may produce hydrofluoric acid, both of them hazardous substances. See Table 14 for properties of plastics.

Puddling process   An obsolete technique for manufacturing *Wrought iron.* Impure *Pig iron* is melted and mixed with mill scale (iron oxide) on the hearth of a *Reverberatory furnace.* The oxide combines with the impurities in the molten iron to form a *Slag.* As the iron becomes more pure its melting point increases and the material on the hearth becomes increasingly pasty, trapping much of the slag. The pasty iron is formed into balls, termed **Balling**, lifted out and then hammer forged or rolled into bar, expelling some of the slag. The bar is cut into lengths, reheated, piled up and forged again. This is repeated a number of times, expelling more slag on each forging operation and also progressively elongating the remaining iron silicate slag inclusions to produce the fibrous structure that is characteristic of classical wrought iron. At an intermediate stage of forging and rolling the bar is termed **Muck rolled** iron and after final rolling, **Refined iron**.

Pull cracking The cracking caused by contraction stresses during *Solidification* of castings.

Pullover mill A non-reversing rolling mill, usually two-high. See Rolling mill. The sheet metal is passed through the rolls a number of times as the roll gap is progressively reduced. Between passes the material is returned to the entry side by being pulled over the top of the mill.

Pulsed arc welding Various electric arc welding techniques in which repeated pulses of high current are superimposed on a steady, lesser background current. In MIG (*Metal inert gas*) welding the background current provides the main heat source and the pulses control the mode of *Metal transfer* from feed wire to weld pool. In TIG (*Tungsten inert gas*) welding the background current maintains arc continuity and the pulses control the heat input into the weld zone.

Punch Various hand or power tools of cylindrical form with an end to strike the workpiece to punch a hole, stamp an imprint, flatten a rivet head, etc.

Purnell process Similar to Austempering.

Push bench See Tube manufacture.

Push fit A tight fit without slack but which can be made by hand. It lies between *Interference fit* and *Sliding fit*.

Push–pull Any system of loading where the load alternates between tension and compression and acts axially rather than in bending or torsion.

Pyrites Generally, any sulphides includ-

ing iron pyrites ('fool's gold') and copper pyrites.

Pyrometry The measurement of high temperature. The term usually implies temperatures above about 300 °C which is about the limit of a simple mercury-in-glass thermometer although more sophisticated mercury/special glass devices can operate at higher levels. A wide variety of devices is available. The simplest technique is merely to judge the colour by eye but refinements of the technique utilize standard colour charts or a series of comparison filters. Such techniques are useful from about 550 °C in dull light, to about 1200 °C although the results are operator sensitive. **Optical pyrometers** superimpose a heated filament on a background of the light from the source being measured; the calibrated filament is then heated until it just disappears, hence the description **Disappearing filament pyrometer**. They are useful from about 700 °C to about 2000 °C. **Radiation pyrometers** focus the radiation on a calibrated thermocouple and have no theoretical upper limit. **Electrical resistance pyrometers** measure the change in resistance of a wire element, commonly platinum, that is exposed to the environment in question. They have a maximum operating temperature of about 1000 °C. **Thermocouples** utilize the *Thermoelectric effects* associated with pairs of metals in contact at their high- and low-temperature junctions and have an operating range up to about 2500 °C. **Pyrometric cones ('Seger' cones)** are tall cones of selected mineral compounds that commence to collapse at the tip at specific temperatures up to about 2000 °C. A series with a range of collapse temperatures is loaded into a furnace with the charge and their progressive collapse observed.

Pyrophoric alloy An alloy which ejects sparks when struck, as in flints for cigarette lighters, e.g. 85% cerium, 15% magnesium.

Q

Quality assurance Systems and procedures in the widest sense for ensuring that appropriate quality levels are identified, working practices defined, records maintained and product quality monitored. They can be considered as the system established in advance to ensure that problems will not arise.

Quality control Systems, including testing procedures and physical measurements, for ensuring that established quality standards are met. They can be considered as the system of checking that confirms that a product is of intended quality.

Quality heat treatment The final heat treatment intended to induce the properties required by the component in service. Previous heat treatments could have intended, for example, to produce *Homogenization, Grain refinement* or softening between stages of cold working.

Quantometer/quantograph (and similar names, some of them proprietary) Analytical instruments, usually spectrographs, in which an arc is struck between an electrode and a sample of the material to be identified, sometimes in a vacuum. The light emitted is presented as a spectrum which is then either displayed on a photographic plate to be measured on associated equipment or, alternatively, selected lines of the spectrum are measured within the instrument and the results printed. In the latter case analysis of an item is a matter of seconds provided that it falls within the standardized range of the instrument.

Quantum The minimum quantity of increment of physical energy.

Quantum mechanics The theoretical concepts necessary to describe and explain activities and reactions on an atomic scale in the same way as Newton's laws and other physical and mechanical laws deal with such matters on a bulk scale.

Quartz One of the three forms of silica, SiO_2. It is hard, No. 7 on *Mohs* scale of hardness, melts at about 1710 °C, has a low coefficient of thermal expansion (5×10^{-7}) and some grades have strong *Piezoelectric* properties. It occurs naturally as **Rock crystal**.

Quasi-cleavage A fracture occurring by a combination of *Cleavage* (brittle) and *Dimpled* (ductile) fracture mechanisms. Often observed in high-strength materials.

Quarternary Comprising four components.

Quench Rapidly cool in various media. Common media include, in increasing order of severity, still air, air blast, oil, water, brine. The process is perhaps most commonly associated with the hardening of *Steel* but it is also encountered in other circumstances such as the preliminary stage of *Precipitation hardening* or it may be utilized to avoid undue oxidation at the high temperature or to induce scale to spall off.

Quench ageing Age hardening following quenching from solution treatment temperature. See Precipitation hardening.

Quench cracks The cracks occurring during or immediately following *Quenching* as a result of *Stresses* induced by uneven thermal contraction or dimensional changes associated with *Phase* changes.

Quicking The deposition of a loosely adherent mercury film on copper alloy components by dipping in a bath of, or swabbing with, various mercury-containing solutions. Normally no electrical current is applied. The mercury inhibits the initial non-electrolytic, non-adherent deposition of silver when the component is subsequently introduced to the silver plating bath. The liquid mercurcy film can cause *Liquid metal attack* at grain boundaries if high levels of *Residual stress* are present and hence variations on the process are sometimes used as a test to confirm freedom from such stresses.

Quicksilver Mercury.

Quill (shaft) A hollow spindle or shaft.

R

R   The stress ratio. See Fatigue.

Rabbling   Stirring molten metal and or scraping *Dross* from the surface.

Rad (1)   The measure of absorbed radiation. 1 rad = 100 erfs/gram = 0.01 J/kg.

Rad (2)   The unit symbol for radian the measure of plane angle.

Radial marks (on a fracture surface) Lines radiating from the fracture origin, particularly on unstable, rapidly propagating fractures.

Radiant tube furnace   A furnace in which combustion of the fuel (oil or gas) occurs in tubes which pass through the furnace chamber. The charge is heated by radiation but not contaminated by the products of combustion.

Radiation damage   Any damage, usually on a microscopic level but possibly affecting bulk properties, resulting from radiation of any sort. However, unless otherwise qualified the term usually implies radiation by sub-atomic particles, particularly neutrons released by radioactive decay in *Nuclear* reactions. Effects may include *Embrittlement*, increase in *Hardness, Tensile strength*, and *Yield strength* and reduction of *Creep* strength. See also Radiation hardening.

Radiation embrittlement   Any damage resulting from any form of radiation of any material, for example sunlight on plastics, but particularly the damage developing in nuclear reactors. In reactors various damage mechanisms affect different metallic materials including the austenitic steels and the nickel base alloys. Until the development of low-copper varieties (< 0.1%) a major concern was the embrittlement of the initially very tough, manganese–nickel–molybdenum bainitic *Steels* often used for pressure vessels. These suffered embrittlement as a result of neutron irradiation causing the formation of epsilon copper, a powerful matrix strengthener.

Radiation hardening   A form of *Radiation damage* resulting from the formation of defects in the *Crystal structure*, such as *Vacant sites* or *Interstitial* atoms which hinder the movement of *Dislocations*.

Radiation pyrometer   See Pyrometry.

Radio tracing   Same as Radioactive tracing.

Radioactive   Having an *Isotope* that spontaneously disintegrates emitting particles or radiation.

Radioactive tracing   The introduction of a radioactive material to a system or another material so that the radiation can be monitored to reveal the progress of reactions or movements in the system.

Radiography   Examination of components by X-rays or gamma rays to reveal sub-surface defects. See Non-destructive testing.

Radium   A radioactive metallic element with a half-life of about 1600 years. In the metallic form its commercial applications usually utilize its radioactive characteristics, for example radiography and radiotherapy. See Table 15 for physical properties.

Radon   An inert gaseous element produced naturally by the decay of radium. It is radioactive with a half-life of about 4 days. See Table 15 for physical properties.

Rag   The burr, or rough edge left on a cut surface.

Rake (of tool)   The angle between the approach surface of a tool and the line perpendicular to the component surface at the cutting edge. An angle greater than 90°, termed **Positive rake**, usually gives a good surface. An angle less than 90°, termed **Negative rake** allows more rapid metal removal at the expense of surface finish.

Ramming   The process of repeatedly impacting a powder with a ram to achieve a compact mass closely moulded to the shape of the container.

Random sequence welding   Deposition of multiple short runs of weld at random along a joint until the full length is complete. This minimizes distortion.

Range of stress   Stress range. See Fatigue.

Rankine   See Absolute temperature.

Rapping   Repeated light tapping, for example to release a pattern from a sand mould or to encourage the flow of dust deposited on the collector plates of an electrostatic precipitator.

Rare earth metals   The lanthanide group of elements comprising Nos 57 to 71 inclusive, scandium No. 21 and yttrium No. 39. See Table 16.

Rare gases   The six inert gases: helium, neon, argon, krypton, xenon and radon. Sometimes termed the Noble gases.

Ratchet marks   Steps on a fracture surface at the intersection between fractures propagating from separate origins. The term is most commonly used in the case of *Fatigue* cracks having multiple origins such as are associated with high stresses and stress concentrations. The term is also used to describe some multi-origin ductile failures such as are observed on threaded fasteners torque-tightened until they fail.

Ratchetting   A progressive change in shape over a series of cycles of loading, usually resulting from, or associated with, a thermal cycle. See Shortening and Thermal fatigue.

Razor steel   Any steel used for razors or razor blades. Open razors and earlier 'safety' blades were usually plain, high-carbon steel (typically 1.5% C); more recently 'stainless' grades with about 12–14% chromium and about 1% carbon have become the norm. All types of steel are used in the hardened and lightly tempered condition.

Reactor   Any vessel or container in which a reaction occurs. Where the reaction is chemical the reactor may be no more than a simple closed vessel, possibly heated and or pressurized, in which the reaction progresses in isolation from the external environment. In a *Nuclear* context the term usually implies the containment vessel, everything within it and also directly associated external equipment such as shielding.

Recalescence   The evolution of heat as a metal undergoes a phase change on cooling. It can cause a transient increase in temperature of *Iron* transforming from austenite to ferrite, the effect being particularly pronounced in *Steels* of *Eutectoid* composition transforming at the eutectoid temperature.

Recarburization   Any process for introducing carbon into *Iron* or *Steel* in the solid state, intentionally or otherwise.

Recovery (1)   The same as *Recrystallization*.

Recovery (2)   The reduction of *Work hardening* effects without necessarily involving recrystallization. One example is the case of some severely work-hardened steels in which heating to a modest temperature, say 250–350 °C, effects an improvement in toughness with little loss of strength.

Recovery (3)   More generally than in the above usages, the return of properties or characteristics, usually desirable, that have been lost or impaired as a result of some prior treatment including radiation as well as working. The recovery may occur naturally or be induced, for example by heat treatment. In the case of *Fatigue*, unloaded rest periods during testing may improve fatigue life so such rests are said to provide some recovery.

Recrystallization The process of forming, in a solid material, a new *Grain* (crystal) structure, usually during heat treatment of a work-hardened material but also when a metal undergoes a *Phase* change in the solid state. Recrystallization commences at individual nuclei, each of which initiates a crystal that grows outwards in three dimensions until it meets its neighbour or the surface. The resultant grains are equiaxed, free from strain and the material is normally in a softened state.

Recrystallization temperature The minimum temperature at which a material will commence to recrystallize.

Red brass Various brasses with more than about 80% copper and hence having a red colour. They typically contain about 15% zinc often with up to 5% tin and, if they are to be machined, possibly up to 5% lead. Depending on the particular grade, they are used for castings or tubing for plumbing and related applications.

Red hardness/strength Hardness and strength of a material at red heat. The term usually implies that the material in question has these characteristics to a useful extent.

Red heat An imprecise term since perceptions vary with the intensity and colour of the background light. In natural, dim light a dark red is readily observed at about 540 °C; above about 1000 °C the colour progressively becomes much more orange/yellow.

Red short Susceptible to cracking during hot working because of poor *Ductility*. The same as hot short except that it obviously only applies to materials that are worked at red heat.

Reducing (1) A broad term applied to many hot or cold working processes for reducing the cross-section of plate, bar or tube.

Reducing (2) (of tube ends) A reduction

in the external and internal diameters without any deliberate change in wall thickness. See Figure 45, located at entry on Tube manipulation.

Reducing (3) An abbreviation for the Cold reducing process, a cold forging process in the final stages of *Tube making*.

Reducing atmosphere/flame (in gas welding) An atmosphere or welding gas flame having an excess of fuel relative to oxygen. It can have a reducing or deoxidizing effect on the workpiece.

Reduction (1) In its chemical sense the removal, from a metal ore, of oxygen and other elements combined with the metal.

Reduction (2) In its *Electrochemical* sense, any chemical process in which a metal gains electrons.

Reduction (3) In manufacturing, the relationship between the original dimensions and the final dimensions, usually stated as a ratio or a percentage.

Reduction of area The amount, usually expressed as a percentage, by which the cross-section area of a tensile test specimen is reduced at the point of failure. See Tensile test.

Redundancy In engineering and metallurgical contexts this does not have quite its usual implication of not being required. Rather it indicates the availability of back-up systems or additional load-bearing features or material to accommodate failure of one component or part thereof.

Redux bonding A proprietary adhesive system.

Reeler straightening A process for straightening round tube. The material is spun between barrel-shaped rolls aligned at a small angle to parallel, similar to the tube piercer shown in Figure 43 located at entry on Tube making. As the tube spirals through the rolls is experiences deformation of increasing and then decreasing severity and emerges circular and straight.

Re-entrant A surface feature that points or projects inwards. A re-entrant angle acts as a notch; a re-entrant feature on a casting prevents its removal from the mould unless the mould has special provision or is expendable.

Reference stress A concept whereby the *Creep* life of a component subject to a complex stress pattern is equal to the life of a simple creep testpiece tested at the reference stress.

Refined In a purified form.

Refined iron Originally, good-quality *Wrought iron (1)*, see Puddling process. The term is now used more vaguely referring either to good-quality *Cast iron*, particularly if it has a fine grain and flake size, or to high-purity iron.

Refinement (of grain size) Any technique for producing a smaller grain size including specialized heat treatments or treatment of the molten metal such as *Modification*.

Refining Purification of impure metal.

Reflowing The brief melting of a surface, often a dip coating, to improve flatness and brightness.

Refractory Historically, the term implied mineral and ceramic materials used in applications such as furnace linings but modern usage includes any material which does not react significantly with its environment and has useful levels of strength for useful periods of time at high temperatures.

Refractory metals/alloys Metals such as tungsten, tantalum and molybdenum which have melting points well above 2000 °C. For comparison, iron melts at about 1530 °C. These metals have applications related to their high strength at elevated temperatures but the term also has implications of difficult manufacturing and working characteristics.

Regenerator A system of heat conservation in which the heat of the process exhaust gas is extracted and returned to the cool ingoing gas. The term usually refers to pairs of honeycomb brick structures, one of which is heated by hot exhaust gas, while at the same time the other has heat extracted from it by the ingoing cold gas stream, the gas flows being alternated from time to time. However, the term may also be applied to other devices such as those with a rotating element passing through the two gas streams.

Regulus metal/regulus of antimony Pure antimony.

Reheat cracking Generally any cracking developing when a component is heated following cooling from casting or other heating cycle. More specifically the term refers to cracking in welds when they are heated either during *Post-weld heat treatment* or when they enter service. This form of damage is most common in the *Heat affected zones* of low-alloy creep resisting steels such as 0.5% chromium, 0.5% molybdenum, 0.25% vanadium. Reheating of such weldments causes precipitation of alloy carbides within the grains which are consequently strong and rigid. The grain boundaries are relatively weak and suffer *Creep* or intergranular hot tensile failure as a result of high levels of *Residual stress* from the welding cycle. Externally imposed stresses exacerbate the problem. Also see Weldability.

Reheat furnace Any *Furnace* in which a part formed item is heated for further processing.

Reheater In a *Boiler*, the system of tubing in which steam is reheated following its first passage through a steam turbine.

Reinforced Containing elements that improve some property. The term usually refers to materials such as *Reinforced concrete* or **Reinforced plastics** which contain, respectively, steel bars and various fibres to improve their tensile and bending strengths.

Reinforced concrete Various forms of concrete containing steel *Reinforcing bars* to improve the material's capability to support tensile and bending loads. The simplest form is produced by pouring concrete over bars that are not restrained other than is necessary to locate them in position. **Prestressed concrete** contains steel bars or rods that are under tension to apply a compressive prestress to the concrete before the service load is imposed. The level of prestress should ensure that even under full service loading the concrete does not develop any tensile stress. If the bars are held in tension during pouring and curing and they require no further system to apply or maintain load during service the material is termed **Pretensioned concrete**. If the tension is applied after the concrete has cured, often during building construction, the material is termed **Post-tensioned concrete**. In this case the reinforcing bars or cable can be inserted through channels cast into the concrete.

Reinforcement (of a weld) Weld metal that stands proud of the straight line joining the *Toes* at the weld to parent metal junctions. See Figures 50 and 51, located at entry on Welding terminology. Some authorities prefer the alternative terms 'excess', but others consider this to have an undesirably pejorative implication bearing in mind that some overfilling is normal, usually fully acceptable and preferable to insufficient filling.

Reinforcing bar Steel bar for *Reinforced concrete*. It is usually hard rolled mild steel or medium-carbon steel with substantial ridges, aligned circumferentially, to improve its 'grip' on the concrete.

Relative atomic mass The strictly correct term for *Atomic mass*.

Relative dielectric constant See Permittivity.

Relative humidity The ratio of the actual water content of air to the saturation content at the same temperature, expressed as a percentage.

Relaxation Strictly, Creep relaxation. The reduction of imposed load as a result of deformation by *Creep*. The mechanism involves the reduction in elastic *Strain* with a corresponding increase in plastic strain, i.e. creep. The phenomenon occurs in components such as high-temperature bolts which are subject to a fixed strain rather than a fixed

stress. See Tensile test and Displacement-controlled loading.

Relaxation time The time taken for some characteristic or property to reduce to a predetermined value. The term is usually used in connection with diffusion-controlled processes such as *Creep*.

Remanence The magnetic flux density remaining following removal of the magnetizing force. See Magnetic.

Renormalizing The *Normalizing* of *Steel* that has undergone some process, for example welding, that has impaired some characteristic. The term obviously implies that the steel was in the normalized condition prior to the process.

Residual stress The *Stress* remaining in a component after all external influences have been removed. Such stresses exist as an internal system with compressive stresses at one location balancing tensile stresses elsewhere. They can be induced in a component by many treatments such as solidification, deformation, machining and uneven heating or cooling (particularly welding). A simple example can be envisaged by considering a tube which is slit longitudinally producing a wide gap. The gap can be closed inducing stress around the tube circumference and if the gap edges are welded together the tube is left with a system of residual hoop stresses, tensile on the exterior, compressive at the bore. In some cases the level of stress may be very high, up to *Yield* magnitude, and often it will cause, or contribute to, many damage mechanisms including *Creep*, *Fatigue*, *Brittle fracture* and *Stress corrosion*. Generalizing, tensile residual stresses are potentially damaging but compressive residual stresses can be beneficial and they may be deliberately induced by *Peening* and similar techniques. See also Stress relief.

Residuals, residual elements Any elements in a metal not added deliberately. They usually remain from the original ore or remelted scrap but may be introduced from other sources such as flux or the furnace environment.

Resilience The capacity of a material to store elastic *Strain* energy.

Resist A coating material, lacquer, etc., applied locally on a component, or generally to equipment to prevent deposition during *Electroplating* or to prevent dissolution during *Etching*. Also called *Stopping*, *Stop-off*, etc.

Resistance butt welding *Resistance welding* processes in which the components are butt jointed, i.e. end face to end face or edge to edge, as opposed to a lap joint. See also Flash welding.

Resistance strain gauge A strip or wire of material whose electrical resistance varies progressively with *Strain*. A calibrated piece of the material is bonded to the component to be monitored and its change in resistance provides a measure of strain.

Resistance welding/brazing/soldering Joining processes in which heat is produced primarily by the passage of an electric current through the contacting surfaces of the two components. The current is delivered via copper base electrodes which also apply a force sufficient to maintain contact and, at least in the final stages, to disrupt surface films allowing formation of a weld. In many cases no significant melting is involved.

Resistivity Resistance to the passage of electricity. It is the reciprocal of conduction measured, in the *SI System*, in microhm centimetres.

Resolution/resolving power Of a *Microscope* or individual lens, it is the size of the smallest feature that can be distinguished. In a conventional light microscope it is about 0.2 μm.

Resonance The excitation of a component or system at its natural frequency of vibration. Such excitation can induce a progressive increase in amplitude of vibration with associated increase in stress and consequent risk of *Fatigue* failure or even simple overload.

Rest periods With reference to *Fatigue*, periods of significant length during which the imposed loading is suspended

and some *Recovery* may occur with consequent benefit to fatigue life.

Restrained weld test A test in which the test piece being welded is firmly secured by some support system of bolts or substantial preliminary welds to minimize movement during the test welding cycle. The restraint imposes high tensile stresses as a result of weld metal contraction during cooling following welding. The test therefore indicates the likelihood of cracking being encountered when welds to a similar procedure are made during production. One example is the *LeHigh test*.

Restrike A repeat of a forging or stamping operation to improve dimensions or quality of surface detail.

Retained austenite See Steel.

Retrogression The phenomenon observed in the precipitation hardening of solution-treated material whereby initial heat treatment at some low ageing temperature subsequently causes, at some higher temperature, a more rapid hardening than normal and a consequent risk of over-ageing. See Precipitation hardening.

Reverberatory furnace A fuel combustion furnace having a wide shallow hearth and a curved roof which is designed to reflect and deflect the flame entering from one side down onto the charge.

Reverse engineering The practice of copying a design by measuring and analysing a sample, preparing drawings, specifications and procedures and producing the copy without infringing the original manufacturer's rights. The legal aspects require great care.

Reverse flow forging Processes in which a billet contained in a die is partially pierced by a ram causing it to expand to fill the die and extrude back around and up the ram. *Impact extrusion* is similar but utilizes a high-velocity ram.

Reverse(d) polarity (welding) An ambiguous term as is its opposite, Straight polarity. In some countries, including the UK, reversed polarity usually means

DC arc welding with the electrode connected to the negative pole of the supply and Straight polarity has the electrode connected to the positive. Unfortunately, American terminology is exactly opposite. Better terms are *Electrode positive* and *Electrode negative*.

Reversed stress   A cyclic stress where the compressive load and the tensile load are of the same value. See Fatigue.

Reversible (electrolytic) cell   A cell in which the two electrodes are immersed in separate electrolytes and in which reversing the current flow results in a reversion of the reactions at the electrodes.

Reversing mill   A *Rolling mill* through which the product passes backwards and forwards a number of times. The direction of rotation of the rolls reverses and a reduction is thickness is produced on each pass.

Reversion   Generally, the return to some previous condition. More specifically, the reversal of a *Martensite* shear transformation in non-ferrous allows to restore precisely the structure of the original pre-quenched *Phase*.

RF   Radio frequency, as in **RF welding** which utilizes a RF power beam for radiation heating.

Rhenium   A metallic element resembling, in some respects, the platinum group metals. It has good resistance to wear and corrosion up to about 300 °C and is readily applied as an electroplate. It has a limited application in the pure form, mainly small plated bearings or contacts, but is used as a hardener for platinum. See Table 15 for properties.

Rheocasting, rheoforming   Die casting or other processes utilizing the thixotropic characteristics of stirred, partially solidified alloys. When a partly solidified alloy is stirred the normal dendritic structure breaks down to form a slurry of rounded particles. In practice the slurry contains little liquid but it becomes increasingly fluid, the more vigorously it is stirred. In the rheocasting process the highly fluid slurry is fed into

the shot sleeve of a pressure die casting machine and injected into the die. **Thixocasting** is a related process in which rapidly cooled slugs of rheocast slurry are held at an intermediate temperature prior to reheating and injection. In **Thixoforging** the rapidly cooled rheocast slugs are forged in closed dies. Benefits claimed for rheocasting include lower injection temperature leading to reductions in die wear, energy costs, process cycle time, shrinkage and porosity, and a capability to maintain high quality in thin sections.

Rheology   The study of the interaction of time, temperature and stress on the properties and characteristics of materials. The term is usually used in connection with non-crystalline materials such as glass, plastic and rubbers but aspects are relevant to the *Creep* of metals and also note the previous entry.

Rhodanizing   Electroplating with rhodium.

Rhodium   A metallic element of the platinum group. It is highly corrosion resistant with a high tensile strength and high elastic modulus as a bulk material. As an electroplated film it can have a hardness of 900 Hv. It is used in its pure form as a wear-resistant plating and as an alloying element with platinum for *Thermocouples*. See Table 15 for properties.

Rifled bore   Spiral ridges, grooves or undulations deliberately introduced along the bore of tubes. In gun barrels they spin and stabilize the trajectory of the shot. In tubes for heat exchangers they caused the contents to swirl giving enhanced heat transfer capacity and reducing the possibility of hot spots and *Steam blanketing*.

Rig testing   Test of structures or components of assemblies intended to simulate service environments, usually on a large scale.

Rightward welding   See Backhand welding.

Rimming steel   A low-carbon *Steel* which, deliberately, is only partially deoxidized before pouring. As a result, during solidification, some of the carbon in the steel reacts with the remaining oxgygen to form carbon dioxide. This occurs fairly early during solidification, so the material forming the outer layers at the sides and bottom of the ingot tends to be lower in carbon and impurites than the interior. Also, a band of voids filled with carbon dioxide develops at about one third the radius measured from the exterior. There is a concentration of impurities at the centre line but, because the volume of voids approximately balances the *Shrinkage*, the central shrinkage pipe is small. During subsequent rolling the voids collapse and fuse. The final product has a high-purity surface layer which is particularly suitable for sheet of good surface quality and corrosion resistance.

Ring tests Various tests of welding characteristics in which a circular disc, hole or groove on or in the material in question is welded and checked for cracking.

Ringing The noise made when a component is struck. It is used as a crude crack detection test.

Riser In casting, a reservoir of molten metal intended to solidify last and hence feed any zones of potential *Shrinkage.*

River lines/pattern Steps on a *Cleavage* fracture surface that radiate from the origin.

Rivet A fastening device having, in its simplest form, a head and a plain shank. The shank is inserted through pre-drilled holes in the two components to be joined and the end of the shank is then forged over, termed clinching, to tighten and maintain the joint. In the case of rivets of substantial size for structural steelwork, the rivet is heated to ease the forging operation and to tighten the joint as the rivet cools. See also Pop rivet.

Rivet joint An overlap joint made between two or more plate or sheet components by a series of *Rivets.*

RMS *Root mean square.*

Roak Same as Roke.

Roasting Heating ores in air to convert them from sulphides (usually) to oxide which is then more readily reduced to the metal.

Robertson test A test to determine the lowest temperature at which a *Brittle fracture* will arrest. A sample of the plate material in question has a notch introduced at one edge as a crack starter. It is then loaded uniformly in tension and subjected to a temperature gradient across its width with the notched edge coolest. A brittle fracture is initiated by impacting the notch zone and the temperature of the point at which the crack arrests identifies the *Crack arrest temperature.*

Rock candy fracture Intergranular fracture surface, particularly when coarse.

Rock crystals See Quartz.

Rockwell test A *Hardness* test in which an indentor is thrust by a known load into the test area, the depth of penetration providing a measure of hardness. The full sequence of testing, performed rapidly by a semi-automatic machine, involves application of a small initial base or minor load with measurement of the resultant penetration. This is followed by the application of the additional major load. The major load is then removed after about 5 seconds, leaving only the base load at which stage the penetration is measured again. The penetration induced by the major load is the difference between the two base load penetration measurements and is presented on a dial gauge as a hardness number. A large number of Rockwell scales, loads and indentors are used to cover the range of materials likely to be tested. Scales A, B and C are the most common.

Rod mill (1) A *Rolling mill* for rod manufacture.

Rod mill (2) A rotating barrel containing short rods to *Peen*, polish or otherwise treat the charge. Similar to a *Ball mill.*

Roke A planar defect penetrating from a surface. It is usually the result of oxide entrapment during casting.

Roll forging A process for producing small items by running precut blanks through a pair of rolls whose contact faces are recessed to the acquired product profile. It is usually a hot-working process.

Roll forming A process in which a series of powered rolls progressively shape material, usually strip, so some relatively complex configuration, for example guttering section or tubes of various shapes which may subsequently be seam welded.

Roll scale The scale formed on hot-rolled material.

Roll welding A welding process in which the components to be joined, usually sheet or plate, are passed through a pair of rolls to effect a joint. The components are usually heated, generally or locally, and the rolls force together the component faces (a lap joint) or edges (a butt joint) producing a *Forge weld* or *Pressure weld* at the interface.

Rolled gold A thin layer of gold, usually of 9 *Carat* or better, applied to a base metal by *Rolling* the two together. See Cladding (2).

Rolled sections Bars rolled to form a wide variety of cross-sections including I, H, L and X.

Rolled steel joist (RSJ) A steel beam of I or H cross-section for load-bearing structural applications.

Roller bearing A *Rolling element bearing* comprising inner and outer races, the gap between them being occupied by rollers. The rollers may be parallel or tapered and they may be in tangential contact or have gaps between them with a cage to maintain their separation.

Roller expansion A technique for expanding the bore of tubes. A ring of hardened rollers supported on a tapered mandrel is inserted into and rotated around the tube bore. As the rolls progress up the taper they expand the bore. The technique is often used to secure the ends of tubes into tube plates or vessels and, with appropriate controls,

strong pressure-tight joints can be achieved.

Roller levelling/straightening Processes for levelling sheet and straightening bar by passing the material through a series of staggered rolls. The first rollers induce a fairly severe bend and subsequent rollers apply bends of progressively reducing severity in alternate directions until the material merges level/straight.

Roller spot welding A *Resistance welding* process in which a series of *Spot welds* is produced in sheets travelling through a pair of narrow rollers (discs would usually be a better description) which apply a driving and clamping force and deliver an intermittent electrical current of high amperage to cause welding. If the spot welds overlap the process may be termed **Roller spot seam welding**. If rotation of the rollers stops during the electrical supply the processes may be termed **Step-by-step roller spot welding**.

Roller threading A process for producing a thread form by forcing grooved roller dies against the bar being threaded. Compared to a cut thread the roll threaded component is usually less a precision product but cheaper, slightly higher bulk strength and contains compressive *Residual stresses* at the surface which can improve *Fatigue* performance.

Rolling (1) A process in which a bar, slab, etc. is reduced in cross-section as it travels between a pair of rolls having a gap between them which is smaller than the original thickness of the material. The rolls are usually driven and tension may be applied to the material on the inlet or exit sides to assist or restrain its passage. See Rolling mill.

Rolling (2) as in **Surface rolling** A process in which a narrow roll bears heavily as it rolls against a component. The intention is to induce beneficial compressive *Residual stresses* which inhibit *Fatigue* crack initiation. It is applied to features such as the radii of crankshafts.

Rolling angle The angle between the line drawn through the working roll centres and the line drawn from one roll centre to the point on the roll surface which is first contacted by the material entering. Also termed Angle of bite, Nip (angle), Contact angle and variations on these.

Rolling element bearing A bearing in which the load between two components in relative motion is carried via rolling elements such as balls, rollers or needles. These rolling elements may bear directly onto the shaft and/or the hous-

ing or they may bear on inner and outer races carried by the respective component. Compare with Plain bearing.

Rolling mill Equipment, or the building containing the equipment, for *Rolling*, i.e. reducing the cross-section of ingots, slab, bar, sheet and foil as it travels through the gap between a pair of rolls. Mills are described by a wide variety of terms, some of which are illustrated in Figure 33. The simplest mills comprise only two rolls, other mills have more, the total number being indicated by terms such as **Two high** or **Four high**. A

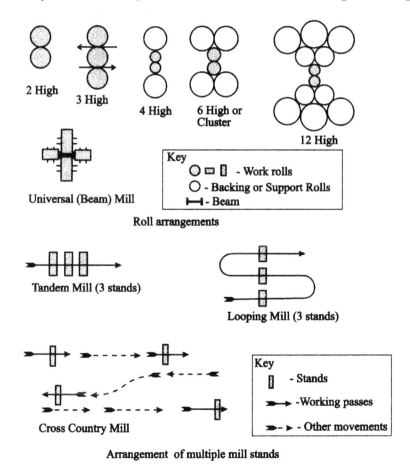

Figure 33   Rolling—arrangement of rolls and stands

Cluster mill or **Sendzimer mill** has six or more rolls. Some mills are **Reversing**, that is, the material is worked as it is passed through first in one direction and then in the other, the gap being reduced between passes. In a **Three high** mill all the rolls are **Working rolls** which rotate continuously in the same direction. The material is reduced as it passes first between one pair and then returns through the second pair. In a Four high and more complex mills only two of the rolls are working rolls, the other **Backing** or **Supporting rolls** minimize flexing of the working rolls. A **Universal mill** has multiple rolls positioned to work both the vertical and horizontal axes of the material. Rolls are carried in **Stands** and material usually passes through a number of stands during a rolling cycle. In a **Tandem mill** two or more parallel stands are grouped close together so that the material enters the second and subsequent rolls before clearing the previous rolls. In a **Looping mill** two or more stands lie side by side so that material, usually bar, emerging from one is directed in a loop to enter the second stand. A single-stand three high bar mill may also be referred to as a looping mill. In a **Cross-country mill** the material completely clears each stand before travelling over roller tables to the next.

Rolling texture Preferred *Orientation* produced by rolling.

Röntgen The unit of radiation dose, $1R = 2.58 \times 10^{-4}$ C/kg.

Root concavity A hollow in the root of a weld. See Figure 52, located at entry on Welding terminology.

Root mean square The formula for calculating the area beneath a waveform. It is the square root of the mean of the squares of each increment of change over a complete cycle.

Root (of weld) For an intended weld it is the area of the parent materials where the first deposit is to be made. For a completed weld it is the face of the first weld run that is remote from the opera-

tor. See Figure 50, located at entry on Welding terminology.

Root of joint (of weld)   Same as Root of weld preparation.

Root of weld preparation The area where two surfaces to be joined are in closest approach and where the first run will be deposited. See Figure 48, located at entry on Welding terminology.

Root rolling The *Rolling* (2) of machined re-entrant radii and notches such as bolt threads to induce favourable *Residual stresses* to resist *Fatigue*.

Root run/pass (of weld) The first run of a multiple run weld. See Figure 50, located at entry on Welding terminology.

Rooting See Oxide rooting.

Rosette (1) In microstructures, a rounded structure as opposed to an elongated form, for example graphite in *Cast iron* as an agglomerate cluster rather than as flakes.

Rosette (2) An array of *Strain gauges* to monitor changes of *Strain* and hence *Stress* in all directions across a surface. Typically three gauges are aligned at $120°$ to each other.

Rotary piecing Processes such as the Mannesman and Stiefel mills. See Tube making.

Rotary forge The Pilger mill. See Tube making.

Rotary hearth furnace A form of continuous furnace with a rotating hearth. The *Charge* is inserted through one door and remains in the hearth for nearly a full circle before being withdrawn through a second door.

Rotary shear A device for cutting sheet and plate utilizing a pair of driven, overlapping disc cutters.

Rotary swaging A process in which a machine applies rapid repeated blows, radially inwards, to reduce the diameter of some components such as bar or tube.

Rotating bending fatigue test Various *Fatigue* tests in which a cylindrical test piece is rotated while under a lateral load so that any point on its periphery undergoes one complete tensile/

compression reverse cycle per revolution. The test piece may be supported as a *Cantilever*, or in *Three-point* or *Four-point bending* depending on the particular machine.

Rough machining Preliminary machining to approximate, slightly oversize dimensions with little regard for surface quality. A subsequent final or finish machining operation achieves the required precision and finish.

Roughing pass/stand The first in a series of *Rolling* passes or stands. Alternatively, any except the final sizing pass or stand.

Roughness See Surface finish.

RSJ *Rolled steel joist.*

Rubert gauges A series of metal plates with machined surfaces of progressively increasing coarseness used for the semi-quantitative assessment of surface finish. A fingernail is scraped across the surface in question and across the gauges to identify the number of the gauge most closely matching it in feel.

Rubidium A metallic element; one of the alkali metals. It is highly volatile and reactive. See Table 15 for properties.

Ruling section The maximum section size of a steel bar that can be fully hardened through to the centre by a particular quenching procedure. See Steel and Hardenability.

Rumbling Treating a batch of components in a rotating barrel so that they rub against each other to remove surface *Flash* or generally improve the surface finish; additional abrasive material may be added to the barrel. The same as *Tumbling* except that larger components and hence more noise may be implied.

Runner A channel or passage for molten metal either entering the mould or between the entrance and the main area of the mould. Also the metal that solidifies in these passages.

Run-off plate (of a weld) Metal attached beyond the finish position of a weld. The plate provides a site for terminal defects and so ensures that full quality weld is

maintained to the end of the main joint. It is removed following completion of welding.

Run-on plate (of a weld) Metal attached to the start position of a weld. The weld is initiated on the plate which becomes the site for start defects and thereby ensure that a full quality weld is established by the time the main joint is reached. The plate is removed following completion of welding.

Runout (1) In casting. Overflow or other excessive leakage from a mould.

Runout (2) Of thread. The tapering of a thread to blend into the plain shank of a bolt.

Runout (3) Of a test specimen. A specimen that has survived the specified test programme without failure. The programme may be, for example, a specified number of *Fatigue* cycles or a period of exposure to an environment capable of inducing *Stress* corrosion.

Rupture The process of breaking or fracture often in a *Creep* context. There are ill-defined subtleties to the use of this term in a metallurgical context (see Break). Normally, when linked directly to a damage mechanism, 'rupture' implies a break involving significant *Ductility* (contrast with *Fracture*). Hence the term 'creep rupture' is common but 'brittle rupture' would usually be regarded as clumsy. However, it is acceptable, when used in a less precise context, as a synonym for 'failure' or 'burst' even when a low-ductility failure mechanism is involved, as in phrases such as 'rupture of a pressure vessel by brittle fracture'.

Rupture data Creep failure data. See Creep, Stress rupture and Larson–Miller parameter.

Rupture strength *Creep* strength.

Rust The corrosion product, ranging in colour through red, brown and yellow, formed on iron and steel in oxygenated moist environments. It is largely ferric hydroxide, hydrated iron oxide, with a variable composition including $Fe(OH)$ and $Fe_2O_3.H_2O$. It is bulky relative to the iron consumed, permeable to oxygen and mositure and is readily detached. It is not protective; attack will continue until the metal is fully consumed provided the environment does not change.

Rust proofing Any surface treatment applied to steel which is claimed to prevent or delay rusting.

Ruthenium A metallic element, one of the platinum group. It is corrosion resistant and, although it oxidizes at high temperatures, the oxide has a high electrical conductivity, hence use of the metal for high-temperature contacts. Apart from this it has little application in the pure form but is an alloy addition to platinum, gold and silver for jewellery. See Table 15 for properties.

Rutile Titanium oxide. The pure form is white and is a common paint pigment.

Rutile electrodes See Electrode (welding).

S

S curve See Isothermal transformation diagram.

S/N curve A graph plotting stress range (S) against the number of cycles to failure (N) to predict the fatigue life of a meterial. See Fatigue.

Sacrificial anode/protection See Cathodic protection.

SAE Society of Automotive Engineers (USA).

Safe working load (SWL) The maximum load which it has been determined a component should carry in normal service. When used in its formal statutory sense the term recognizes that there is an adequate and defined margin between the safe working load and the *Proof load* and a further, less well-defined, margin between the proof load and the expected breaking load. See Safety factor.

Safety critical Identifies a component or system whose function is vital to the safety of plant, operatives or the population at large.

Safety factor An allowance made by a designer to accommodate unknowns. The factor can be applied to one or more of the properties on which the design is based, including yield, tensile, fatigue or creep strength, corrosion or erosion rate, see respective entries on these topics. For example, a component could'be designed on the basis that the expected maximum load will induce a stress of only a third the *Ultimate tensile stress.* This safety factor of 3 allows for unpredictable factors such as inadvertent excessive loading, variation in materials properties, minor unrecognized manufacturing defects or limited undetected deterioration in service. Depending on the particular application, the safety factor may be calculated on the relationship between the expected breaking load and either the *Safe working load* or the *Proof load.* Generalizing, considerations of economy encourage reduction in the safety factor towards unity but this requires increasingly tight *Quality assurance* with accurate prediction and monitoring of service conditions, possibly supported by regular inspection of critical items.

Sal-ammoniac Ammonium chloride, NH_4Cl, sometimes used as a soldering *Flux.*

Salt bath A vessel containing molten salts of various compositions. When used for heat treatment or quenching the salt offers rapid and uniform heat transfer with protection against oxidation. Some salts are selected to be neutral towards the component being treated but others are not. In particular, salt baths are ofen used to carburize steel products as the first stage of *Case hardening.*

Salt spray test Various test of corrosion rate and *Stress corrosion* susceptibility. Many such tests have been devised, the common factor being that a specified salt solution is sprayed onto the test specimens for a defined period at regular intervals. The solution is usually sodium chloride up to 20%, and the environment may by an open-air or an enclosed cabinet with environment control. The test specimens may be unstressed or stressed in tension or bending and may be welds or other joints or bimetallics. The test period may be predetermined or continued until the test criterion is met. The possible criteria include failure, cracking, corrosion rate or merely surface appearance.

Saltpetre   Potassium nitrate, KNO_3, a constituent of gunpowder.

Samarium   A metallic element, one of the *Rare earth metals*. It has little commercial application. See Table 15 for properties.

Sand blasting   Similar to *Shot blasting* except that sand or another sharp-edged mineral grit is used.

Sand blister/buckle/scab   A bulge on the surface of a *Sand casting*. It contains a large proportion of sand mixed with the metal and is easily scraped or chipped off. It results from an area of soft sand being permeated by molten metal.

Sand casting   Any casting process in which the mould is formed from compacted sand with a binder to maintain its shape. The compacted mould may be used 'green', i.e. without further treatment or it may be dried.

Sandpaper   A strong paper backing carrying a layer of graded sharp sand for use as an abrasive.

Sandwich   Any triple- or even multilayer construction. Typically it might comprise outer faces of high-strength durable material and a thick inner layer of low-density material with sufficient strength to maintain the bond and spacing between the outers and hence the rigidity of the assembly.

SAP   Sintered aluminium powder, usually with reference to material that has been *Sintered* then *Extruded* to provide a material with better strength at ambient and elevated temperatures compared with conventionally cast and extruded aluminium. The improved properties result from the *Dispersion hardening* effect of particles of aluminium oxide originating from the surface of the powder particles.

Sapphire   A naturally occurring form of aluminium oxide, Al_2O_3, containing small quantities of oxides of other metals such as chromium, cobalt and titanium which give a characteristic blue colour. Non-gem quality can be manufactured.

Saturated (1)   A *Solution* containing the maximum solute content at equilibrium.

Saturated (2)   A magnet that is magnetized to the maximum level achievable.

Sauveur's diagram   A means for estimating the carbon content of annealed carbon *Steels* (maximum of 0.5% manganese) by examining their microstructure to determine the amount of pearlite. The diagram (Figure 34) is a straight-line plot between zero and 100% pearlite at eutectoid composition, usually taken as 0.87% carbon, followed by a further straight line falling to 85% pearlite at 1.5% carbon.

Scaffold   In a metallurgical context it is the fused mass of coke and other materials which builds up in the shaft of a *Blast furnace* and impedes normal downflow of the charge.

Scale (1)   The thick corrosion film, usually oxide, formed at high temperature on metal surfaces.

Scale (2)   The material deposited on a surface exposed to impure water. Its primary constituent is mechanically deposited impurities but it may include a percentage of corrosion product.

Scale (3)   Of a drawing, the ratio of the size as drawn to the size in reality.

Scalping   The removal, by mechanical or flame cutting techniques, of surface defects prior to working a newly cast *Ingot*.

Scandium   A metallic element, one of the *Rare earths*. It has little commercial application. See Table 15 for properties.

Scanning electron microscope   See Electron microscope.

Scantlings   The dimensions of the elements of a structure.

Scarf joint   A joint formed at overlapping tapers. See Figure 47(b), located at entry on Welding Terminology.

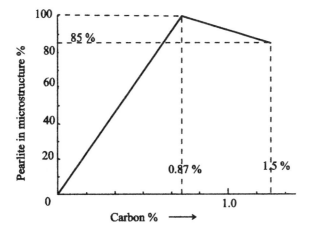

Figure 34 Relationship between pearlite % in microstructure and carbon content in annealed plain carbon steel (Sauveur's diagram)

Scatter band See Creep.

Scavengers Materials in a *Flux* above a molten metal that combine with contaminants to form *Slag*.

Scavenging Processes in which some substance, particularly a gas, is added to or bubbled through molten metal to remove dissolved gases or other impurities. The gas may be inert and act purely physically by allowing the contaminating gas to diffuse into it or, alternatively, it may be active and react chemically with the contaminant.

SCC *Stress corrosion cracking.*

SCF *Stress concentration factor.*

Schaeffler diagram A diagram predicting the microstructure of rapidly cooled high-chromium nickel *Stainless steels*, in particular welds. The various alloying elements are categorized as austenite formers or ferrite formers and ascribed a factor according to their relative potency compared with nickel or chromium respectively. The total austenite-forming effect is plotted on the vertical axis and the ferrite-forming effect on the horizontal. Zones in the diagram then indicate whether the structure will be austenite, martensite, ferrite or some mixture. See Figure 35.

Scleroscope See Shore scleroscope.

Scorification An initial stage in the refining of gold, silver and other precious metals. The **Scorifier** is a shallow crucible of fireclay or bone ash in which the precious metals are dissolved in lead while the impurities form a slag on the crucible. The lead alloy is then subjected to *Cupellation* to extract the precious metal.

Scragging The deliberate, pre-service straining of components, particularly springs, beyond *Yield* to improve *Fatigue* life. See Autofrettage which is a similar technique applied to hollow products.

Scrap (1) Material discarded or spoilt and not suitable for continued processing to a finished component.

Scrap (2) Previously used metal that is suitable for reprocessing, usually remelting, into a product of adequate quality. Material rejected or removed during production and returned for remelting is ofen termed **Process scrap**. Also see Secondary metal.

Scratch brush A wire bristle brush.

Scratch test Various fairly crude tests in which the material in question is scratched. The two concepts are either

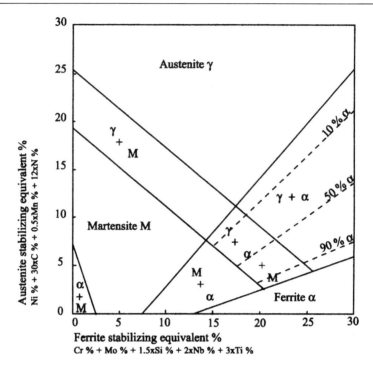

Figure 35   Schaeffler diagram. The microstructure developed in alloy steel welds

that the component is scratched by a series of calibrated implements (see Mohs' test and File test) or it is scratched by a single implement to find variations in hardness.

Screw   A rod or bar with a one or more helical fins. Some versions, **Wood screws** or **Self-tapping screws**, taper so that they can cut their way into un-threaded holes. A **Machine screw** is similar to a bolt in having no taper and requiring a pre-threaded hole to enter but it is threaded for the full length of the shank. A **Drive screw** either has a driven member mounted on it or it runs in a fluid to which it imparts motion, termed an **Archimedes screw**.

Screw dislocation   A *Dislocation* that travels in a direction normal to the direction of formation of the slip step.

Scuffing   Light surface damage by a mechanical rather than a corrosion process. The term is used for the early stage of both *Adhesive wear* and *Abrasive wear*.

Seal weld   A weld that forms a continuous impervious barrier to fluid but which makes no significant contribution to joint strength. Compare with Sealing run.

Sealing processes   Various processes intended to improve the condition of a surface, particularly its resistance to fluid penetration but also its resistance to staining or tarnishing. Examples include processes for coating with oil, grease, silicone or paraffin/lanolin mixtures, the application of phosphate *Conversion coatings*, the immersion of newly *Anodized* coatings in boiling water to block the natural pores with the resultant aluminium hydroxide and the pressure or vacuum impregnation of porous casings with resins.

Sealing run (of a weld)   The run laid on the remote side of a previously welded root. Compare with Seal weld.

Seam (1)   A defect in the form of an elongated, surface emergent fissure produced during working operations.

Seam (2)   Any linear joint formed by soldering, welding, folding, etc.

Seam weld   A joint between overlapping or butting edges of sheet material. See also Roller spot welding.

Seamless tubing   Any tubing that does not contain a longitudinal or helical seam weld. It may be made by various processes. See Tube making and Extrusion.

Season cracking   An obsolete term for *Stress corrosion* cracking in brass. The term is said to derive from the spontaneous cracking, during the Indian monsoon season, of brass cartridges containing significant *Residual stress* and stored in the ammonia-rich environment of stables.

Seasoning   Various circumstances in which a product is left to develop some characteristic over a period of time.

Secant modulus   The ratio of stress to strain, analogous to the elastic modulus, for cases where the stress–strain relationship is not linear (see Tensile test). The secant modulus ignores non-linearity by calculating a specific value for each measured combination of stress and strain.

Second moment of area   The measure of the capability of a section to resist flexing.

Secondary bonds   The weaker *Interatomic bonds* such as the van der Waals' force.

Secondary creep   The stage of *Creep* during which the rate of creep deformation is approximately constant, assuming constant stress.

Secondary hardening   The hardness increase in some alloy steels produced by *Tempering*. It results from carbide precipitation effects plus, in some cases,

the transformation to martensite of retained austenite following the reduction of its carbon content by the carbide precipitation. See Steel.

Secondary metal   Remelted scrap, which may be further refined.

Secondary phases   The *Phases* in alloy systems that do not include a pure metal in their range, i.e. all phases in a system other than the two primary phases.

Secondary stresses (1)   Bending stresses as opposed to direct tensile stresses.

Secondary stresses (2)   Additional stresses superimposed on the main stress.

Seebeck effect   The electromotive force (emf) due to the difference in temperature between two junctions of dissimilar conductors in a circuit. See also Thermoelectric effect.

Seed crystal   A small single crystal from which a much larger one is grown.

Seger cone   Tall, narrow pyramids (about 8–10° slope) of fireclay and *Refractory* mixtures having compositions selected to cause the pyramid to collapse at some specific temperature. They are used usually in multiples covering a range of temperature to monitor the temperature in furnaces.

Segregation   Variations in impurity and alloy content across a grain or casting. The effect arises because the first metal to solidify is usually relatively pure so impurities and alloying elements tend to be rejected into the band of material immediately ahead of the advancing *Solidification* front. Within a grain, significant alloy composition gradients can occur if there is a wide composition band and a large solidification temperature range; this effect is termed **Coring**. On a larger scale, when advancing grains meet, impurities pushed ahead of the two grains concentrate at the grain boundaries—termed **Minor segregation**. If solidification continues progressively from the periphery to the centre of the casting, impurities and the alloy constituents tend to become concentrated near the centre line. This is called **Normal segregation, Major segrega-**

tion or **'V' segregation**, the last reflecting the characteristic distribution of the impurities revealed by sectioning the cast ingot. In large ingots, the impurity level in the band of molten metal just ahead of the solidification front may depress its freezing point to such an extent that the purer metal at the centre of the ingot solidifies first, leaving a band of impurities at about mid-radius. This is termed **Inverted 'V' segregation**, again because of its appearance on sectioning. In extreme cases material rich in the low melting point alloy can be squeezed back along unsolidified grain boundaries towards or onto the exterior surface, termed **Inverse segregation**. Eruptions of this low melting point, alloy-rich material onto the surface are termed **Blebs** or **Blebbing** or **Cauliflowers**.

Seizing/seizure/seize-up An extreme case of *Adhesive wear* when the two contacting surfaces become permanently locked together. The engagement mechanism may be purely mechanical, resulting from mutual distortion and embedding of the two surfaces or there may be a measure of welding, perhaps considerable, resulting from the high local temperature generated by friction. See also Solder lock.

Séjournet process The use of glass as lubricant during *Extrusion* at high temperature, particularly of steels. Immediately prior to entering the extrusion chamber the heated *Billet* is coated with glass, typically by rolling it down a tray of powder, and a bonded pad of powder or fibre is placed in the die end of the chamber.

Selective freezing When an alloy phase with a wide composition range commences to solidify the first crystals are considerably enriched with the higher melting point constituent. These may be selectively removed for use or to be discarded.

Selective leaching The dissolution and removal of one component from a mixture which may be an ore or other mix

of powders or it may be a solid alloy. The term is sometimes used as a generic term for corrosion processes such as *Dezincification* in which one element of an alloy is removed. However, purists may argue that dezincification and similar processes involve solution of both elements followed by redeposition of the survivor.

Selenium An allotropic element, one of whose forms is metallic or metalloid. It is used in electrical rectifiers and about 0.2% can be added to *Stainless steel* to form selenides in the form of fine, well-distributed globular particles that do not significantly damage mechanical properties but improve *Machinability*. See Table 15 for properties.

Self-adjusting arc welding This term usually refers to *Metal inert gas welding*. It is so called because the electrical characteristics of the equipment are designed so that the wire feed speed, the arc current and the wire burn rate mutually interact to maintain a constant arc gap.

Self-annealing materials Those materials which require no heat input to *Recrystallize* during working at ambient temperature, for example tin and lead.

Self-locking fastener Bolts and similar threaded fasteners with anti slackening features as described for Self-locking nuts.

Self-locking nut A nut with features which resist slackening resulting from in-service vibration and other inadvertent loads. Apart from the obvious risk of a joint completely parting if the nut is lost, even a limited reduction in *Preload* raises the risk of *Fatigue* failure. There are numerous locking features although they tend to fall into two categories. One general type, often termed **Stiff nuts**, relies on some form of distortion of the nut thread to introduce a high level of friction against, and possibly indentation into, the mating thread. A second category has inserts of some resilient material, typically a hard plastic, set into the nut thread. With some types of locking features it is good practice to discard the component after a single use as the locking effect can be much reduced on second tightening. Also see Lock nut.

Self-hardening (steel) Same as Air hardening.

Self-lubricated bearing A plain bearing formed by *Sintering* and having an considerable quantity of interconnected porosity which is impregnated with a lubricant such as grease or graphite. They are installed in locations where lubrication in service is not practical.

Self-tapping screw See screw.

Selvyt A proprietary deep nap cloth used, with a suitable abrasive agent, for *Metallographic* polishing.

SEM Scanning electron microscope. See electron microscope.

Semiconductor A material having electrical conducting characteristics that are intermediate between a conductor and an insulator. In a solid material the electrons can exist only in certain bands of energy level that are separated by regions normally forbidden to electrons. Three general relationships between filled, unfilled and forbidden bands can be considered. First, certain solids have some allowed bands completely filled with electrons and others empty. If the forbidden regions between the allowed bands are large the electrons cannot move from one band to the next and the material is an **Insulator**. Second, in materials where bands are not completely filled or their energy levels overlap the electrons can move freely between bands. The electrons moving in this manner are termed **Conductivity electrons** and the material is a metallic **Conductor**. Finally, in cases where the energy gap between a filled and an empty gap is small then a small input of energy can cause an electron to jump the gap. Silicon is one example. At very low temperatures the four valence electrons of each atom are associated in a strong **Covalent** bond with another four electrons of an adjacent atom. As the temperature increases, i.e. energy is

added, electrons can move to another band leaving a site for an incoming electron and conductivity is achieved. Pure metals with this characteristic are termed **Intrinsic semiconductors**. The effect can also be induced by adding particular impurity elements that assist an electron to bridge the forbidden gaps. The impurity elements are termed **Dopants** and the alloyed materials are **Extrinsic semiconductors**. The dopants can work in two ways. Silicon, a Group IV element with four valence electrons, is an intrinsic semiconductor in its pure state, as described above. However, if an atom of a Group V element with five valence electrons is added to form a substitutional solid *Solution* one of the electrons is in excess of the number required for bonding. This electron takes up a position in the normally forbidden gap from where it is easily dislodged by a small energy input to contribute to conduction. Dopant elements contributing an excess electron are termed **Donors**. In a similar manner a Group III element atom in solid solution will have only three valence electrons required for bonding with the four required by the adjacent silicon atom. This effectively produces a hole in the valence band which can accept electrons jumping from a silicon bond to another, again contributing to conduction. Such elements are termed **Acceptor** dopants. Conventionally, conductivity in semiconductors is considered on the basis of the movement of holes into which electrons can jump. The movement of a hole is regarded as a 'positive' movement and hence the materials with an acceptor dopant providing holes are referred to as producing *p*-type **semiconductors**. Similarly, the movement of an electron is regarded as a negative movement of a hole and hence materials with a donor dopant are *n*-type **semiconductors**.

Semi-killed steel A *Steel* which has been partially deoxidized during melting. The steel poured into the mould retains just sufficient oxygen to react with some of the carbon during solidification to form the quantity of carbon monoxide necessary to compensate for *Shrinkage*. The voids filled with carbon dioxide are closed during subsequent hot working.

Semi-metal An element having many but not all of the characteristics of a *Metal*. Also termed a Metalloid.

Semi-permanent mould A metal mould used repeatedly with fresh sand cores.

Semi-permeable membrane A membrane which is permeable to the solvent but impervious to the solute.

Semi-steel Low-carbon *Cast iron* having graphite flakes that are small in size and quantity and well distributed. The material consequently has relatively good *Ductility* compared with normal grey cast iron. It is usually made by melting scrap steel with the pig iron.

Sendzimir mill A cold *Rolling mill* for producing precision sheet in which the two small work rolls are supported by multiple layers of backing rolls.

Sensitization Usually, this term refers specifically to the heating of certain *Austenitic stainless steel* in the approximate temperature range 450–850 °C which causes chromium carbide precipitation in the grain boundaries. This leaves the immediately adjacent zones depleted of chromium causing them to be susceptible, i.e. sensitive, to corrosion. The intergranular corrosion is often termed *Weld decay* even when the heating was not the result of welding. The term is also used more generally where some treatment causes the material involved to be sensitive to some subsequent treatment or susceptible to a damage mechanism.

Sensor Any detection device, with or without a measuring capability.

Sessile dislocation A *Dislocation* formed by a plane of vacancies surrounded by normal lattice. This type of dislocation cannot move.

Set (1) Deformation.

Set(t) (2) A substantial chisel for cutting metal.

Shadowing A technique for increasing the contrast of *Replicas* for *Electron microscopy*. The process is carried out in a vacuum chamber in which a suitable metal is evaporated and projected at an acute angle onto the replica surface to exaggerate topographical irregularities.

Shaft furnace A vertical cylindrical furnace packed nearly to the top with the charge. The materials are fed intermittently into the top to react as they progress downwards. The products, usually liquid, are periodically tapped from the base. See Blast furnace.

Shank The plain shaft which carries the active end of a component, for example the unthreaded central length of a *Bolt* or the plain end of a drill bit.

Shape memory effect (SME) A phenomenon, observed in some alloy systems, which is associated with a reversible and progressive transformation to *Martensite* as temperature falls. The unusual characteristic is that if the material is deformed (within limits) while in one phase it will, on changing temperature to the other phase, revert to the original shape. The dimensional change can be much larger than normal thermal expansion.

Shaping A machining process in which the cutting tool, fixed in a substantial arm, reciprocates across the stationary component. The tool cuts on the forward stroke and pivots clear on the return stroke. The table supporting the component moves incrementally sideways between cutting strokes.

Shatter cracking Low-ductility cracking, usually subsurface, by various mechanisms including *Brittle fracture* and *Hydrogen embrittlement*. May also be termed *Fisheyes* or *Flakes*.

Shear (1) The cutting action in which a pair of blades or similar abutting faces slide across each other cutting the entrapped material as with scissors. **Shear strength**, in this context, refers to the capacity to resist this form of loading.

Shear (2) The deformation mechanism in which layers of atoms on the *Crystal* *lattice* slide across one another like a pack of playing cards. The layers involved, termed the **Shear planes,** are usually those at an angle of 45° to the principal stress. **Shear strength** in this context refers to the stress necessary to induce permanent deformation by such a mechanism.

Shear bands Markings observed on polished and etched samples of *Cold worked* material, particularly rolled sheet. They extend over groups of grains and are caused by variations in the severity of deformation.

Shear ledges Steps between areas of *Crystalline* fracture on a *Brittle fracture* surface. They are normally aligned along the line of crack propagation and hence point to the origin.

Shear lip An area, at the edge of a *Flat fracture* surface, where the plane of fracture is at about 45° to the direction of loading. It occurs by ductile shear at the final stage of crack propagation leaving, usually, a sharp fracture edge. See Shear (2).

Shear modulus The ratio shear stress to shear strain. Same as Bulk modulus.

Shear plane See Shear (2).

Shear strength See Shear, definitions (1) and (2).

Sheared-off This term is used loosely with respect to virtually any failure of a bolt or similar fastener. It is probably best limited to those cases where the bolt has been cut by a shearing action of one bolted surface sliding across the other, a relatively rare occurrence.

Shed A feature intended to shed, i.e. discard, unwanted deposits.

Sheffield plate Decorative silver *Electroplating* on copper or *Nickel silver.*

Shell The orbits in which electrons are arranged on the *Atomic structure* of elements.

Shell marks Concentric marks on the surface of a *Fatigue* fracture. They are visible to the unaided eye and mark the crack front at irregularities in the load cycling such as interruptions in cycling or an abnormal load. Also termed *Con-choidal marks, Beach marks* or *Arrest marks.* Compare with *Striations.*

Shell mould A mould formed in a sand plus thermosetting binder mixture by a metal master die. The sand mixture may be sprinkled or sprayed onto the die prior to baking or alternatively a heated die may be pressed into the mixture. Multi-piece moulds, with cores if required, can be made for complex castings in a wide range of sizes and metals. See also Investment casting.

Sheradizing A process in which a steel component is heated in zinc powder at 400 °C, i.e. just below the melting point of the zinc, to form a protective zinc–iron alloy diffusion coating.

Shielded (metal) arc welding Processes in which the electric arc and weld vicinity are protected from reaction with the environment, by, for example, a *Flux* layer, or a gas shroud produced by decomposition of the *Electrode* coating or supplied via the torch.

Shim(ming) Thin sheet material used as a spacer to correct the gap between two components.

Shore hardness (1) A hardness test for relatively soft materials which measures the depth to which a standard indentor penetrates in standard conditions. Sometimes termed **Shore 'A'** hardness.

Shore (scleroscope) hardness (2) A *Hardness* test for hard materials including metals in which an indentor, in a glass tube, falls from a set height onto the test surface, the height of rebound providing a measure of hardness.

Short As in 'hot short', and 'cold short', this term indicates low *Ductility* and a susceptibility to cracking during working operations at the indicated temperature.

Shortening as in **copper shortening** The phenomenon by which some components become progressively smaller in one dimension when subject to a cycle of operation involving both a change in temperature and a change in *Stress.* In a typical case a copper conductor bar is contained within a longitudinal cavity just below the periphery of the rotor of an electric motor. When the motor is started the rotor rapidly accelerates to speed, centrifuging the bar against the rotor body. The bar then rises in temperature but its expansion is constrained by friction against the body. The resultant compressive stress may immediately cause *Yield* of the low-strength annealed copper or it may cause *Creep* over the period of time the motor is hot. When the motor is switched off the rotor speed drops rapidly, removing the frictional constraint, and the copper bar can contract normally as its temperature drops. However, because of the yield or creep in compression it will be shorter than originally. This process is repeated on every cycle of operation with the result that the bar becomes progressively shorter. This may in turn impose tensile loads within the bar or in attachments if the ends are constrained. See also Thermal fatigue.

Shorterizing   A process for *Case hardening* appropriate *Steel* or *Cast iron* in which the component surface is heated by a flame and immediately quenched by a following water jet. The term is usually used where the process is semi-automated to cover an extended surface and the flame and jet traverse a predefined path.

Shot   Metal of approximately ball shape. The term was originally used for any size or form of material fired from guns, for example large pieces fired individually (i.e. cannon balls), medium-size pieces, loosely packed together, termed grapeshot, or smaller particles packed in the cartridge case of a hand **Shot gun**. Larger balls, usually cast iron, were cast individually, smaller particles, typically lead for shot guns, were formed by pouring from a height in a **Shot tower** into water.

Shot blasting   A process in which hard steel or cast iron *Shot*, typically a few millimetres in diameter, is projected at a surface to remove scale and other con-

taminants or to produce a roughened surface providing a good key for subsequent surface treatments such as painting. The term usually implies that the shot is projected by an air blast but in a *Wheelabrator* it is flung from a rapidly rotating wheel. The shot initially is rounded but broken shot is of no great concern as the sharp edges may improve the descaling process. Compare with Peening. Grit blasting is similar except that mineral grit is used as the abrasive.

Shot peening   See Peening.

Shottky defect   A *Vacancy* in the *Crystal lattice*.

Shrink fit   See Interference fit.

Shrinkage   Reduction in dimensions resulting from contraction during *Solidification* and cooling. Also termed **Patternmaker's shrinkage**. Dimensional changes are unavoidable but the contraction may lead to defects such as internal **Cavities, Voids** and **Porosity**. If contraction is constrained the resultant stresses can lead to **Shrinkage cracking** or **(Hot) tearing**. Such damage is characteristically *Interdendritic* or *Intergranular*. Damage can be minimized by careful designing of the casting and the system for introducing and feeding the molten metal.

Shrinkage allowance   The amount by which a mould is made oversize to allow for *Shrinkage* of the casting. Also termed **Patternmaker's allowance**. The allowance for particular metals is influenced by various factors but typical examples are:

Aluminium:	up to about 1 m, open mould,	1 in 77
	over 1.8 m open mould	1 in 96
Brass	open mould	1 in 64
Steel	up to about 0.6 m, open mould	1 in 48
	over 1.8 m, open mould	1 in 77
Grey iron	up to about 0.6 m, open mould	1 in 96
	over about 1.2 m, open mould	1 in 144

Shrinkage groove (of weld) Shallow groove along the edge of the *Penetration bead* of a weld caused by *Shrinkage* during solidification. See Figure 52, located at entry on Welding terminology.

Shrinking-on Same as Shrink fit. See Interference fit.

SI system Système International d'Unités. The International System of Units adopted by the General Conference of Weights and Measures and endorsed by the International Organization for Standardization. The system is used internationally in scientific fields and in most countries for industrial, commercial and domestic activities. It differs in significant respects from the older metric system, referred to as the CGS (centimetre, gram, second) system. The seven **Base units** of the SI system are:

Quantity	Unit name	Symbol
length	metre*	m
mass	kilogram*	kg
time	second	s
electric current	ampere	A
thermodynamic temperature	kelvin	K
luminous intensity	candela	cd
amount of substance	mole	mol

*Agreed English translations.

Sialons See Silicon nitride.

Sieve analysis The determination of the particle size and proportionate distribution of a powder by passing it through a series of sieves having decreasing mesh sizes. The mesh size, termed the **Sieve mesh number** is defined as the number of apertures per inch.

Sigma (phase) An intermetallic compound, particularly iron–chromium, FeCr, but also other compounds having similar complex tetragonal crystallographic structures. FeCr precipitates in *Stainless steels* having more than about 18% chromium which are heated between about 410 °C and 870 °C for prolonged periods. The maximum rate of formation is at about the middle of this range (see Figure 16, located at entry on

Gamma loop). Sigma formation is favoured by *Ferrite* stabilizers such as molybdenum, niobium, titanium, aluminium and silicon and also by *Austenite* stabilizers such as nickel if it is in excess of about 8% and manganese in excess of about 4%. Lesser quantities of the latter two have neutral or inhibiting effects. Sigma does not have a major effect on *Hardness* or *Tensile* strength but it reduces both *Tensile ductility* and *Creep ductility* by a significant amount. As an example, in a hot tensile test at 700 °C, a typical 18% chromium 10% nickel austenitic steel would have about 50% ductility but a few per cent of sigma is sufficient to reduce this to about 6%. Sigma also reduces corrosion resistance particularly at the grain boundaries and this can cause grain boundary fissuring that acts as a stress raiser causing premature low-ductility creep failure.

Silal A high-silicon, low-carbon *Cast iron* which resists *Growth*. Typically, it contains 6% silicon, 3% carbon, 0.7% manganese, 0.3% phosphorus.

Silane See Silicon nitride.

Silica Silicon oxide, SiO_2, a mineral which occurs naturally as quatz. It retains good mechanical properties close to its melting point of 1710 °C, has a low coefficient of thermal expansion (5×10^{-7} per °C) and hence is widely used as a *Refractory*. It is *Allotropic*, the **Quartz** phase being stable to approximately 870 °C, **Tridymite** stable from 870 °C to 1470 °C and **Crystobalite** stable from 1470 °C to 1710 °C.

Silica gel A hydrated form of silica. It is strongly hygroscopic but absorbed water is readily driven off by heating so it is a common desiccant.

Silicate A compound in which silicon is combined with other elements.

Silicate process Same as CO_2 process.

Silicon A *Metalloid* element. It is a *Semiconductor* and an important alloying element in many metals. See entries below, Killed steel and Table 15. Compare with Silicone.

Silicon brass   An alloy of nominally 60% copper 40% zinc to which up to about 1% silicon is added to give improved strength with increased resistance to oxidation and *Wear*.

Silicon bronze   An alloy of copper with 1–5% silicon plus, possible, small amounts of iron, manganese and zinc but not usually any tin. These alloys have good strength and corrosion resistance.

Silicon carbide   SiC. A hard refractory material, used as an abrasive.

Silicon nitride Si_3N_4. A hard abrasion-resistant ceramic capable of taking and retaining a sharp cutting edge. It has good high temperature properties with, for a ceramic, a good toughness. The high-temperature properties can be further improved by additions of aluminium and oxygen to form the **Silanes**. These can be regarded as β silicon nitride, Si_6N_8, in which the aluminium substitutes for some of the silicon and the oxygen substitutes for some of the nitrogen.

Silicon steel   Silicon is added to many steels in small quantities, up to about 0.4%, as a deoxidizer (see Killed steel) or to improve fluidity in casting but these amounts do not normally merit the description 'silicon steel'. This term usually indicates larger amounts of silicon added for more specific effects on mechanical or electrical characteristics. Silicon improves *Hardenability*, allowing oil rather than water quenching so some spring steels contain up to 2% silicon. It also improves high-temperature corrosion resistance and hence some valve steels contain about 3.5%. A low-carbon (0.07%) steel with about 4.5% silicon, suitably manufactured, combines high magnetic permeability with low hysteresis loss and hence is used for the laminated core of electrical transformers.

Silicone   A material containing a silicate, i.e. silicon combined with another element (commonly oxygen), and an organic material (i.e. a hydrocarbon), the whole forming a long-chain molecular compound. Relative to other organic materials, siicones tend to be chemically stable, i.e. resistant ot oxidation or other corrosion, thermally stable over a wide temperature range above and below ambient, not readily 'wetted' by water and not affected by immersion.

Siliconizing   A process for developing a silicon-rich surface layer on steel by packing a component in ferro silicon and heating at about 900 °C for a few hours. The layer has good wear and oxidation resistance.

Silky fracture   A fracture surface that is fine textured and dull but clean. It is characteristic of *Ductile* failure.

Silver   A metallic element. It is resistant to oxidation and so is usually regarded as a noble metal although it tarnishes particularly in polluted environments. It has the highest thermal and electrical conductivity of any material, apart from *Superconductors*. It is widely used in pure or alloyed form for decorative items and industrial applications particularly photographic emulsions. See Sterling silver and Table 15.

Silver solder   *Brazing* alloys, also termed **Hard solders**, based mainly on silver, copper, zinc and cadmium, for example 50% silver, 19% cadmium, 16% zinc and 15% copper. The various alloys have melting ranges commencing at about 620 °C which is the *Eutectic* for the alloy quoted. Compared with the tin lead *Soft solders* the silver brazes give considerably higher strength but cannot be applied with a simple soldering iron. Compared with the brass and high-copper brazes they offer similar strength and require similar equipment for application but their lower melting point is generally beneficial, particularly in minimizing distortion. However, they are more expensive than any of the other solders or brazes and the grades containing cadmium present a potential health hazard unless precautions are taken to deal with the toxic cadmium fumes released during brazing. Some grades of silver solder have a colour closely matching silver but the alloys generally are not confined to joining silver components. They have numerous applications for steel and copper items where the requirements of strength and relatively low melting point justify the cost.

Silver steel A high-carbon steel supplied as softened, bright precision ground bar or strip for machining purposes. It has a silvery sheen compared with most engineering steels but it does not contain silver. See Steel and Table 2 for composition and properties.

Silvering The application of a reflecting surface to glass. Techniques including chemical deposition in various solutions and sputtering or *Vapour deposition*.

Single crystal Material which throughout its volume conforms to one continuous *Crystal lattice*. In addition, by implication, the crystal is deliberately grown by specialized techniques to a size sufficiently large to be of commercial or engineering interest.

Sinking (1) In the context of tube manufacture the term refers to drawing without the use of an internal mandrel or plug to control the bore size. See Drawing.

Sinking (2) *Die sinking.*

Sintering A process, below the melting point but usually at elevated temperature, in which contacting particles mutually diffuse and bond. The process may be used to agglomerate fine mineral ore dust to make it more easily processed and handled or it may be used to make substantial objects of metal powders for further processing or to directly enter service. If the atmosphere is intended to react with the powder during the operation to promote bonding the process is termed **Activated sintering**.

Situ See *In situ*.

Size factor The relationship between the sizes of atoms which determines, in part, how they form solutions and compounds. If atoms of different elements, or more strictly their positive ions, are similar in size then, if other factors are correct, they can form substitutional solid solutions. If they are widely different and

other factors are appropriate they may form interstitial solid solutions or compounds. See Solution and next entry.

Size factor compounds These are intermetallic phases formed when the *Size factor* and other aspects are suitable and the elements are present in specific simple proportions. The **Laves** phases are based on the relationship AB_2, hence $MgCu_2$ or $TiCr_2$, which are formed when the constituent atoms differ in size by about 22.5% allowing a particular form of close packing. The **Interstitial compounds** form between certain metals and non-metals having widely differing atomic sizes and so able to adopt an interstitial structure. Examples include carbides such as Fe_3C, WC and Mo_2C or nitrides such as TiN.

Sizing pass The final light working pass in *Rolling* or *Drawing* to provide a high-quality surface with precise dimensions.

Skelp Slit strip with any necessary edge preparation that is to be formed into a tube by longitudinally rolling and or drawing through a die prior to *Seam welding*.

Skin effect The tendency of high-frequency electrical current to concentrate at the surface of the conductor.

Skip sequence weld A welding technique in which a long run of weld is built up by a number of short deposits laid in a pre-planned sequence such that the first series of welds are made with gaps between them and subsequent welds fill the gaps. See Figure 49, located at entry on Welding terminology.

Slab An imprecise term referring to part-rolled material with a thickness about half its width. It is implicit that it will subsequently be rolled to plate or sheet.

Slack quenching Cooling steel at a rate insufficient to form a fully martensitic structure but fast enough to form bainite. See Steel.

Slag The mixture of *Flux* and impurities extracted from the metal which forms during various melting processes. It floats on, and remains separate from, the underlying metal and has the additional benefit in some cases that it protects the underlying metal from the environment.

Slag fibre/wool Fine-stranded fibres formed from various mineral *Slags* and similar in texture to glass wool or fibre. They are often used for insulation purposes but, compared with glass, some varieties may have low softening temperatures or may slump and powder relatively rapidly particularly if subjected to vibration.

Slag trap (of weld joint) Any joint geometry or other feature which could retain molten slag and impede its subesquent removal.

Sliding fit A fit with just sufficient slack to allow the mating components to move axially relative to each other without *Galling* but without perceptible lateral play. This usually implies an interfacial gap just sufficient to accommodate a film of lubricant. Contrast with Push fit.

Slip (1) Deformation by planes of atoms in the *Crystal lattice* sliding over each other. The sliding action is facilitated by the movement of *Dislocations*.

Slip (2) The thick water-based paste used in *Slip casting*.

Slip casting A technique in which a water and powder paste is poured into an absorbent plaster mould. The mould absorbs most of the water leaving a fragile replica of the mould interior.

Slip lines/bands Marking produced on deformed surfaces that were polished prior to deformation. They arise because neighbouring areas have deformed along differently aligned *Slip planes* giving rise to surface irregularities.

Slip planes The planes in the *Crystal lattices* across which slip can occur.

Slow bend test A bend test to determine the *Ductility/Brittle* characteristics of metals, usually steel. The term contrasts with the fast loading rate of *Impact tests* for determining such characteristics. The test specimen may be notched or plain and may be loaded in either *Three-point* or *Four-point bending*.

Slow butt welding Same as Resistance butt welding.

Slug test Same as Peel test.

Slugging The illicit insertion of solid bulk material into a weld joint. Such material partly fills the joint and is hidden by subsequently deposited weld metal but is not fully fused and forms a serious weakness.

Slurry (1) A thick suspension of solids in water or other liquid, less viscous than a paste and easily poured.

Slurry (2) Same as Suds.

Slush Material in a intermediate stage between fully molten and fully solid.

Slush casting/moulding (1) A technique for producing hollow castings. Molten metal is poured into a mould, usually of metal and often of complex shape. The molten metal is usually swirled round and, after a solid skin has formed over the mould interior, the remaining molten metal is poured out. In a variation on the process a measured amount of molten metal is introduced through an orifice which is then sealed prior to swirling. In this way a fully enclose hollow is produced.

Slush casting/moulding (2) Processes in which the alloy is poured or injected into a mould while it is in a pasty, part-solidified, state. Also see Rheocasting.

Smart material A material which alters some significant characteristic when subjected to some external stimulus, for example piezoelectric crystals which generate a voltage when stressed.

SME *Shape memory effect.*

Smelting Processes involving chemical reaction at high temperature to reduce *Ore* to molten metal or to some intermediate product. Where the final product is to be molten metal a *Flux* is usually added to combine with the unwanted oxide and other impurities forming a *Slag*.

Smoke Airborne, finely particulate materials usually products of combustion.

Smut (1) Finely particulate material released into the atmosphere usually as a result of combustion.

Smut (2) Surface blackening produced

on some metals by pickling in caustic soda solution.

S/N curve See Fatigue.

Soaking Prolonged heating.

Soaking pit A chamber in which newly cast ingots, particularly steel, are stored to allow the temperature to equalize prior to the first hot-rolling operation. There may also be some *Homogenization* of the structure. The facility may be no more than an insulated but unheated hole in the ground usually with some form of cover but supplementary heating may be applied.

Soapstone The bulk form of talc, also termed **Steatite**, hydrated magnesium silcate, $3MgO.4SiO_2.H_2O$. A very soft mineral, it is No. 1 on the *Mohs* scale of hardness. It has a high electrical resistance and hence is used for insulators. It is readily carved for ornamental applications. In powdered form it is termed **French chalk** or, if perfumed, **Talcum powder**.

Soderberg diagram See Fatigue and associated figure.

Sodium A metallic element that oxidizes readily in air and reacts violently with water to release hydrogen with the heat of the reaction then igniting the hydrogen. It is a good thermal conductor and is used within hollow exhaust valves of internal combustion engines and as a coolant in certain nuclear reactors. Small quantities are added to '*Modify*' aluminium–silicon casting alloys. See Table 15.

Sodium silicate process Same as CO_2 process.

Soft The opposite of *Hard* in its various technical senses.

Soft solder Alloys primarily of lead and tin. They melt at low temperatures to provide a fairly strong joint and seal between metals such as lead, steel (usually pre-tinned), copper and copper alloys. The term soft solder is in contra-distinction to **Hard solder**, the term applied to *Braze fillers* and *Silver solders* which, apart from higher strength and hardness, have melting ranges commencing above 450 °C, the usually accepted watershed. The soft solders offer a range of compositions with a useful variation in freezing ranges (see Figure 36). **Plumber's solder**, 70% lead, 30% tin, has a wide freezing range, 250–183 °C, with a long pasty stage allowing the joint to be 'wiped' to a smooth profile. **Electrician's solder**, also termed **Tinman's solder**, is of approximately *Eutectic* composition, 62% tin, 38% lead, freezing at 183 °C. The soldering processes and techniques are largely similar to *Brazing* apart from the lower tempera-

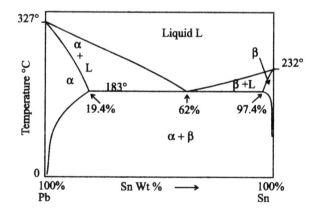

Figure 36 A eutectic. The lead–tin system, the soft solders

tures and lower strength. However, because of the lower temperature, small soldered joints can be made with the heated tip or 'bit' of a **Soldering iron**. Another common process is **Wave soldering** in which the components to be soldered, for example circuit boards, are suspended joint face down, just above a bath of solder. A wave is then induced in the surface of the bath and the component passed over skimming the wave. In most cases a *Flux* is applied prior to, or during, soldering to clean the surfaces of oxide and other contaminants which would inhibit wetting and bonding.

Softening   Various processes of reducing Hardness and Tensile strength and, usually increasing Ductility, Malleability and Toughness. See entries on these topic plus Annealing, Tempering, and Normalizing.

Solder   See Soft solder.

Solder lock   The phenomenon whereby if a *Soft soldered* joint between copper components is heated for too long a period, including remelting, the copper and solder mutually diffuse causing the melting point of the solder to increase to such an extent that it cannot be melted by the usual heating system. Plumbers sometimes use the term 'seized joint' for the phenomenon.

Solder paint   A suspension of powdered *Soft solder* and *Flux* which can be painted onto the surface to be soldered. After heating to fuse the solder and form the joint the flux is cleaned off.

Solid particle erosion (SPE)   Erosion by solid particles entrained in a fluid. The term is usually used in the context of damage observed on the early stages blades of a steam turbine. Such damage is caused by particles of oxide or other debris entrained in the steam from earlier parts of the system. Oxides may be released by *Spalling* in the superheater and reheater stages and debris may be produced by poor practice during welding. This contrasts with the more common erosion observed on the final stages of blades that are exposed to steam

cooled to a stage where large quantities of water droplets are produced. See Steam erosion.

Solid phase diffusion welding   See Diffusion welding.

Solid solubility   See Solution and Phase

Solid solution   See Solution and Phase.

Solid solution hardening/strengthening   See Solution hardening.

Solid state   In the context of electronics this term implies electronic devices that comprise only solid materials such as transistors rather than devices containing gases such as thermionic valves.

Solid state welding   Any welding process that does not involve melting of either the parent materials or a *Filler*. See Diffusion welding.

Solidification   The process whereby a metal changes, on cooling, from the fully liquid state to the fully solid state. Solidification commences when a small group of atoms forms a nucleus from which a crystal (grain) grows by additional atoms attaching themselves at specific locations to build up the *Crystal lattice*. Each crystal develops by forming, in three dimensions, a system of angular branches termed a *Dendrite*. As a dendrite grows the gaps between main branches are filled by further branches. Eventually, in most practical cases, the dendrite will meet a neighbouring dendrite growing to form another crystal. The line at which they meet is termed the *Crystal boundary* or *Grain boundary*. Provided sufficient molten metal is available in the immediate vicinity all gaps will be filled with solid material. If not, porosity will remain. This is termed **Interdendritic porosity** when located within a grain at gaps in the dendrite branches or **Intercrystalline** or **Intergranular porosity** if it is located at the boundaries. All pure metals solidify at a fixed temperature while alloys, except in special cases, solidify progressively over a temperature range. If an alloy is held at a temperature within the range it will comprise a pasty mixture, sometimes described as a mush or slush, of solid and liquid components in fixed proportions and specific compositions defined by the bulk alloy composition and the temperature. See also Shrinkage, Segregation and Lever rule.

Solidification cracking Any cracking occurring during *Solidification*, usually as a result of *Stress* imposed by constraint of thermal contraction. Such cracking is normally located at grain boundaries as these are the last material to solidify and hence are relatively weak and unable to withstand the loads imposed by the surrounding material as it cools. Also see Weldability.

Solidus The line defining the lower limit of the melting range of an alloy. A pure metal melts and solidifies at a specific temperature but, apart from special cases, alloys melt and solidify progressively over a range of temperature. Within the range the alloy will exist as a pasty mixture of solid plus liquid. When represented graphically the line defining the upper limit of the range is referred to as the Liquidus and that defining the lower limit is the Solidus.

Soluble anode process Various processes of refining metals by making impure material the anode of an *Electrolytic* cell and depositing pure material onto a high-purity starter plate forming the cathode.

Solute The materal taken into *Solution* by the *Solvent*.

Solution A *Phase* containing more than one component. It is commonly recognized that a liquid can form a solution, i.e. water can dissolve salt to form a salt solution. Similarly, metals in the solid state can form solid solutions. As with liquids the principal constituent will be termed the **Solvent** and the secondary constituent the **Solute**. When the solvent contains the maximum possible amount of solute it is termed **Saturated**. Treatment which precludes *Equilibrium*, for example fast cooling, can cause excess quantities of solute to be retained in the solvent. Such unstable solutions are termed **Supersaturated**. A solid solu-

tion, in equilibrium, will have a uniform composition at all positions and, under the microscope, will appear as a featureless single phase. In the cases of some pairs of metals, such as copper and nickel, the two are completely soluble in each other at all compositions but in most other alloy systems solubility is limited and multiple phases occur. Where the composition range of a phase includes the pure metal it is termed a **Primary** solid solution, otherwise it is termed a **Secondary** solid solution or **Intermediate phase**. Where the range of the intermediate phase is narrow and based on a simple ratio of the atoms of the two elements it may be termed an **Intermetallic compound**. In a **Substitutional** solid solution the solute atoms take positions on the crystal lattice normally occupied by a solvent atom. In an **Interstitial solid solution** the relatively small solute atoms fit into the spaces between solvent metal atoms.

Solution anneal A heat treatment intended to effect *Annealing*, i.e. softening and *Recrystallization* of work-hardened material, and to take into *Solution* all (or most) precipitates. It is also implied that there will be no significant reprecipitation during subsequent cooling. The term is often used in the context of *Austenitic stainless steels* and where such material is intended for high-temperature service there may also be an implication that some *Grain growth* is intended so the *Creep* properties will be improved.

Solution hardening The hardening of an alloy as a result of one or more elements being in solid solution in another. The effect results from the different size of the *Solute* atoms distorting the crystal lattice of the *Solvent* and thereby impeding *Dislocation* movement. The term is sometimes used casually to refer to the hardening process involving *Solution treatment* plus *Precipitation hardening*, but this is usually regarded as erroneous.

Solution treatment Heating to take

much, if not all, of a secondary phase into solution in the primary phase. See Solution.

Solvent The material that dissolves the *Solute* material to form a *Solution*.

Solvus The line, in a *Phase* diagram, defining the limit of a *Solid solution*.

Sonic fatigue Cracking by *Fatigue* induced by atmospheric vibration associated with noise, air turbulence or *Vortex shedding*. Also called Acoustic fatigue.

Sonic testing Any testing using audible sound waves. Usually, this merely involves striking the component in question and listening to the 'ring' emitted. A long pure tone implies freedom from gross cracks while a brief dull note indicates. See Ringing and Ultrasonic testing.

Sonotrode The vibration-emitting head of an *Ultrasonic welding* unit.

Sorbite A term, largely obsolete, referring to a microstructure in steel comprising fine spheroidal carbide in a ferritic matrix, for example that produced by tempering martensitic or bainitic structures above about 400 °C. Such structures are now usually referred to as fine spheroidized carbide, tempered martensite or similar phases. See Steel.

Space frame A load-bearing structure of struts and stays.

Space lattice Either an alternative term for *Crystal lattice* or a notation system for defining the location of atoms on the crystal lattice. See Bravais lattice.

Spalling Loss of material from a surface as flakes or sheets.

Spangle The bright crystalline pattern on *Galvanized* steel.

Spark machining Metal removal by repeatedly striking an electric arc against the component to remove material by melting and vaporization. Usually, the component is submerged in paraffin or a similar medium during machining and individual arc strikes are very small. The technique is useful for very hard materials or for holes which are of complex shape. The component is usually the anode and the tool the cathode.

Spark test A crude check of *Steel* composition in which the component is abraded with a powered grinding wheel. The quantity and form of the sparks can, within limits and to experienced eyes, give a useful indication of composition, particularly carbon.

Spatter Metal particles unintentionally sprayed out during welding. They may become strongly bonded to the component and, at best, will be unsightly or, at worst, actual or potential damage sites.

Specific gravity The ratio of the weight of a substance to the weight of the same volume of water. The *SI* system prefers the term 'relative density'.

Specific heat The heat required to raise unit mass of material through one temperature unit.

Specific resistence Resistivity. The electrical resistance of unit length of unit cross-section of material. Measured in microhm centimetres.

Spectrograph An instrument for producing and displaying the spectrum of radiation emitted by a material when excited by, for example, an electric arc. It can be used for the analysis of metals. Where the instrument measures the intensity of the emission it may be termed a **Spectrometer**.

Speculum (1) Alloys of copper with tin in the range 25–45%, possibly with zinc, antimony, etc. The various alloys are typically hard, brittle, brilliant white and corrosion resistant and have been used since antiquity for mirrors and ornaments. Many of the alloys can be applied as an electroplate to give a hard silver-like deposit.

Speculum (2) A mirror usually formed from *Speculum (1)* metal and possibly incised with a graticule.

Spelter (1) Zinc, particularly in impure form or as a casting material for sculptures as a cheap alternative to bronze.

Spelter (2) Brass for *Brazing filler* or, less commonly and as in the previous entry, as a cheap alternative to bronze for sculptures.

Spent uranium See Uranium.

Sphere models Physical or graphical models of crystals which uses spheres to represent atoms. See Crystal structure for examples.

Spheroidal/spherulitic graphite The graphite in *Cast iron* having a spherical rather than a flake shape.

Spheroidized carbide The carbide in steel that has become spherical in shape as a result of heating. See Steel.

Spheroidizing treatment The heat treatment of steel at just below the lower critical temperature allowing the formation, by diffusion, of relatively coarse sperical carbides. See Steel.

Spiegel, spiegeleisen Ferromanganese master alloys used in the final stages of steel production. Where they are differentiated the former has about 20–30% manganese and 5% carbon, remainder iron, and the latter about 5–20% manganese and 5% carbon.

Spin welding The same as *Friction welding* but more common in the plastics field.

Spinel A simple cubic *Crystal lattice* found in various minerals. The term is sometimes used specifically for MgAl$_2$O$_4$.

Spinning A manufacturing process in which sheet or plate material is rapidly rotated and forced by a non-cutting tool against a shaped former. The variations range from the simplest case of a plain, hand-held, shaping tool to high-powered systems where a roller head tool is carried on a powered carriage; the latter may be termed **Flow spinning**.

Spiral welded tube A tube formed from strip rolled to a helix and welded on the spiral interface between the edges.

Splash lines Damaging or unsightly material on the surface or in the interface of *Spot* or *Seam welds* caused by the ejection of material during welding.

Splat casting Droping small quantities of molten metal onto a substantial cold metal surface to achieve very rapid rates of cooling. A variation of the technique runs a thin stream of molten metal onto a cooled metal wheel.

Splicing The joining of two multistranded ropes or cables by intertwining the individual strands of each into the other.

Sponge iron/metal A porous iron or other metal produced by reduction from its ore without melting.

Spot test A test in which spots of chemical reagents applied to the test surface produce reactions indicative of the composition.

Spot welding (1) A *Resistance welding process* in which the components, usually sheet, are clamped between two electrodes supplying heating current. The weld formed is approximately the size of the electrodes, or the smaller of them if they differ. See Figure 47(b), located at entry on Welding terminology.

Spot welding (2) Any localized weld formed by any process.

Spray transfer (in welding) See Metal transfer.

Spraywelding Various processes for producing coatings on metals. The basic characteristic is that the material is initially applied by some metal powder spraying technique and the component is then heated by a flame or other means to cause the deposited powder to fuse to the component.

Sprigs Pins or rods inserted to strengthen local weak areas of sand moulds.

Spring brass An imprecise term applied to various hard-rolled brass alloys but particularly 70% copper, 30% zinc. See Table 8.

Spring steel An imprecise term indicating *Steels* capable of being hardened and tempered to give high yield strength and low mechanical hysteresis. Depending on the application, compositions vary considerably, ranging from simple 0.5% carbon, 0.8% manganese steel to low-alloy steels with 0.6% carbon, 2% silicon, 0.85% manganese, 0.3% chromium and 0.25% molybdenum. Hardness levels also vary with the application but are usually within the range 350–620 Hv. For specialist applications other more complex steels may be employed as springs, for example the 18% tungsten, 4% chromium, 1% vanadium steel, better known as a *High-speed* cutting steel, is often used for springs in high-temperature service but such steels would not normally be included in the term 'spring steel'.

Sprue   The parts of a sand casting which comprise the *Feeder* and *Riser* systems and which are normally cut off and discarded.

Sputtering   The ejection of surface atoms as a result of ion bombardment. No heat is involved. The ejected atoms are deposited on surrounding surfaces.

Stable   Not liable to change physically or chemically.

Stabilized stainless steel   Austenitic stainless steel containing sufficient quantities of elements such as titanium, or niobium (columbium) with a strong affinity for carbon. These combine preferentially with carbon which, in their absence would, upon heating the steel in the 400–900 °C range, combine with the chromium. This would cause local chromium depletion, particularly at *Grain boundaries*, leading to corrosion problems, for example, after welding. To achieve immunity the alloy addition needs to exceed some level related to the carbon content; typical figures are 10 × carbon for niobium (columbium) and 5 × carbon for titanium. Immunity from this form of damage can also achieved by limiting the carbon to less than 0.03%, by quenching the steel from about 1050 °C or by adding about 3% molybdenum to induce a small percentage of ferrite which localizes the carbide precipitation. However, although these latter three treatments confer immunity the term 'stabilized' is usually confined to steel varieties with the strong carbide formers. See also Steel, Sensitization and Weld decay.

Stablizing treatment   Any process intended to stablize the microstructure, dimensions or other feature of a component so that undesirable changes do not occur in subsequent service or treat-

ment. For example, some complex components intended for high-temperature service are subjected to a pre-service heat treatment at a slightly higher temperature so that distortion in service is minimized. See Thermal instability.

Stacking fault   A defect in the *Crystal lattice* in which the normal stacking arrangement is disrupted with partial *Dislocations* at the perimeter.

Staggered intermittent weld   A weld made intermittently along the two sides of a joint, for example a T-fillet weld, with the welds on one sides lying opposite the gaps on the other. Where the welds on one side lie opposite the welds on the other the joint is termed a **Chain intermittent weld**. See Figure 49, located at entry on Welding terminology.

Stainless (steel/iron)   Any iron or steel containing more than about 12% chromium can be termed stainless. With about 12% or more of chromium the material, when exposed to the normal atmosphere, rapidly forms a thin, impervious, chromium-rich, oxide film which protects the underlying steel from further attack. There are many forms of stainless steel (see entry on Steel). The term **Stainless iron** has been used for widely differing alloys ranging from 12% chromium, low-carbon steel to 30% chromium, high-carbon *Cast iron*; it should therefore be viewed with caution. See Steel for further commentary on stainless steels.

Stamping   Pressing and forging sheet and plate in *Closed dies.*

Stand   Two or more rolls in a housing. See Rolling mill.

Standard (1)   A basis of comparison on which units are based. For example, units of measurement were traditionally based on a solid metal standard bar retained by some national or international body.

Standard (2)   A specification covering all relevant aspects of a material, procedure, system, etc. Aspects defined include composition, properties, dimensions and performance.

Standard gold In the UK 91.66% gold, remainder copper. In the USA 90% gold, remainder copper.

Standard silver In the UK 92.5% silver, remainder copper. In the USA 90% silver, remainder copper.

Standard temperature and pressure A reference environment, conventionally, 0 °C and 101 325 Pa. This is the term now preferred to the nominally identical Normal temperature and pressure.

Standard wire gauge A series of standard diameters.

Static fatigue A term sometimes applied to various forms of delayed failure where the loading is static rather than cyclic, in particular *Hydrogen damage* in metals and *Creep* in plastics. Contrast with Fatigue.

Static plate See Explosive welding.

Staving A process for increasing the external diameter of a tube end while maintaining the bore constant. Also termed *Upsetting*. See Figure 45, located at entry on Tube manipulation.

Stay A structural member of slender proportions, such as a bar, beam or wire, carrying tensile loads.

Steadite The *Eutectic* between austenite and iron phosphide in *Cast iron*. It has a melting point of about 980 °C.

Steady-state creep See Creep.

Steam blanketing Many boilers generate steam by circulating water through tubes heated on their exterior. Normally, the steam is generated as discrete bubbles (termed 'nucleate boiling') which rise by gravity or are swept by the pumped circulation. However, in some adverse circumstances, such as an excessive heat flux, the steam can form a persistent layer along the bore. This layer is termed a 'steam blanket'. It can lead to various adverse effects, including *Caustic attack*, excessive thickness and fissuring of the normally protective magnetite film, overheating of the tube and distortion.

Steam erosion The erosion of a surface by high-velocity steam impinging on a component. In practice, erosion by pure gaseous steam is rare and the term is often a misnomer (although see Wire drawing). Usually, the damage is caused either by entrained solids (see Solid particle erosion) or by water droplets. The term is commonly used regarding the damage observed on the last stage of moving blades of a steam turbine where the steam is very 'wet' containing a considerable number of condensed water droplets. However, even in this case the steam velocity is not responsible for the metal loss since the damaging droplets are those that drip off the stationary blades to be impacted by the moving blades. Once droplets have been accelerated to steam velocity they follow the streamline flow and cause little damage. Strictly, also, this form of **Water droplet** damage is not a form of abrasion by a cutting action, rather it is the result of repeated high-velocity impact and hence is more akin to a surface *Fatigue* process. The surface produced is not smooth or grooved but develops a *'Cat's tongue'* texture of sloping sharp spikes and pits which become progressively coarser as damage progresses.

Steam side See Fire side.

Steatite See Soapstone.

Steckel mill A rolling mill in which the material is pulled backwards and forwards through rolls that are not driven.

Steel This entry offers a brief outline of the metallurgy of steel. For examples of properties to be expected from various steels see Tables 1 to 5. For an outline of the system followed in BS 970 to designate steels see entry on Engineering steels. Steel is iron with another element intentionally present. However, alloys containing more than about 2.5% of carbon are termed **Cast iron** (see entry on this topic). The following section provides a broad overview of the metallurgy of steel in three subsections: (1) Steel types, (2) The physical metallurgy and heat treatment of steel and (3) Corrosion-resistant steels.

(1) Steel types
Steels normally contains less than about

1.5% carbon and are categorized, rather imprecisely, on the basis of their carbon and alloy element content. Some of the more common terms and their usual meanings are as follows.

Plain carbon steels contain no significant quantity of alloying elements other than carbon although they may contain many impurities from the ore (particularly sulphur and phosphorus up to about 0.05% each), from scrap (including copper or tin perhaps up to 0.5% each) or from materials added to assist the steelmaking process (particularly manganese up to about 0.8% and silicon up to about 0.3%). The plain carbon steels can be sub-categorized as:

Mild steel or, less commonly, **Low-carbon steel**, which has less than about 0.2% carbon. It has only a limited capability to harden by heat treatment (as described below) and is usually employed in the *Normalized* or *Cold worked* condition.

Carbon steel, which contains more than about 0.2% carbon. These steels can be hardened by heat treatment and may be used in the normalized, cold worked or hardened conditions. They are often further subdivided into:

Medium-carbon steel containing about 0.25 to about 0.5% carbon.

High-carbon steel containing about 0.5 to about 1.2% or, rarely, up to about 2% carbon.

Carbon manganese steels contain carbon above about 0.15% plus manganese of about 1–2% and the usual impurities. In these steels the increased level of manganese improves hardenability (defined below) and tensile strength in the normalized and cold worked conditions.

Alloy steel has one or more elements, other than carbon (and manganese below about 2%), deliberately added in sufficient quantity to improve some property such as hardenability or high-temperature strength. Such steels are often designated by to their principal alloy contents, for example, 'nickel steel' or 'chrome(ium) moly(bdenum)

steel'. Almost invariably, carbon is also present and, unless the other alloying elements are present in large quantities, the steel responds to heat treatment broadly as summarized below. Such steels are obviously expensive and are normally used in the hardened condition. In a few specialized alloy steels, carbon may be deliberately excluded. Above about 5% total alloy content, again excluding carbon and manganese, the term **High-alloy steel** is common.

Stainless steels are high-alloy steels but are dealt with in some detail in subsection (3) Corrosion-resistant steels

(2) The physical metallurgy and heat treatment of steel

An important characteristic of iron is that it is *Allotropic*, that is, it undergoes *Phase* changes as its temperature varies. At temperatures up to about 910 °C pure iron is in the form of **Ferrite** or alpha (α) phase, with a *Body centred cubic* (BCC) *crystal lattice* structure. From 910 °C to about 1400 °C the stable phase is **Austenite** or gamma (γ) with a *Face centred cubic* (FCC) arrangement. Between 1400 °C and the melting point at about 1530 °C the structure returns to the BCC form but is referred to as delta (δ) phase or **Delta ferrite**. Note, there is no **Beta** phase in iron. It used to be thought that a beta phase existed between 910 °C and the Curie point, that is, the change from *Magnetic* to non-magnetic, at 770 °C.

In the case of pure iron, the change from one phase to another is not readily detectable outside the laboratory and has no great commercial significance. However, the addition of small amounts of carbon to form steel introduces a series of useful effects. These are illustrated in Figure 37, usually termed the **Iron carbon (equilibrium) diagram**.

The *Solubility* of carbon in iron varies considerably. In the low-temperature ferritic phase solubility is very low, about 0.03% maximum. However, the austenitic phase can hold in solid solution all the carbon present in conventional

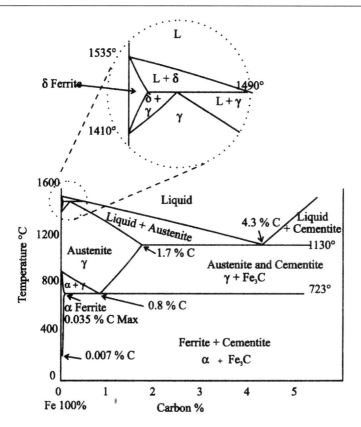

Figure 37 The iron–carbon phase diagram. More strictly, the phase diagram for iron–metastable Fe₃C

steels. Furthermore, the addition of carbon extends the temperatures range over which austenite can exist under equilibrium conditions. Thus, on slow cooling, the austenite does not transform fully to ferrite at 910 °C but commences to transform at some lower temperature, termed the **Upper critical temperature**, which depends on the carbon content. Transformation then continues progressively as the temperature falls further but the ferrite which is formed contains very little carbon so the remaining austenite becomes progressively richer in carbon. Ultimately, at a specific temperature, all the remaining carbon-rich

austenite will transform to a mixture of ferrite and **Iron carbide, Fe₃C**, also termed **Cementite**. This temperature, 723 °C in plain carbon steels, is termed the **Eutectoid temperature** or the **Lower critical temperature**. The upper and lower critical temperatures are also referred to as, respectively, the **A₃** and the **A₁**. The rarely used term A₂ refers to the change from magnetic to non-magnetic at about 770 °C. Often, the 'A' temperatures are further defined as in Ar₁ or Ac₁. The 'r' suffix denotes the temperature measured during cooling (from the French *refroidissement*) and 'c' the slightly higher temperature measured during heating (*chauffage*). The range of temperature between A_1 and A_3 is variously described as the **Critical range, Transition range** or **Transformation range** although it should be noted that these terms can have different meanings in other contexts.

The mixture of ferrite and cementite formed at 723 °C is usually precipitated as the *Eutectoid* form comprising alternate plates, or lamellae, of carbide and ferrite. This can have an appearance under the microscope similar to mother of pearl, hence its name **Pearlite**. Microstructures formed in this way usually comprise grains of ferrite plus grains of pearlite, the amount of pearlite being roughly proportional to the quantity of carbon in the steel. About 0.8% carbon, termed the **Eutectoid composition**, gives a fully pearlitic structure (see entry on Sauveur's diagram). The terms **Hypo-** and **Hyper-eutectoid** indicate, respectively, compositions of lower or higher carbon content. Hyper-eutectoid steels have a structure composed predominantly of pearlite with a small amount of cementite as films or discrete areas. In isolation, cementite is hard and brittle but pearlite, with its alternate layers of cementite and relatively soft, tough ferrite, provides a good combinations of hardness, strength and toughness. Generalizing, the hardness of the steel increases with the proportion of pearlite in the microstructure.

All the treatments considered so far have involved slow cooling rates so that appreciable diffusion can occur and transformation effectively reaches the stable equilibrium state. The process of very slow cooling from above the transition range, often in a furnace, is termed **Annealing** or **Full annealing**. This produces a coarse pearlite and ferrite structure with the steel in a soft, low-strength, condition. A slightly faster cooling rate, which is usually achieved by removing the steel from the furnace to cool in still air, is termed **Normalizing**. This produces a finer pearlite with

slightly higher strength and hardness but, for practical purposes, normalized steel is still usually regarded as being in a softened condition and a near-equilibrium state. However, if a steel is cooled more rapidly by **Quenching** into oil, water or brine, etc., insufficient time is available for diffusion. This forces transformation to take place below the A_1, resulting in non-equilibrium transformation. With only a slight increase in cooling rate the effect is to precipitate a finer form of pearlite but as the rate increases the lamellar form is suppressed and the carbide is deposited as a dispersion of particles in an acicular, i.e. needle-like, ferrite matrix. This structure, termed **Bainite**, can form over a wide range of temperature below about 550 °C. Transformation high in this range produces **Upper bainite** with a microscopical appearance often described as 'feathery'. Faster cooling delays transformation to a lower temperature, producing the more obviously acicular structure of **Lower bainite**. Bainitic structures have a higher hardness than the ferritic/pearlitic structures.

At even higher cooling rates, when the **Critical cooling rate** is exceeded, **Martensite** is formed by a diffusionless shear process which commences at the **Ms** (martensite start) temperature and continues until the **Mf** (martensite finish) temperature. Transformation is not time dependent but is a function of the temperature in the Ms–Mf range. In terms such as 'M_{10} temperature' the subscript indicates the percentage transformation to martensite at the temperature in question. If the Mf temperature is below ambient, the transformation to martensite will remain incomplete and the structure will contain **Retained austenite**. This may be transformed to martensite by cooling below ambient. The steel will remain fully martensitic on return to ambient. Fully martensitic structures offer the maximum hardness that can be developed by heat treatment of plain carbon steels. The crystal structure of martensite is usually described as body centred tetragonal and under the microscope it has a fine accicular appearance.

The process of heating a steel into the austenitic range and then cooling at a rate sufficient to cause transformation to bainite or martensite is usually referred to as **Hardening**. The martensitic structures in particular are hard and of high tensile strength but they tend to be brittle. Consequently, quenched steels are usually reheated for about an hour to temperatures in the range 100–600 °C to produce progressively softer but more *Tough* steels. This treatment, termed **Tempering**, causes the *Metastable* bainite and martensite phases to transform progressively to ferrite and iron carbide. The carbide particles produced by low-temperature tempering are extremely fine and difficult to resolve in a light microscope but become increasingly coarse, the higher the tempering temperature. Modern practice is to refer to these structures simply as tempered martensite or tempered bainite. Earlier practice used other terms. **Troostite** was defined as a dark etching, unresolvable structure developed by light tempering, i.e. up to about 300 °C, and **Sorbite** was defined as a fine, but resolvable, iron carbide in a ferrite matrix produced by tempering above about 400 °C. When the tempering temperature exceeds about 500 °C the carbides become increasingly coarse and are now usually described as **Spheroidized carbide** although the older terms **Sorbite** or **Sorbitic carbide** are still occasionally used. Cooling rate is not critical for tempering treatments but the time at temperature does affect the structure particularly at higher temperatures where longer times produce coarser carbides.

Up to this point interest has centred on plain carbon steels, that is, ones that contain no deliberate alloying additions apart from carbon and up to about 2% manganese. These steels can develop

very high strengths but to do so they need to be cooled very rapidly during the hardening process. This can be achieved in the case of small components but with larger sizes the material, particularly at the centre of the section, will not reach the critical cooling rate to transform to martensite. A further problem associated with rapid cooling is that it can cause distortion or even cracking, particularly of complex shaped items. This latter problem arises partly because of the thermal contraction that occurs as the temperature falls and partly because of the volume changes associated with the various phase changes. These difficulties are overcome by adding to the steel further alloying elements that delay transformation so that slower cooling rates can develop the martensite or bainite structures. Many elements have this effect, including manganese, nickel, chromium, molybdenum, vanadium and, of course, increased carbon. They are said to confer **Hardenability**. This term, therefore, refers to the ease with which a steel may be hardened; it does not refer to the level of hardness that can be achieved. Hardenability is measured in terms of the maximum section size, or **Ruling section** that can be fully hardened by a given cooling rate. Some of the elements that are added remain in solid solution but others can combine with the carbon to form alloy carbides which can confer further beneficial effects or, in other circumstances, present problems. When large amounts of some alloying elements, such as nickel and manganese, are added to the steel the transformation characteristics are changed to such an extent that the steel remains partially or fully austenitic down to ambient temperature even when slowly cooled. Such steels are termed **Austenitic steel**, or, if they are partly austenitic, they may be termed **Duplex**. Fully austenitic steels are non-magnetic and, since they remain austenitic they cannot be hardened by rapid cooling. If, in addition to the austenite stabilizing elements (nickel and manganese), the steel contains more than about 12% chromium it can be termed **Austenitic stainless steel** which is considered further below. It will be recognized, however, that not all austenitic steels are stainless.

(3) Corrosion-resistant steels

An adverse characteristic of iron and steel that reduces their usefulness is the lack of corrosion resistance. Iron reacts with oxygen in the atmosphere to form a range of iron oxides. The rate of attack in dramatically increased in the presence of moisture and the oxide formed in these circumstances is often combined with water to form **Rust**. The unfortunate characteristics of rust are that it is very bulky compared with the iron that it replaces, it is readily detached and, most critically, it does not form a barrier to further contact with the environment. Consequently, rusting continues progressively in appropriate atmospheres until all the iron is consumed. Most alloying elements have little effect on the corrosion resistance of iron. Elements such as copper, aluminium and silicon can confer a small benefit in certain environments but only chromium has a major effect. Chromium reacts with atmospheric oxygen very strongly and rapidly but the oxide formed is thin, transparent, tightly adherent and acts as a barrier to further attack. This benefit is conferred on iron if the chromium content of the alloy exceeds about 12%. Thus, steels with more than this quantity of chromium are referred to as **Stainless steels**. If chromium is the only significant element the alloy is sometimes referred to as **Stainless iron** but this term has been applied to various steels and cast irons and it should be used and interpreted with caution. The term **Ferritic stainless steel** usually indicates an iron–chromium alloy containing insufficient carbon to undergo a hardening heat treatment as described above, while higher-carbon alloys that can be hardened are termed **Martensitic stainless**

steel. If more than about 8% nickel is added to a high-chromium steel the austenite-to-ferrite transformation is suppressed to such an extent that the material remains austenitic at ambient temperatures and the alloy is referred to as **Austenitic stainless steel**. A common alloy contains 18% chromium plus 8% nickel, hence the popular designation '**18/8**' **stainless**. The **Stabilized stainless steels** (see entry on this topic) contain elements, such as titanium and niobium (columbium), to avoid susceptibility to intergranular corrosion. Although austenitic steels cannot be hardened by conventional heat treatment as described in the previous sub-section certain more complex stainless steels can be hardened by **Precipitation hardening** processes.

Stellite   A range of proprietary alloys mostly based on cobalt and having excellent corrosion resistance, hardness and strength.

Step (fatigue) test   A *Fatigue* test programme in which a specimen is subjected to a prescribed number of cycles at each of a series of progressively increasing *Stresses*. The fatigue limit is then taken as the penultimate stress, i.e. the highest that did not cause failure or, alternatively, the average of the penultimate and the final stresses.

Stereographic projection   A technique for presenting, in two dimensions, information regarding the three-dimensional orientation of crystallographic planes. The unit cell of a *Crystal* comprises a small number of atoms arranged in a simple geometric pattern, for example at the corners of a cube. The crystallographic planes are the planes on which the atoms lie, for example the cube faces or the cube diagonals. The basic stereographic projection usually deals with a single crystal which, for the purpose of the exercise, is considered to be located at the centre of a large sphere. Figure 38 illustrates the technique for a single plane, C. A line is projected perpendicularly from plane C to intersect the sphere

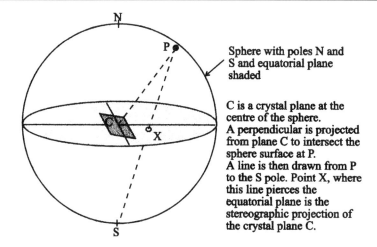

Sphere with poles N and S and equatorial plane shaded

C is a crystal plane at the centre of the sphere. A perpendicular is projected from plane C to intersect the sphere surface at P. A line is then drawn from P to the S pole. Point X, where this line pierces the equatorial plane is the stereographic projection of the crystal plane C.

Figure 38   A stereographic projection of a single plane of a crystal

surface at point P. A further line is then drawn from P to the pole of the other hemisphere, point S. The point X at which this line pierces the equatorial plane is the stereographic projection of the plane C. This procedure is repeated for all crystallographic planes of interest and the pattern of the series of X points forms the stereographic projection of the crystal. The technique can also be applied to multiple crystals such as a small piece of sheet. If the crystals of this sheet are aligned at random then the X points will be randomly distributed on the stereographic projection. However, a grouping of points will be evidence of preferred *Orientation*. The term **Pole figure** is often used interchangeably with the term 'stereographic projection'. Where they are differentiated, pole figure usually refers to projections of multiple crystals.

Stereotype metals   Tin antimony alloys similar to type metals. See White metals and Babbit metals.

Sterling silver   Alloy of 92.5% silver minimum and 7.5% copper (usually), maximum.

Stick electrode (welding)   The coated

rod *Electrode* used for *Manual metal arc welding.*

Stiction   A term used to describe the force necessary to overcome static friction.

Stiff nut   See Self-locking nut.

Stiffness   The ability to resist elastic deformation.

Stitch weld   A line of overlapping *Spot welds.*

Stitching   See metal stitching.

Stopp(ing) off (agent/coating etc.)   Material applied to a surface to prevent, locally, some reaction or effect, for example to prevent local plating or etching during such processes or to prevent adhesion of weld *Spatter.*

STP   See Standard temperature and pressure.

Straight polarity   See Reverse polarity.

Strain   Dimensional change resulting from an applied stress and normally expressed as a ratio of the change to the original dimension. See Tensile test.

Strain ageing or strain age hardening   A spontaneous increase in *Hardness* occurring in some materials, particularly certain steels, after they have been deformed during manufacture or service. The hardness increase is usually small but it can cause serious *Embrittlement*. It may occur over days or even years and can be accelerated by moderate heating as occurs, for example, during hot-dip *Galvanizing*. The effect results when dissolved *Interstitial* elements such as carbon and particularly nitrogen diffuse through the *Crystal lattice* until they become engaged with, and hence impede the movement of, *Dislocations*. Contrast with Strain hardening.

Strain energy The energy associated with elastic deformation, for example the energy in a compressed spring. Its measure is the area under the elastic portion of the stress strain curve obtained in the *Tensile test*.

Strain energy release rate The elastic strain energy released in unit propagation of a crack.

Strain gauge Any mechanism or device for measuring dimensional changes and hence *Strain*. However, the term usually refers to **Electrical resistance strain gauges**, devices comprising a length of resistance wire, typically laid in a tortuous path, which is bonded to some backing sheet which in turn is bonded to the test piece or component being investigated. The change in electrical resistance is monitored as the component is deformed and, by reference to calibration data, the strain deduced.

Strain hardening The increase in *Hardness*, together with increases in *Tensile* and yield strength, that results from plastic deformation. Also termed work hardening.

Strain lines/markings Lines of severe deformation revealed by polishing and etching.

Strain annealing A technique for developing a very large grain size by annealing a material after it has received a small but critical amount of cold working. See Critical strain.

Strap joint A joint formed by butting two plate edges together and laying a strap along the butt line. The strap is then joined to the individual plates by techniques such as bolting, riveting, welding or brazing.

Strauss test A test for *Weld decay* in which a specimen is boiled in a copper sulphate/sulphuric acid solution for 72 hours and then bent to check for cracking.

Stray arc/flash In arc welding, the unintentional striking of an arc against the component, including previously deposited weld metal, and the surface damage so caused.

Stray current Unexpected electrical currents arising in processes or encountered in the environment. For example, earth return or other currents associated with tramways can produce currents in other buried components.

Stray current corrosion *Electrolytic* corrosion caused by *Stray currents* in the environment. Electrical currents are found in soil as a result of its deliberate use as an earth return or as a result of accidental leakage. Such currents can cause severe corrosion of equipment at considerable distances from the location where the current enters the soil.

Stress Force per unit area. See Tensile test. The three basic stresses are *Tension*, *Compression* and *Shear*.

Stress analysis Any technique for determining the level and distribution of *Stress* in complex components and structures. Examples include scale solid modelling or mathematical and computer modelling including *Finite element analysis* and *Photo-elastic* techniques.

Stress assisted/accelerated corrosion Forms of corrosion in which the rate of attack is significantly increased by *Stress*. Where the attack causes cracking it would normally be termed **Stress corrosion cracking**.

Stress concentration The local increase in *Stress* at a crack, notch or other section change, also termed a **Stress raiser**. A component containing a crack is obviously weaker than one which is defect free but the reduction in strength is often more than would be expected simply from the reduction in cross-section area. The cause can be visualized by considering a cracked bar in which the load is evenly distributed over many individual strands. Remote from the crack the strands and the load will be evenly spread across the section but at the crack the load in the severed strands will be transferred to the immediately adjacent, intact strands. The strands at the zone at the crack tip will, therefore, be subjected to their normal share of the load plus the load transferred from the severed strands. In ductile materials this zone will *Yield* transferring load deeper into the surrounding intact material. However, non-ductile material will not yield and locally the stresses remain high, leading to an extension of cracking at the crack tip. Stress concentrations can also be caused by holes, steps and section changes. Generalizing, the sharper the step or the smaller the crack tip radius, the more severe the stress concentration. See also Notch strengthening and Fracture toughness.

Stress concentration factor (SCF) The factor by which stress is increased by a *Stress concentration.* Not the same as *Stress intensity factor.*

Stress corrosion (cracking) (SCC) The cracking resulting from the combined and concurrent effect of corrosion and a *Stress.* Some metals are highly susceptible to cracking when stressed while exposed to specific corrodants. In the absence of the corrodant the stress has no adverse effect and, even when both stress and the corrodant are present, the volume of metal lost to the corrosion process is usually negligible. The attack normally takes the form of cracking commencing at the surface and penetrating along a multiple branching path. The stress, almost invariably tensile, can be externally imposed or *Residual* and the path will be either *Intergranular* or *Transgranular* depending upon the particular material and environment. Susceptible combinations include *Brass* in ammonia solutions and austenitic stain-

less steel (see Steel) in warm chloride solutions.

Stress cracking or alternatively, **Environmental stress cracking** Any cracking occurring as a result of the combined and concurrent effect of a hostile environment and stress. This term is commonly applied in the context of polymeric (plastic) materials in which case the responsible environmental factor may be liquid, gaseous or radiation including light. However, the terms are occasionally, perhaps increasingly, used in the case of metals as an alternative to *Stress corrosion.*

Stress intensification (factor) This term has been used to refer to both *Stress concentration* and the quite different *Stress Intensity* potentially leading to confusion.

Stress intensity factor The measure of the combined effect of stress and crack size at the crack tip, usually designated by the K symbol. See Fracture toughness. It is not the same as *Stress concentration factor.*

Stress–number curve The S/N curve depicting fatigue properties. See Fatigue and the associated Figure 12.

Stress raiser Any feature such as a section change, crack or inclusion which causes an increase in the stress in its vicinity, i.e. a *Stress concentration.*

Stress ratio See Fatigue.

Stress relaxation The reduction in *Stress,* either applied or residual, by *Creep.*

Stress relief Any process for reducing *Residual stress* induced by processes such as cold working or welding. The term usually implies processes which involve heating the component to some moderate temperature for a short period, typically an hour or two. As examples, stress relief of welds in normalized plain carbon steel would be at about 550 °C and in low-alloy creep-resisting alloy steels at about 700 °C. At such temperature the *Yield strength* is greatly reduced allowing a combination of *Plastic deformation* and *Creep* to convert much of the *Elastic strain* to *Plastic strain*. Consequently, stress relief may cause some deformation of the component and, even after stress relief, some stress will remain. The treatment may also produce some beneficial softening of hardened areas of the heat affected zones of welds but it is usually implicit that the *Tensile* properties of the bulk component will not be adversely affected to any significant extent. In the case of cold-worked material there may be useful improvements in *Ductility* and *Toughness*. See also Vibratory stress relief.

Stress rupture (properties)   *Creep* failure (properties). There are various conventions for presenting graphically the interaction between stress, temperature and time to failure for a particular material. The term 'Rupture data' usually refers to a plot of stress on the vertical scale against temperature on the horizontal with lines on the graph identifying time to failure. This system is useful as it can also present Proof stress data as shown in Figure 5, located at entry on Creep. See also Larson–Miller parameter.

Stress strain diagram   See Tensile test.

Stretch forming   Processes in which sheet material is stretched, or at least restrained, in one direction at the same time as a former is pressed into its surface.

Stretcher strains/s/Lines/Markings and similar terms   Same as Lüders lines.

Striation   A linear surface mark, in particular the line on a *Fatigue* fracture surface marking the crack front at the end of each cycle. Individual striations are not visible to the unaided eye although the shading or texture they collectively produce may be. Compare with Beach mark.

Strike (1)   A thin *Electroplated* coating intended to provide a good bond for a subsequent coating.

Strike (2)   A defect produced during welding by the accidental, brief contact of an arc welding electrode. It may be

merely adherent metal but there may also be a crack.

Striker/striking plate   In electric arc welding, a piece of metal, not part of the component to be welded, on which the arc is initiated immediately prior to its transfer to the joint being made. Its function is to avoid arc initiation defects on the joint.

Stringer bead (of a weld)   A weld run deposited in a straight line with no significant side-to-side deflection of the *Electrode, Filler* rod, arc or flame relative to the main line of progress. See Figure 49, located at entry on Welding terminology.

Stringers   Discontinuous lines of non-metallic *Inclusions* in the microstructure. Their alignment results from, and provides information on, deformation of the material during manufacture.

Strontium   A metallic element, one of the *Alkaline earth* series. It has strong deoxidizing characteristics and has been used as deoxidant in the manufacture of copper alloys. It burns with a strong red flame and hence is used in flares and fireworks. See Table 15 for properties.

Strut   A column or beam carrying a compressive load.

Stub   The discarded end of some consumable such as a manual metallic arc welding *Electrode*.

Stub-in   In welding, the entrapment of the electrode or filler wire in the solidifying weld metal.

Stud (1)   Various components, usually fitments to some larger assembly, which act as connectors, retainers or locating devices.

Stud (2)   A threaded fastener with both ends threaded, one for insertion into the main body of a component and the other end to carry a nut. For example, a set of studs would be screwed into a cylinder block prior to positioning the cylinder head and installing the nuts. Compare with Studbolt.

Stud weld   A joint formed between a stud, or similar item, and the main component. The stud is normally carried in a purpose-built gun or other device and, after an initial heating process is applied to the stud and, usually, the target area, the stud is pressed hard against the component to form a *Pressure weld*. Heating processes include *Electric Arc*, *Electric Resistance*, *Capacitor discharge*, and *Friction*.

Studbolt A threaded fastener with both ends threaded and carrying nuts at both ends. For example, a studbolt would pass through a pair of flanges to hold them together. Compare with Stud.

Studding Bar threaded over its full length, often supplied in standard lengths to be cut to size as required. Also multiple *Studs (1)* or *(2)*.

Sub-atomic The particles forming an atom or processes occurring on a smaller scale than an atom.

Sub-critical annealing Softening steels at a temperature just below the transition range, typically about 650 °C for about an hour. Also termed **Process annealing**. Such treatment is normally applied to heavily cold worked material so it causes the ferrite to recrystallize but does not take the carbide into solution. The resultant microstructure therefore comprises equiaxed grains of ferrite superimposed on elongated stringers or bands of spheroidized carbide. The benefits of the process compared with normalizing or full annealing include lower heating costs, less scaling and, in some cases, good machining characteristics. See Steel and Annealing.

Sub-grain Areas within a *Grain* having lattice *Orientations* differing by a detectable amount but not sufficient to constitute separate grains.

Sublimation Evaporation directly from the solid to vapour without an intermediate liquid *Phase*.

Submerged arc welding *Arc welding* with the arc and weld area covered for protection by a substantial layer of granular *Flux*. It is normally an automatic process with the *Filler* and *Flux* delivery and the electrical conditions controlled by the machine.

Substitutional Occupying a position on the *Crystal lattice* normally filled by atoms of another element. See Solution.

Substrate The material underlying the surface film or outer layer.

Sub-zero treatment Any treatment below 0 °C (or 0 °F depending on the context) particularly chilling of steel to cause transformation of retained austenite. See Steel.

Suds Water or emulsion containing corrosion inhibitors and other additives used to cool and lubricate components and tooling during severe metal cutting operations.

Sulphating The development of a layer of lead sulphate on discharged lead acid cells. It causes permanent damage as it is not reduced by subsequent charging.

Sulfinuz process A proprietary *Case hardening* treatment for steel carried out in a salt bath which produces a case rich in nitrogen, sulphur and carbon having superior *Fatigue* and *Fretting* resistance.

Sulphur A non-metallic element. It is often encountered in metallic ores, it is usually undesirable even in small quantities and is not easily removed totally. In *Steel* it can form low melting point iron sulphide which is very damaging but it combines preferentially with manganese additions to form the relatively innocuous manganese sulphide inclusions observed in most steels. See Table 15 for properties.

Sulphur print A print formed by placing a suitable photographic paper, previously soaked in 3% sulphuric acid, against a prepared steel surface to reveal the presence and distribution of sulphur compounds particularly manganese sulphide *Inclusions*. These inclusions reveal the *Flowlines*, i.e. the pattern of deformation, of the material during the various working operations since the material was cast.

Superalloys An imprecise term usually meaning materials significantly superior to high-alloy steels. This includes any of a wide range of complex alloys based mainly on metals such as nickel and cobalt but with additions of many other metals including aluminium, iron, molybdenum, titanium and tungsten. Such alloys were, and still are being, developed primarily for aerospace applications requiring high strength and corrosion resistance at high temperatures.

Superconductor A material having no resistance to the passage of direct current electricity below some critical temperature. A few elements and many intermetallic compounds have critical temperatures close to absolute zero but there are a few specialized alloys with critical temperatures close to ambient. In the superconducting state a material will act as a perfect diamagnet in a weak magnetic field. Type I superconductors lose this characteristic at a specific field strength; Type II lose it progressively over a range of field strength.

Supercooling Cooling, without a *Phase* transformation occurring, to a temperature below that at which transformation would occur under equilibrium conditions. The term is often used with particular reference to the rapid cooling of molten metal to below its normal freezing temperature or range. This promotes the formation of abnormally large numbers of the nuclei that initiate *Solidification* and hence leads to a very fine grain size.

Supercritical Any action that takes a system or material past some critical point.

Supercritical fluid Water or other material that is pressurized to such an extent, i.e. beyond the *Supercritical pressure*, that on heating the fluid does not boil in the conventional sense. It does not change from an identifiable liquid phase to an identifiable gas phase, there is no change in density, no latent heat is involved and there is no physical interface such as exists normally between water and steam.

Supercritical pressure The pressure at which a substance becomes a *Supercri-*

tical fluid. For water it is 22.1 MPa (3208 lb/in^2) absolute.

Superfinish A fine abrasive polishing procedure to produce a high-quality finish particularly on races, rollers and balls of bearings.

Superheating (1) Heating steam to a temperature greater than the boiling point.

Superheating (2) Heating, without a phase transformation occurring, above a temperature at which transformation could occur under equilibrium conditions.

Superheating (3) Heating molten metal well above its melting point to improve fluidity, etc. during casting.

Superlattice See Order.

Superplasticity The ability of a metal to accept very large amounts of deformation, often greater than 2000% elongation, without cracking or *Necking*. The phenomenon typically occurs in very fine-grained material deformed at temperatures and strain rates such that progressive deformation is matched by continuous *Recrystallization* and *Recovery* at a rate sufficiently rapid to maintain the fine grain size. The temperature is typically about half the melting point on the absolute scale and the strain rate is usually low compared with industrial manufacturing and testing practice. If the circumstances are correct then any neck which tends to form will become a site for an increase in strain rate with consequent increase in stress and work hardening. Deformation will therefore cease at this location and be transferred to material elsewhere.

Supersaturated See Solution.

Surface effects This term usually refers to the influence on *Fatigue* properties of surface features such as polish, scratches, machining, pitting, local *Residual stresses* and *Fretting*.

Surface engineering Matters concerning any aspect of surface condition or treatment including machining, peening, residual stress, polishing, plating, case hardening, diffusion treatment, conversion coatings and key for painting.

Surface finish The form and quality of a surface. It may be defined in qualitative terms such as *Polished*, or *Mill finished* or, particularly in the case of machined surfaces, it may be measured quantitatively or semi-quantitatively. Older semi-quantitative techniques compare the **Roughness** of the surface in question with a series of standard machined surfaces, the comparison being made by eye or feel (see Rubert gauge). The more modern quantitative techniques use devices which measure the surface texture and profile by optical methods or, in the case of instruments such as the **Talysurf**, by running a stylus over the surface and measuring the height and pitch of the undulations. The irregularities are rarely simple and the electronics in the equipment integrate the various long- and short-range deviations to give a roughness figure in micro-metres, microns or micro-inches.

Surface hardening See Case hardening.

Surface rolling See Rolling.

Surface tension The phenomenon by virtue of which a liquid appears to have a skin and forms a meniscus or stands as globules on a non-wetted surface. It arises from the mutual attraction between molecules exposed at the surface and the consequent tendency of a surface to contract to the smallest possible area.

Surfacing The deposition of material, particularly by welding or similar processes, to provide a surface coating having properties such as high hardness or corrosion resistance superior to the substrate. Where the intention is merely to restore dimensions the term *Building-up* is often considered more appropriate.

Suspension A liquid containing a solid material in finely divided particulate form and which is evenly dispersed rather than settled as a deposit.

Susceptibility See Magnetic.

Sustained load failure A form of failure in *Steel*, usually a high-tensile grade, that occurs at *Stress* levels that would normally be acceptable and possibly after a considerable period of time. It results from excessive hydrogen content and is accompanied by negligible ductility. See Hydrogen damage.

Swaging Reducing the diameter of bar, tube or similar sections by processes such as forging or hammering by hand or machine as in *Rotary swaging*.

Swarf The strands or chippings of scrap material cut from components during machining operations such as drilling and *Turning*. Also the sludge formed by grinding and comprising mainly metal and grit.

Sweat (1) Low melting point material exuding from the surface of castings during the final stages of solidification. See Segregation.

Sweat (2) Moisture condensing on cold components heated with a gas flame.

Sweated joint One made by *Soft soldering* or, less commonly, *Brazing* or *Silver soldering*, particularly when the mating components are pre-tinned and then held together while heating them.

Swedish iron An imprecise term usually implying high-purity iron, particularly one made from high-grade ores smelted with charcoal and refined to minimize contamination.

Sweeps/sweepings Dust and other debris collected from the floors and other surfaces of premises handling precious metals. It is collected for extraction of the valuable metal.

SWG *Standard wire gauge.*

SWL *Safe working load.*

Système International d'Unités The International System of Units adopted by the General Conference of Weights and Measures and endorsed by the International Organization for Standardization. See SI system.

T

T joint A joint between the end or edge of one component and the face of another. Some usage of the term is confined to joints where the two components are at approximately right angles but other usage includes joints where the angle is much more acute and the weld is a *Fillet*. See Figure 47(b), located at entry on Welding terminology.

Tack welds Small, well-spaced welds applied to maintain components in alignment during the main welding operation.

Tagging Reducing the diameter of the end of a bar to allow its initial insertion through a drawing *Die* or into a *Swaging* machine.

Tailings The material left after a low-grade ore has been concentrated. It will contain some of the desired metal but not in economically recoverable quantities.

Talbot process The process of rolling newly cast ingots while their interiors are still molten to improve the closure of internal shrinkage cavities. After the initial rolling the ingots pass to a *Soaking pit* to allow any final solidification and temperature equalization prior to hot rolling.

Talc A naturally occurring mineral, hydrated magnesium silicate, $3MgO.4SiO_2.H_2O$, also termed **Soapstone** or **Steatite**. A very soft mineral, it is No 1 on the *Mohs* scale of hardness and is readily carved for ornamental applications. A high electrical resistance leads to its use as an insulator. In powdered form it is termed French chalk or, if perfumed, talcum powder.

Talysurf See Surface finish.

Tandem mill A *Rolling mill* in which two or more successive stands are close together with their roll axes aligned parallel so that the product travels in a straight line and enters the second stand before leaving the first. The rotational speeds of the individual stands are co-ordinated to maintain a steady movement of the product without interstage kinking or excessive tension.

Tangent modulus The elastic modulus measured over a small change in stress. It is relevant for those cases where stress is not proportional to strain so the stress–strain line in the elastic range is a curve rather than a straight line. As the term indicates, it is measured as the tangent to the line at the stress of interest. See Tensile test.

Tantalum A metallic element having a strong affinity to oxygen and forming a tenacious oxide film to provide excellent corrosion resistance in oxidizing environments. It is occasionally added as a carbide stabilizer in *Austenitic steels* and forms a hard carbide for cutting tools. See Table 15.

Tap A tool for cutting threads on the surface of a hole.

Taper section A *Metallographic* technique in which the specimen is cut so that the surface examined is at a shallow angle to the original surface. The edge of the prepared specimen thereby provides an exaggerating view of features at the original surface.

Tapping (1)   Drawing molten material from a furnace by piercing the sealed **Tapping hole** or opening the appropriate mechanism.

Tapping (2)   Forming a thread with a *Tap*.

Tarnish   Light surface discoloration due to thin corrosion films.

Taylor process   A technique for producing very fine diameter wire by hot drawing through a die with glass as a lubricant. The wire may be coated by being inserted into a glass tube or by being drawn through a bath of molten glass.

TD nickel   A form of nickel that is *Dispersion strengthened* by thoria (thorium oxide).

Technetium   A metallic element having little commercial application. See Table 15.

Teeming   Pouring molten metal, particularly steel, from a melting or holding furnace and usually via a *Ladle*, into ingot moulds.

Teeming line/height   The level in an ingot mould to which the molten metal is initially poured.

Tellurium   A toxic metalloid element. Up to about 1% may be added to *Stainless steel* and copper (termed tellurium copper), to improve machining characteristics. See Free machining and Table 15 for physical properties.

TEM   Transmission electron microscope. See Electron microscope.

Temper (1)   In many metals, but not usually steels, the various strength ranges are referred to as tempers. For example, *Aluminium* has **Temper designations** such as T4 or T6 with the higher number having higher strength. Also, terms such as half hard, quarter hard and full temper are referred to as tempers or **Temper grades**.

Temper (2)   In *Steels* this usually refers to the various meanings of *Tempering* or, less commonly, to the carbon content.

Temper bead technique   A welding technique, usually involving *Manual metal-*

lic arc welding of *Steel*, in which the *Electrode* is manipulated and other variables controlled so that as one *Bead* is deposited it tempers, i.e. softens, the underlying bead and reduces *Residual stress*.

Temper brittleness   The same as *Temper embrittlement*.

Temper carbon   Graphite nodules formed by heating certain *Cast irons*.

Temper colours   The colour of the oxide formed on steel during *Tempering*. It provides a guide to the metal temperature.

Temper embrittlement   The reduced impact strength and low *Fracture toughness* of some low-alloy martensitic or bainitic *Steels* after heating into or cooling slowly through the 300–600 °C range. The temper embrittling effect is recognized first, by the shift to a higher level of the temperature range over which the steel changes from brittle to ductile behaviour and second, by the intergranular cracking mode as opposed to the cleavage mode that is characteristic of other forms of brittle fracture in steel. The embrittlement rate is at its maximum at about 450 °C and the effect can be reversed by fast cooling from about 600 °C. However, in practice, large components cannot be cooled fast enough to avoid embrittlement without risking cracking or distortion. The embrittling mechanism is associated with the segregation of impurity elements such as phosphorus, tin, arsenic and antimony to the *Prior austenite grain boundaries*. Other elements such as manganese and silicon exacerbate the embrittling effect of the segregating elements.

Temper hardening (1)   Artificial ageing as part of a *Precipitation hardening* process.

Temper hardening (2)   *Secondary hardening*.

Tempering   In *Steels* this usually refers to the practice of reheating steel, previously hardened by heat treatment, in the approximate range 100–700 °C.

This alters the microstructure, reduces *Hardness* and *Tensile strength* but increases *Ductility* and *Toughness*. This is the most widely accepted meaning among metallurgists. However, the term is also casually in reference to the whole process of hardening steel by heat treatment.

Tempilstiks A proprietary brand of crayon available in a wide range of melting points. A material or component is marked with one or more crayons and observed during, or examined after, processing or testing to determine the temperature reached.

Temporary instability See Thermal instability.

Tenacity Tensile strength.

Tensile modulus Elastic modulus, see next entry.

Tensile strength The strength in tension. See Tensile test.

Tensile test A mechanical test in which material gripped at its two ends is pulled apart. A basic *Testpiece* comprises two ends to be gripped in the machine plus a central portion, the **Gauge length**, on which the various measurements are taken for calculation of the mechanical properties. Loading the testpiece induces a stress, an extension and a strain. These terms have specific meanings in a metallurgical context but can be confused. **Stress** is defined as the force per unit area and is calculated by dividing the applied load by the cross-section of the gauge length. The *SI* unit of stress is the Pascal (P), or the more practical MegaPascal (MPa); typical Imperial units are tons per square inch (T/in^2) or, particularly in the USA, pounds per square inch (psi) or thousands of pounds per square inch. Non-SI metricated units include kilograms per square millimetre (kg/mm^2) and Newtons per square millimetre (N/mm^2). The **Extension** is the total amount by which the gauge length of the testpiece is stretched. **Strain** is defined as the extension per unit length and is calculated by dividing the extension by the original gauge

length. It is usually expressed as a percentage although terms such as inches per inch may be used.

For most tensile tests it is conventional to plot a graph of stress against strain, often referred to as a **Stress/strain curve**, to allow calculation of mechanical properties (see Figure 39). As the load is increased progressively during a typical test the strain increases in direct proportion to the increase in stress so that, on the graph, a straight-line relationship is maintained. During this stage the specimen behaves in an **Elastic** manner such that when the load is removed all deformation disappears and the specimen returns to its original dimensions (however, see Elastic after-effect). The slope of this straight-line stage, mathematically the stress divided by the strain, is variously termed the **Elastic modulus**, the **Tensile modulus** or **Young's modulus**. At some point, the straight-line relationship will cease. This point, and the associated stress, are referred to as the **Limit of proportionality (LOP)**. With many materials a further small increase in load may induce further elastic deformation but the material soon reaches its **Elastic limit**. Increasing the stress above this level then produces a combination of elastic deformation and **Plastic deformation**, predominantly the latter. Plastic deformation, also called **Permanent deformation** or **Permanent set**, is defined as that deformation not recovered by removal of the load and is measured as **Plastic strain**. The position on the stress/strain curve at which plastic deformation commences is also loosely termed the **Yield point** but the stricter use of this term refers to the phenomenon in some materials, particularly some steels, whereby at the yield point an incremental increase in strain occurs without any increase in stress. The stress at the yield point is termed the **Yield stress** which gives the **Yield strength**. As further load is applied beyond the elastic limit and yield point there will be

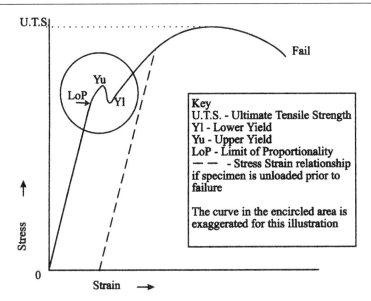

Figure 39   Tensile stress–strain curve for mild steel (annealed or normalized)

a further period during which deformation is uniform over the full gauge length of the specimen but, ultimately, at some **Maximum load**, the specimen will commence to thin locally, termed **Necking**. This marks the onset of **Plastic instability** following which the cross-section of the neck reduces rapidly, the load registered by the test equipment falls steadily and the test-piece breaks.

Engineering designs are often based on the stress at which plastic deformation will commence. In the case of many steels and some other metals the yield point is conveniently identifiable as a distinct inflection, or kink, on the stress/strain graph, indeed, in some test circumstances the line exhibits two inflections, termed the **Lower** and **Upper** yield points as illustrated in Figure 39. However, for most metals the elastic limit, i.e. the transition from elastic to plastic deformation, is indistinct as illustrated in Figure 40. It is then normal practice to determine, from the stress/strain graph, the stress required to cause a specified small amount of plastic deformation, usually referred to as a **Percentage strain** or **Offset**. This is determined from the graph by identifying on the strain axis the required level of strain and, from it, drawing a line parallel with the straight-line portion of the stress/strain curve. The point at which this parallel line intersects the curve then identifies the required stress. The strain specified is often 0.2% and the associated stress is then referred to as the 0.2% **Proof stress (PS)** or 0.2% **Offset**. Other percentages may be quoted but the differences, in practical terms, between the various levels of proof stress are small. The maximum load, divided by the original cross-section of the specimen, gives the most common measure of the material tensile strength, known by various terms including **Maximum stress (MS), Tensile strength (TS)** or **Ultimate tensile strength (UTS)**. After testing, the broken specimen is reassembled for the

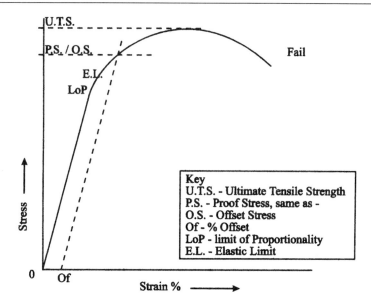

Figure 40   Tensile stress–strain curve typical of most metals, i.e., those not having a pronounced or double yield

gauge length to be remeasured. The increase in gauge length is calculated as a percentage of the original gauge length and reported by terms such as **% Elongation to failure** or, simply, **% Elongation**. Finally, the size of the area at the point of failure is measured to calculate the **% Reduction in area**. The elongation and reduction in area are measures of **Ductility**. The terminology summarized above is by far the most common in commercial practice but reference is occasionally made to terms such as **True** stress/strain or true maximum stress. The term 'true' recognizes that in the conventional tensile test all calculations are based on the original cross-section area, despite the progressive reduction in cross-section following initial yielding. Calculations which take account of this progressive reduction will provide higher values of UTS than those conventionally calculated. These are termed the 'true' values.

Tensile testpiece   Any material prepared for testing, usually until it breaks, in a tensile testing machine. The basic characteristics of the testpiece are that it is representative of the material in question and that it is of accurately measured dimensions so that the tensile strength and related properties can be calculated. Figure 41 illustrates a typical test piece. It is characteristically reduced in section over the parallel length, which includes the gauge length, to establish the necessary accuracy and to ensure that failure occurs in the gauge length rather than within the machine grips.

Tensometer   A small testing machine capable of small-scale tensile, compression and bend testing.

Terminal phase   The same as Primary phase.

Ternary   Containing three components, in particular alloy systems with three elements.

Terne metal   Alloy of, typically 80% lead, 20% tin, 0.2% antimony, see next entry.

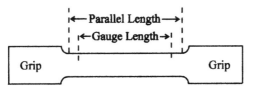

Gauge length usually: 5.65√A or 4√A
Where A = Original cross sectional area

The parallel length may be full thickness sheet, tube
wall, etc., or machined from bulk.

The grip options include:
Plain for friction or claw grips, threaded if bar
or drilled for a pin to assist friction grip

Figure 41   Tensile testpiece

Terne plate   *Steel* sheet with a coating of
Terne metal applied by dipping in the
molten alloy.

Tertiary   Third. See entry on Creep for
Tertiary creep.

Tessellated stress   The stress that develops within a material because of local
variations from point to point in some
physical property. Examples include
differences in thermal expansion or
Elastic modulus between neighbouring
phases.

Testpiece   A sample of some larger item
that has been prepared for some form of
testing to determine the properties of the
larger item.

Tetra-   Four-.

Tetrahedron   A solid object with four
faces, in particular a pyramid with three
sides and a base.

Texture (1)   The characteristics of a surface such as roughness, alignment and
sharpness of topographical features.

Texture (2)   Preferred orientation. See
Orientation.

Textural stress   Same as Tessellated
stress.

Thallium   A metallic element of high
toxicity and little commercial application. See Table 15.

Thermal activation   The increase in
temperature that provides the additional

energy input necessary to initiate and
maintain a process.

Thermal analysis   Techniques for closely monitoring the rate of change of temperature during the cooling or heating of
metals. The rate is usually plotted graphically and inflections in the curve
indicate *Phase* changes.

Thermal centre   The position within a
component such as a casting or item
being hardened which experiences the
slowest cooling rate.

Thermal conductivity   The ability to
transmit heat across a temperature gradient. All materials conduct heat by
the transmission of vibration between
neighbouring atoms and molecules. Metals also conduct energy via the cloud of
free electrons that surround the atoms in
the *Crystal lattice*, a much more effi-
cient process, hence the better thermal
conductivity of metals.

Thermal cutting/gouging etc.   Cutting
etc. by means of thermal processes
such as *Electric Arc, oxyacetylene*
which heat and melt the material. The
term includes processes in which
excess oxygen is injected to promote
cutting by combustion of the material
and by blasting away molten material
and oxide. **Powder cutting** is a variant
in which a powder is injected to

assist the process largely by a scouring action.

Thermal etching Usually the same as *Heat tinting* but where the terms are differentiated thermal etching is carried out in a vacuum, oxidation is prevented and some diffusion effects are subsequently observable.

Thermal expansion The increase in dimensions associated with an increase in temperature.

Thermal fatigue Damage, particularly cracking, resulting from thermal cycling. This form of damage is differentiated from repeated *Creep* loading which is seen as the simple loading and unloading of a component at a high temperature. The critical feature of thermal fatigue is that at some position the normal thermal expansion (less commonly, contraction) is constrained by neighbouring material that has not expanded to the same extent because of a difference in temperature or, less commonly, a difference in coefficient of expansion. This constraint induces localized stresses. In the simplest case the stresses remain within the elastic range and disappear when the temperature differential is removed. The result is a simple *Fatigue* cycle leading to fatigue cracking; some damage may occur because of the high temperature but, assuming a short duration, it can be regarded as hot tensile damage rather than creep in its usual sense. In more severe circumstances the temperature differential is sufficient to induce stresses that exceed *Yield*. In many practical cases the situation is one where a small volume of material, perhaps an edge or corner, rises rapidly to its high operating temperature but its expansion is constrained by the large underlying volume which is slower to heat up. The yield strength falls as the temperature rises and, as a result of these two effects, the high-temperature surface material yields in compression. When the underlying material eventually reaches the operating temperature, the crushed surface reverts to a

tensile stress. This tensile stress remains active at the operating temperature, causing *Creep*. The time spent under load and at temperature during each cycle is termed the **Dwell period**. More severe temperature gradients or combinations of thermal and imposed service loads can cause more complex cycles of stress and plastic deformation. In some cases material may experience both plastic compression and plastic extension during each thermal cycle; in other cases the component may progressively extend or compress over a series of cycles, sometimes termed **Ratcheting** or **Shortening** (see Copper shortening). In addition to the effects of the thermomechanical cycle, thermal fatigue is often accelerated by corrosion and oxidation effects. Such attack may help both crack initiation and propagation by, for example, preferential oxidation at grain boundaries or by forming stress raising, sharp tipped cracks in oxide films. Depending upon the relative contributions of the thermomechanical and the corrosion effects the attack may become a form of high-temperature *Corrosion fatigue*.

Thermal instability The tendency to distort as the temperature is raised. The phenomenon results from various effects and can have various consequences. For example, small variations in thermal expansion coefficient associated with differences in composition or microstructure within a component may induce distortion as the temperature changes. Typically, the distortion will repeatedly develop and disappear as the temperature rises and returns. The phenomenon is difficult to eliminate so it is sometimes termed **Permanent instability**. In other cases a component will contain significant levels of *Residual stress* so that on heating it experiences *Creep relaxation* or even *Yield* if the yield strength falls with the temperature rise. Some or all of the resultant dimensional change does not disappear on cooling, it becomes **Permanent distor-**

tion. However, when the component is heated again there will be less, or even no, further distortion so the phenomenon may be termed **Temporary** or **Transient instability**. In such cases stability at the normal operating temperature can often be achieved by appropriate heat treatment at a temperature above that expected in service, a process referred to by various terms including **Stabilizing heat treatment**, **Thermal stability soaking** or *Stress relief*.

Thermal paint/crayons/strips Materials that show some physical change, usually a colour change, as they reach some predetermined temperature. Series of such materials, covering a wide range of temperature, are available for use as a form of temperature measurement. One proprietary form is the *Tempilstik*.

Thermal severity number (TSN) A means for quantifying the cooling capacity of a joint to be welded. Satisfactory welding of *Steel* is critically dependent on avoiding excessive cooling rates otherwise hardening and cracking will occur. The cooling rate is directly dependent on the cross-section of the material conducting the heat away and the TSN is a measure of the total of the available conduction paths in units of some arbitrary but convenient size, often $\frac{1}{4}$ in. For example a *Butt* joint between two pieces of plate $\frac{1}{4}$ in. thick would have a TSN of 2 (two conduction paths each one unit), and a *Tee* joint between 1 in. plates would have a TSN of 12 (three conduction paths each four units). The TSN is correlated with other inputs to indicate any requirements for preheating the joint (see Weldability). Also see Controlled thermal severity test.

Thermal shock Effects, particularly high stress and resultant cracking or, less commonly, *Spalling*, resulting from severe temperature gradients with constraint of the associated thermal expansion or contraction. The term is often used interchangeably with *Thermal fatigue*, which results from broadly similar factors. Where they are differentiated,

thermal shock usually implies a relatively small number of thermal cycles and the development of damage immediately the temperature change is imposed.

Thermal stability See Thermal instability.

Thermal straightening Same as Hot spot straightening.

Thermal stress A stress produced by restraint of thermal expansion or contraction.

Thermic lance A form of *Oxyacetelene cutting* in which the oxy-acetylene flame is formed at the tip of a steel tube and directed at the item to be cut. The tube is progressively burned away and the heat from this reaction adds to the normal heat from the oxy-acetylene flame and from any combustion of the material being cut. In some cases further heat may be developed by inserting steel rods or similar material into the tube.

Thermionic Concerned with the emission of electrons by material at high temperature.

Thermistor A contraction of **Thermal resistor**, a semiconductor, the electrical resistance of which varies considerably and progressively with temperature.

Thermit(e) reaction/welding The Thermit process utilizes the powerful exothermic reaction resulting from ignition of finely divided aluminium powder mixed with the oxide of a metal such as iron, chromium, molybdenum or manganese. The reaction oxidizes the aluminium, reduces the ore to its metal and the heat released is sufficient to melt even high melting point metals. The process can be used to produce such metals but it is better known for its use in Thermit welding. In a typical application, joining railtrack, the reaction between the aluminium powder and the iron oxide is initiated in the upper chamber of a refractory mould and superheated molten iron flows down into a lower chamber to fill the gap between, and fuse together, the rail faces.

Thermocouple A temperature-measuring device comprising wires of two different metals joined to each other at both ends. It produces an electromotive force approximately proportional to the difference in temperature between the two joints. Calibrated and with circuitry to measure the emf it is widely used for temperature measurement in industry and technology. See Thermoelectric effect.

Thermodynamics The study of the interaction between heat and mechanical energy in any process.

Thermoelectric cooling The extraction of heat by the Peltier effect (see next entry).

Thermoelectric effect(s) This term is used as a synonym for the Seebeck effect or as a collective term for all the following effects. The **Seebeck effect** is the electromotive force (emf) produced by the difference in temperature between two junctions of dissimilar conductors in a circuit. The **Thomson** (alternatively the **Kelvin**) **effect** is the emf produced by a difference of temperature between two sites of a conductor or, alternatively, the release or absorption of heat associated with a electrical current flowing between two sites on the conductor which are at different temperatures. The **Peltier effect** is the release or absorption of heat as an electric current crosses the junction from one conductor to a dissimilar one.

Thermogalvanic corrosion *Electrochemical* corrosion mechanisms in which a primary factor is the difference in temperature between the anode and the cathode. The colder is usually the anode and corrodes.

Thermometal An alternative term for *Bimetallic* strips.

Thermometer Any instrument for measuring temperature.

Thermonuclear Usually a nuclear reaction in which a dominant factor is the impact of particles having a high kinetic energy such as in the hydrogen bomb.

Thermoplastic Becoming plastic upon

heating and returning to a rigid state on cooling. The transition usually develops in the range 80–120 °C.

Thermosetting Materials which, when first heated, can be moulded and then 'cure', undergoing a chemical change to become rigid. Subsequent heating does not cause softening.

Thermospray A form of metal spraying in which powdered metal or other material is injected into a high-temperature flame impinging on the surface to be coated.

Thick film lubrication The formation of a substantial film of lubricant between the faces of a plain bearing. This is the usual design intent. See Oil wedge.

Thin film examination The examination in the transmission *Electron microscope* of thin films of material. The films are prepared by cutting, grinding and then *Electropolishing* the material to be examined until it is thin enough to transmit the electron beam.

Thin film lubrication The formation of a thin, perhaps monomolecular, oil film. This is, in practice, not normally regarded as satisfactory for long-term operation as intermittent metal-to-metal contact is likely.

Thin nut See Lock nut.

Thixocasting, thixoforging See Rheocasting.

Thixotropic The characteristic of a material whereby its viscosity is dependent on the shear rate. In practice the more rapidly the material is stirred, the more fluid it becomes. The process is reversible so that when stirring ceases the material becomes more viscous as in the case of non-drip paint. See also Rheocasting.

Thomas converter Similar to Bessemer converter.

Thomson effect (alternatively, the **Kelvin** effect) The emf produced by a difference of temperature between two sites of a conductor and, conversely, the release or absorption of heat associated with a electrical current flowing between two sites on the conductor which are at different temperatures. See also Thermoelectric effect.

Thoria Thorium oxide, ThO$_2$. Small amounts are added to tungsten for lamp filaments to restrain grain growth during long-term high-temperature service.

Thorium A radioactive metallic element. About 1–2% is added to tungsten to improve electron emission in applications such as TIG welding electrodes and thermionic valve filaments. See Table 15.

Thread The helical ridge and groove around a *Threaded fastener.*

Thread rolling The process of forming a *Thread* by rolling the component between dies. The result is less accurate than machining but may introduce favourable compressive *Residual stresses.*

Threaded fasteners Devices such as bolts, studs and screws and the associated internally threaded nuts or holes which allow components to be clamped together. See Preload.

Three-high mill See *Rolling mill.*

Three-point bending A bend test in which the testpiece is supported at two points and loaded at a point midway between them. The stress peaks at the central loading point unlike *Four-point bending.*

Throat thickness (of fillet weld) The thickness of weld metal measured from the point of maximum penetration at the root to the outer surface. See Figure 51, located at entry on Welding terminology.

Through weld A weld made through a previously intact component to another component beneath.

Throwing power The ability of a electroplating solution to apply a uniform deposit on irregular and recessed shapes. Compare with Covering power.

Thulium A metallic element, one of the *Rare earth* group, with little commercial application. See Table 15.

Tide marks Same as Beach marks.

Ties, tie rods Bars, beams, tubes or similar components carrying significant tensile loads in a construction or assembly.

Tie line See lever rule.

TIG Tungsten inert gas.

Tilt boundary An interface between two grains or the sub-grains of the same grain where the angular misalignment is very small so that the boundary is little more than a single plane of *Dislocations.*

Tilt casting/moulding Casting processes in which the mould is initially tilted to allow the molten metal to be poured gently onto a sloping side. As the mould fills it is progressively rotated to its vertical position. The process is intended to eliminate turbulence which might entrain *Dross.*

Tilt hammer A primitive forge in which a centrally pivoted beam carries the tup (hammer) at one end and the other end is struck repeatedly by an eccentric on a rotating shaft. The eccentric raises the hammer which then falls freely to strike the workpiece on the anvil. The drive could be provided by various sources— often water wheel or steam but even animal power.

Time delay (of *Yield* in some *Steels*) The phenomenon whereby rapid application of a load above the yield stress does not induce immediate yield. This is because a period, typically a few milliseconds, is required for *Dislocations* to break away from the lattice features, such as interstitial atoms, to which they are pinned.

Time, temperature, transformation diagrams/curves Same as Isothermal transformation diagram.

Time quenching Same as Martempering.

Tin A soft corrosion-resistant metallic element. It is allotropic, changing from a fairly ductile body centred tetragonal structure to a low-ductility diamond structure on cooling below about minus 20 °C, although 13.2 °C is the actual equilibrium temperature. The higher-temperature allotrope is sometimes termed **White tin** and the lower-temperature allotrope **Grey tin**. The damaging formation of the low-temperature phase during service is referred to as **Tin pest**. Tin has good corrosion resistance and is used as a protective coating on steel for canning foodstuffs and other purposes. The tin offers protection only by excluding the environment. Unlike zinc, it does not offer sacrificial protection so if a tin coating is scratched the exposed steel corrodes. See Electrochemistry and Galvanizing. It is a major constituent in *White metals* and *Soft solders* and is a minor alloying element in some *Brasses* and *Bronzes*. See Table 15 for properties.

Tin can   A can made from tin-plated *Steel*.

Tin cry   The creaking noise produced by tin and its alloys during deformation.

Tin foil   Very thin sheet of tin or high-tin alloy. The term is also casually but incorrectly applied to aluminium foil which is commonly used for wrapping foodstuffs and similar applications.

Tin–nickel plating   An electroplated coating of nickel and tin, usually on a brass substrate, giving a highly reflective, corrosion-resistant and low-friction surface.

Tin pest   See Tin.

Tin plate   Steel sheet with a coating of tin usually applied by running the steel, rolled to final size, through a bath of molten tin beneath a suitable *Flux*.

Tin sweat   The formation of tin-rich beads on the surface of cast **Bronze** due to inverse *Segregation*. It may also occur if bronze is heated close to its melting point.

Tin–zinc plating   An electroplated coating of an alloy of tin and zinc, typically about 75% tin, 25% zinc. The coating on steel offers corrosion protection close to that of zinc alone (see Galvanizing) but is more readily *Soft soldered*.

Tinman's solder   A *Soft solder* of eutectic composition.

Tinning   Applying a (thin) *Tin* or *Solder* coating, usually prior to making a soldered joint.

Tinsel   A decorative lustrous material usually as thin strands or foil. It was originally an alloy of about 60% zinc

and 40% lead but is now usually some plastic material.

Titania   Titanium oxide, TiO_2. It is used as white pigment for paints and is a common constituent in the coating of arc welding *Electrodes*, in which case it is more usually termed 'rutile'.

Titanium   A corrosion-resistant metallic element with a density between the light metals, aluminium, etc. and steel. It reacts readily with oxygen but forms an adherent impervious oxide which confers excellent corrosion resistance. It forms various alloys, some of considerable strength (see Table 12).

Toe   (of a weld)   Usually this term refers to the line on an exposed surface where the boundary of the deposited weld metal meets parent metal. Less commonly, it can refer to any line where the surface boundary of a weld run meets parent metal or a previous weld run. It will be seen that this includes positions where the toe has been buried by a subsequent deposit. In most cases references to 'toes' will imply the first definition but sometimes the terms **Primary toes** and **Secondary toes** are used to refer, respectively, to the first and second definitions. See Figure 50, located at entry on Welding terminology.

Toe crack   A crack commencing at the *Toe* of a weld and running into parent metal.

Tolerance   The permissible variation in dimensions.

Tool steel   Any steel capable of being hardened for use as a tool. Such steels range from plain carbon steels used for hand files to very highly alloyed steel used for high-temperature cutting and other working operations.

Top hat furnace   See Bell furnace.

Top pouring   The normal process of pouring molten metal into the mould from the top. This allows solidification to commence at the bottom and has the advantage of simplicity. However, it has the disadvantage that it may induce turbulence, splashing, erosion of the mould and entrainment of dross.

Topaz   The naturally occurring mineral, aluminium fluosilicate, $Al_2F_2SiO_4$ having various colours depending on the individual impurities. It hard, No. 8 on the *Mohs'* scale.

Torch   See Blowpipe.

Torque   The force of rotation, the product of force and the perpendicular distance between the axis of rotation and the direction of the force.

Torr   A non-*SI* unit of pressure used for measuring near vacuum conditions. 1 torr = 1 mm Hg = 133.322 Pa.

Torsion   Twisting. The state of *Strain* induced by *Torque*.

Total carbon   In *Cast iron*, the sum of combined carbon and free carbon.

Touch welding   *Metal arc welding* in which the tip of the coating on the consumable *Electrode* remains in contact with the parent metal during the welding operation. The electrode has a coating formulated to form a cup standing proud of the metal core during welding to avoid core to parent contact.

Touch pitch copper   See Copper.

Toughness   An imprecise term encompassing the capacities to absorb energy prior to and during the cracking process, to deform plastically without cracking and to resist crack growth. It implies resistance to impact and hence the results of *Impact* tests may be referred to as 'toughness'. A measure of toughness is also provided by the area under the curve plotted from a *Tensile test*. Also see Fracture toughness.

Tracer techniques   The use of small quantities of radioactive materials to monitor processes such as corrosion, wear, diffusion, pollution, etc.

Tracking   The breakdown of electrical resistance via a narrow track across a surface of an insulator. It usually results from some external contamination.

Tramp elements   Impurity elements in a metal, usually undesirable, introduced at the casting stage by *Scrap* metal.

Tramp materials   Any contaminating materials entrained in a process and undesirable because they might enter the product or damage the production equipment.

Transcrystalline/granular   Running across the crystal/*Grain*, for example a crack running fully or partly across a grain rather than around the grain boundary. Contrast with Int*ra*granular which refers to features such as precipitates within a grain.

Transducer   A material capable of converting electrical energy to mechanical energy or vice versa.

Transferred arc process   See Plasma welding.

Transformation   The change from one *Phase* to another at the **Transformation temperature** or over the *Transformation range*.

Transformation range   The range over which a *Phase* change occurs, in particular the temperature range over which austenite changes to ferrite structures on cooling and vice versa on heating. See Steel.

Transformer steel   A steel having low magnetic hysteresis and hence used as sheet for the laminations of electrical transformers and generator stators. Such steels are typically very low carbon with 2–5% silicon and are carefully cold rolled to maximize their magnetic characteristics.

Transient (stage) creep   The primary stage of *Creep*.

Transistor   A semiconductor that is capable of amplifying electrical current.

Transition lattice   The localized, distorted and unstable *Crystal lattice* formed at an intermediate stage in a phase change.

Transition joint (weld)   A weld formed between two different metals.

Transition metals   The elements, all metals, usually shown as a block in the Periodic Table between Groups II and III. They arise because their atomic structure includes incompletely 'filled' inner sub-shells, See Atomic structure and Periodic Table.

Transition temperature   Any temperature at which a transition occurs but in particular the temperature at which steel changes from *Ductile* to *Brittle* behaviour. See also Fracture appearance transition temperature.

Translation   The relative movement as one block of atoms slides across another.

Transmission electron microscope   See Electron microscope.

Transuranic/transuranium elements   The man-made elements having an atomic number greater than the 92 of uranium.

Transverse test (bend, tension, etc.)   A *Test* across the primary axis of the material or across the line of a weld.

Treeing   Localized excessive deposition of electroplate due to high current density in the area.

Trepanning   Boring a hole by cutting away only a narrow circumferential band rather than the full cross-section. This releases a central core when the process is completed to the far surface.

Triaxial stress   The state in which none of the three *Principal stresses* is zero.

Tribology   The study of surface contact phenomena, including friction, wear and lubrication.

Trichloroethylene   A volatile, nonflammable hydrocarbon used as a solvent and degreasant.

Tridymite   An *Allotrope* of quartz. See Quartz.

Triple point (1)   The point in a system where three phases are in equilibrium.

Triple point (2)   The point, observed on the surface of a *Metallographically* prepared section, where the boundaries of three grains meet.

Tritium   Hydrogen with three neutrons in the nucleas. See Atomic structure.

Troostite   A largely obsolete term. It was previously used in reference to a structure of fine, dark etching unresolvable (in the light microscope) carbide and ferrite. See Steel.

Tropenas converter   A *Converter* in which air, possibly oxygen enriched, is blown from the side across the charge of previously desulphurized molten pig

iron and scrap. The process is used for steel castings of relatively low quality.

True stress/strain   See Tensile test.

Trumpet alloy   A brass with about 20% zinc, 1% tin, remainder copper, used for musical instruments.

TSN   See Thermal severity number.

TTT diagram   Time, temperature, transformation diagram. See Isothermal transformation diagram.

Tube   A pipe or other hollow section. The various terms are largely interchangeable and usage is dependent on the particular industry but 'tube' would usually indicate a fairly small diameter and a simple cross-section.

Tube making   The basic tube making sequences can be categorized as:
(1) Piercing and subsequent processes
(2) Welding of rolled sheet or plate
(3) Extrusion.

Piercing sequences commences with hot piercing, following by hot reducing and cold reducing. In the **Press piercing** process a heated cylindrical *Billet*, usually set in a die, has a tool forced along its central axis to fully penetrate it or nearly so. Alternatively, the billet may be fully pierced in a **Barrel** or **Mannesman rotary piercer**. This comprises a pair of barrel-shaped rolls set not quite parallel and rotating in the same direction, see Figure 42. The heated billet is introduced, end-on, into the end gap of the rolls. The rotation of the rolls spins the billet and their angled alignment draws it in. At the same time the barrel taper compresses the spinning billet across its diameter which tends to open up a zone of weakness at the billet centre line. A **Piercing point** carried at the tip of a *Mandrel* is located part-way along the roll gap to coincide with the zone of weakness and the forward motion of the billet thrusts the weakened centre over the piercing point to form a crude, thick-wall tube. A freshly preheated piercing point is usually fitted for each billet. The mandrel serves only to locate the point, it does not support or control the bore.

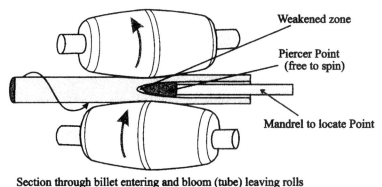

Weakened zone

Piercer Point
(free to spin)

Mandrel to locate Point

Section through billet entering and bloom (tube) leaving rolls

Figure 42   The rotary (Mannesman) piercing process

The **Disc** or **Stiefel piercer** processes are similar in concept and action except that the rolls are mushroom-shaped and set with their axes near-parallel, their slight angular alignment inducing billet progress. The **Three-roll piecing** process is again similar to the Mannesman mill except that three rolls are used, the central weak zone is less pronounced and the billet is effectively forced over the piercing point. Fully pierced billets are usually termed **Blooms** or **Hollows** and near-pierced billets are termed **Bottles**.

After piercing the bloom is **Hot reduced** by one of various processes including hot rolling, the push bench, the Pilger mill and the Assel mill. In **Hot rolling**, the blooms or bottles, sometimes reheated, are passed through grooved, driven rollers with a *Mandrel* set in the bore to provide support and to control its size. The **Push bench** process is applied to bottles. Immediately after piecing and without further heating the bottle is set on a mandrel to be pushed through a series of dies and/or non-driven rolls of progressively reducing diameter. The **Pilger mill**, also termed the **Rotary forge**, comprises a pair of rolls with deep tapered grooves (see Figure 43). The rolls have parallel axes

and run in opposite directions with their faces in contact so that as the rolls rotate the pair of grooves form, at the contact line, a orifice of varying diameter. The rolls rotate continuously and, when the orifice diameter is at its maximum, the heated bloom, carried on a mandrel, is thrust forward into the rolls against the direction of rotation. The rolls then force the tube backwards and the taper on the rolls forges a small length of the tube, reducing its exterior diameter. This sequence is repeated with the tube being rotated 90° on each forward movement and the tube moves progressively through the mill. The **Assel mill** is similar to the Mannesman piercer except that three rolls are deployed at 120° and they usually have a well-radiused step at mid-length to forge the heated bloom as it passes through on a mandrel. The Assel mill offers better dimensional control, in particular less variation in wall thickness, than the Pilger mill. Cold reducing is commonly achieved by a drawing process in which the tube is pulled through a die. Usually the bore is controlled by a mandrel or by a *Plug* located in the tube bore at a position coinciding with the die. Drawing without use of a mandrel or plug is termed **Tube sinking**, or, in the industry, just

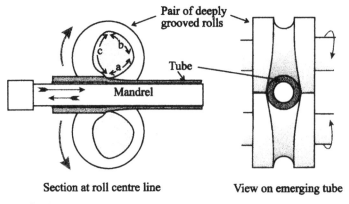

Section at roll centre line   View on emerging tube

a- forging taper
b - parallel length
c - clearance

The rolls rotate continuously. The tube bloom, on a mandrel, is pushed
forwards incrementally but can be thrust backwards against a spring.
At the position shown the rolls have just engaged the bloom.
They thrust it backwards as the taper on the rolls forges the tube.
The parallel portion of the rolls then smooths the previously forged surface.
When the rolls reach the clearance position the tube is rotated 90° and
advanced more than the back thrust.
The sequence repeats until the full length has been forged.

Figure 43   The Pilger process for the hot forging of tube

Sinking. Multiple drawing passes may
be made with, if necessary, interstage
annealing. An alternative is the **Cold
reducing process**, also termed the
Rockrite process, which is similar in
concept to the Pilger described above in
that it uses rolls with tapered grooves to
produce a forging action. However, in
this case (see Figure 44) the groove
diameter is much smaller in proportion
to the roll size and the pair of rolls
reciprocates along a track changing rotation direction at the extremities of each
stroke. The tube, on a tapered mandrel,
is fed in at the point of maximum opening at the end of a stroke. It is not
pushed backwards by the rolls but is
rotated slightly at each forward step to
produce an even product. Large reductions of external diameter, bore and wall
thickness can be achieved with good

control of dimensions including eccen-
tricity. The Rockrite process may, therefore, follow the Assel mill in the
production sequence for tubes requiring
tight dimensional control.

Welded tube is produced from rolled
sheet or plate, cut to width and rolled
into the tubular form and then welded.
The seam is usually longitudinal but
may be spiral. Virtually any forms of
welding can be employed depending
upon the tube size and application but
for higher-quality tubing in smaller size
ranges **Electrical resistance welding** is
popular.

See the entry of Extrusion for details
of this common route for tube production. Often extruded tube receives no
further working but in some cases one
or more drawing passes may be required
for sizing or other reasons.

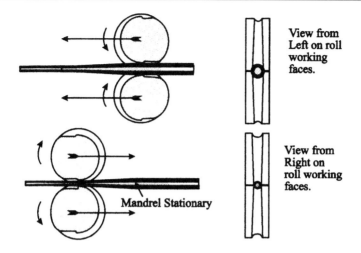

View from
Left on roll
working
faces.

View from
Right on
roll working
faces.

Mandrel Stationary

The rolls reciprocate along a toothed track, revolving about half a
revolution.
The diagrams depict the rolls at the extremities of the strokes.
The tube moves forward incrementally and turns at the ends of the strokes.
The tapered mandrel remains stationary.
On forward strokes the tapered roll working faces forge the tube.

Figure 44   The cold reducing process for tubes

Tube manipulation   Various process applied to tubes ends are illustrated in Figure 45.

Tube plate   The perforated plate into which the tubes of a heat exchanger are sealed by some process such as *Expanding*, *Welding* or *Soft soldering*.

Tuberculation   Forms of corrosion in which the corrosion product forms prominent adherent growths or tubercles on the affected surface.

Tumbling   A process in which components are enclosed in a rotating barrel to be treated by some other material enclosed with them. The other material may be an abrasive, polishing or plating agent or peening shot—hence terms such as **Tumble polishing**, **Tumble peening**, **Tumble plating**, etc. If both plating and peening agents are included the process may be termed *Peen plating*.

Tungsten   A dense metallic element still occasionally referred to as **Wolfram**. It has a high melting point and hence is used in a near-pure form for the filaments of electric lamps and similar applications. In steels, it forms carbides stable to high temperatures and hence the steels are used for high-temperature cutting and hot working applications. See Table 15 for properties.

Tungsten inert gas welding (TIG)   Welding in which the *Weld zone* is shielded by a gas that is inert during the welding operation and the electric arc is struck between the component being welded and a non-consumable tungsten electrode.

Tup   The moving head of a forge, in particular a drop forge.

Turning   A machining operation in which the workpiece is gripped in the chuck of a lathe and rotated against a tool bit which cuts away material.

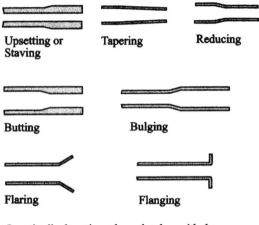

Upsetting or Staving   **Tapering**   **Reducing**

Butting   **Bulging**

Flaring   **Flanging**

Longitudinal sections through tubes with the manipulation on the right hand end

Figure 45   Tube end manipulations

Turnings   The *Swarf*, i.e. strands or chips of metal removed by *Turning*.

Tutenag   Either commercially pure zinc originating in China and neighbouring countries, or alternatively an alloy of copper, zinc and nickel similar in composition and appearance to *Nickel silver* and used for semi-decorative domestic ware such as firegrates.

Twin   An arrangement of the *Crystal* *lattice* in which the planes of atoms on adjacent blocks in a crystal are aligned in a mirror image of each other (see Figure 46). Under the microscope the intersection between the blocks appears as a line. **Annealing twins** can occur during annealing of metals having a *Face centred cubic* structure. In this case the individual twin lines in a pair are straight, run from grain

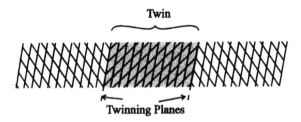

Twin

Twinning Planes

On this wireframe model the atoms lie at the grid intersections. The crystal structure is the same throughout but lies in mirror image either side of the twinning planes.

Figure 46   Wireframe model of a section through a twin

boundary to grain boundary and may be well separated. **Mechnical twins** can be formed during working of *Body centred cubic* and *Close packed hexagonal* structures. Microscopically these twins appear within a grain as a pair of closely spaced, curved lines joined at their ends in a lenticular, ◊, formation. Mechanical twins, formed in ferritic *Steel* failing in a brittle manner, usually as a result of high rates of loading, are termed **Neuman bands** or **Neuman lamellae**.

Twist drill A *Drill bit* with a pair of cutting edges formed by the junction of the helical flutes and the conical tip. The function of the flutes is to lead away the *Swarf* as the drill penetrates.

Type metals A range of low melting temperature alloys of tin, lead and antimony for casting printing type. See White metal and Babbitt metal.

Ugine–Séjournet process Same as Séjournet process.

Ultimate load (in *Brinell* test) The load that will just force the Brinell ball to a depth of half its diameter.

Ultimate tensile stress/strength The maximum *Stress* a material can support. See Tensile test.

Ultra-High tensile steel An imprecise term usually taken to mean steels with tensile strengths in excess of 100 tons/in^2 or 1500 MPa.

Ultra-light metals See Light metals.

Ultrasonic Sound waves at frequencies above those audible to the human ear. They are readily transmitted through solids and can be used to detect sub-surface defects. See Non-Destructive Testing and also the following entries.

Ultrasonic cleaning A process for cleaning components by immersing them in a bath of liquid which is subjected to ultrasonic energy to assist release of surface contamination.

Ultrasonic machining Processes in which abrasive material is ultrasonically activated by a tool or former to abrade the workpiece.

Ultrasonic soldering The use of an *Ultrasonic* beam to disrupt the normally tenacious oxide film, such as that on aluminium, which would otherwise impede bonding between the underlying metal and the *Soft solder.*

Ultrasonic welding A welding process utilizing the heat energy derived from the high-frequency mechanical excitation of the intended joint interface. The vibration, typically in the range 20–

40 kHz, is generated electrically with available power of 200–3000 W and is delivered to the weld zone by the machine head, sometimes termed a **Sonotrode**. An additional static interfacial force is normally applied and additional external heating is an option.

Under-annealing Generally, in a pejorative sense, annealing at a temperature or for a time insufficient to induce full softening. The term is also used in a more specific sense when referring to the deliberate annealing of *Steel* in the transition range so that the pearlite areas and a proportion of the ferrite areas are recrystallized.

Underbead crack A sub-surface crack in the *Heat affected zone* of a weld. See Figure 52, located at entry on Welding terminology.

Undercooling Same as Supercooling.

Undercut (1) A *Re-entrant* detail on a casting.

Undercut (2) A groove at the *Root* or *Toe* of the weld. See Figure 52, located at entry on Welding terminology.

Undercut (3) A localized reduction in some machined dimension, intentional or otherwise. If unintentional, the term may be used in the sense of a *Stress raiser.*

Underfilling (of a weld) An insufficiency of weld metal in a joint. See Figure 52, located at entry on Welding terminology.

Underflushing (of a weld) See flushing.

Understressing In *Fatigue*, the deliberate application of one or more cycles at a stress below, but usually close to, the

Fatigue limit. This can have a *Strain ageing* effect improving the fatigue properties and life.

UNE, UNO, UNQ, etc. In an attempt to avoid disputes over who or what should be honoured it has been proposed that man-made elements, particularly those discovered in future, be named in terms of their atomic number in pseudo-Latin. Hence, element 104 is un–nil–quadum i.e. '**Unnilquadum**' shortened to UNQ and so on.

Uniaxial Acting or aligned along one axis of a crystal or a component.

Unit cell The arrangement of the smallest number of atoms that, repeated, forms a *Crystal lattice.*

Universal mill A rolling mill having pairs of both horizontal and vertical rollers and hence capable of working all surfaces of the section.

Unsoundness This term usually implies volumetric defects such as internal cavities, voids and porosity as opposed to planar defects such as crack or *Laps.* Nevertheless, a product with a crack would still be termed 'unsound'.

Upper shelf See Brittle.

Upper yield See Tensile test.

Upset/upsetting A local increase in cross-section area resulting from a longitudinal force. The term is commonly used in operations where the deformation is deliberate and a die is used to shape the upset, for example the formation of a bolt head. The process is beneficial as it produces a desirable flow, i.e. pattern of deformation, and it reduces the amount of machining. For upsetting of tube ends see Figure 45 located under Tube manipulation.

Upset butt welding Same as Resistance butt welding.

Uranium A toxic metallic element having the highest atomic number, 92, of all the naturally occurring elements. It is *Allotropic*, having an alpha orthorhombic phase stable to 660 °C, a beta complex tetragonal phase stable from 660 °C to 722 °C and a gamma body centred cubic phase stable from 722 °C to the melting point at 1130 °C. In its naturally occurring form it is composed of three isotopes mainly U_{238} with 0.7% of U_{235} and 0.008% of U_{234}. In a nuclear reactor the U_{235} undergoes *Fission*, i.e. the atoms are 'split' as a result of neutron bombardment, releasing heat and the U_{238} accepts neutrons to form plutonium. See Table 15 for properties. Apart from its nuclear applications uranium has some commercial applications as **Spent uranium**. This is material with little remaining radioactivity used for shielding purposes or for its high density, for example the keel of racing yachts.

UTS Ultimate tensile strength. See Tensile test.

V segregation See Segregation.

Vacancy An unoccupied atomic site in a *Crystal lattice*.

Vacancy jump The movement of atoms in substitutional solid *Solution* into a vacant site. This can be regarded as a movement of the vacancy in the opposite direction.

Vacant site Same as Vacancy.

Vacuum A space that is devoid of matter. Loosely, a pressure significantly less than atmospheric.

Vacuum arc melting/remelting (VAM/VAR) A purification process for metals such as *Steel* and *Titanium* in which a DC electric arc is struck between an electrode of the impure metal and the water-cooled copper mould over which it is suspended. The impure electrode is progressively melted and a new ingot solidifies in the mould. A high vacuum avoids oxidation and removes volatile elements. The process is deliberately slow with minimum turbulence so impurities float to the surface of the molten metal.

Vacu-blasting A *Grit blasting* process in which the grit or shot, after impacting the target component, is recovered by vacuuming for re-use.

Vacuum deposition A process in which material, in a vacuum, is evaporated and then condenses on a component forming decorative or functional coatings.

Vacuum furnace Any furnace in which the charge is treated under *Vacuum*. In melting furnaces the vacuum extracts dissolved gasses as well as protecting from contamination. In heat treatment furnaces it prevents oxidation, carburization and contamination.

Vacuum plating A form of *Vacuum deposition* in which some reaction is involved such as between the surface being treated and some vapour introduced into the vacuum chamber.

Vacuum treatment/melting Treatment of solid or molten metal under vacuum. Vacuum treatment of solid metals, usually at high temperature, assists removal of dissolved gases. Vacuum treatment of molten metal removes gases and, depending on the vacuum, other contaminants. Bismuth, calcium, magnesium and zinc and lead are removed from molten steel at about 10 kPa, antimony at about 100 Pa and phosphorus sulphur and arsenic at 1 Pa.

Valence electrons The electrons in the atom's outer shell, the number of which establishes the *Valency*.

Valency A measure of the proportions in which atoms combine. The valency of any element is that number of atoms which will combine with or replace one atom of hydrogen. It is controlled by the number of electrons in the outer shell of the atom, the so-called *Valence electrons*, these being the only electrons available for interactions with other atoms. See Interatomic bonding.

Vanadium A metallic element used mainly as an alloying element. In *Steel* it acts as strong deoxidizer, improves *Hardenability*, forms carbides stable to high temperatures and promotes *Secondary hardening*. See Table 15.

Van der Waal's bonding/forces See Interatomic bonding.

Vapour blasting A cleaning process in which components are scoured by an aqueous suspension of abrasive entrained in a high-velocity airstream.

Vapour degreasing The exposure of components to a vapour released by degreasants, in particular chlorinated hydrocarbons.

Vapour phase inhibitors Substances which release vapours that inhibit corrosion on neighbouring metal surfaces in moist conditions. Various materials, often impregnated into paper, are used but they tend to be material-specific, protecting some metals but not protecting others or even promoting their attack.

Vapour pressure The measure of the tendency of a material to release molecules to its surroundings. In liquids the phenomenon is readily recognized as evaporation but similar effects occur more slowly in solid materials.

VAR Vacuum arc remelting. See entry.

Veins, veining The fine lines on a *Metallographic* specimen formed at the boundaries of *Sub-grains*.

Verdigris The green corrosion product formed on *Copper* exposed to the atmosphere. It is usually copper sulphate but may also contain chlorides and carbonates depending on local pollution. It is sometimes termed 'patina'. Historically, the term was used for green pigments, particularly basic copper acetate.

Vermiculite A mixture of hydrated silicates of aluminium, iron and magnesium. On heating it expands and fragments producing the low-density material commonly used as thermal insulator and for other applications.

Vertical welding position Any position in which the weld run is approximately vertical. This is usually taken to mean a *Weld slope* between 80° and 90°. See also Weld position.

Vibratory stress relief The practice of submitting structures containing residual stresses, in particular welds, to controlled programs of vibration. It is claimed, not without some opposition, that this induces a reduction in residual stress without causing significant damage.

Vickers (hardness) test A *Hardness* test utilizing a diamond indentor which is thrust into the component by a known load and the size of the indentation provides a measure of the hardness. The standard indentor is a square-based pyramid with a 136° included angle and, on the standard machine, weights are hung on the machine to induce, via a lever system, loads of up to 120 kg at the indentor. After a few seconds application the load is removed and the indentation measured by a suitable microscope usually attached to the machine. The hardness number can then be calculated as load per unit area of impression in kg/mm from

$$H = P \times \frac{1.8544}{d^2}$$

where P is load in kg and d is the diagonal of the impression in millimetres. The hardness is independent of the load but, in practice, to keep the impression size within practical limits and to maximize accuracy, the load is matched to the material, typically 30 kg for steels and 10 kg for many non-ferrous metals. Tables for standard loads relate diametral measurements to hardness. Vickers Hardness results are commonly reported as Hardness-Vickers, abbreviated to **HV**, or as Hardness Diamond (**HD**) or Vickers Pyramid Number (**VPN**). For example, $212 \, HV_{30}$ (or HD_{30} or VPN_{30}) i.e. indicates a hardness on the Vickers scale of 212. The subscript, $_{30}$, merely indicates the load in kilograms applied during the test and, in most cases, it can be ignored as the tables used for the calculation make due allowance for the load applied. Variations on the basic theme include microhardness devices where extremely small loads are used to measure microscopical features and the *Knoop* tester which utilizes a pyramid with an elongated base to allow the testing of thin sheet on edge. Also, various portable devices which utilize spring or hydraulic systems to apply the load have been developed for site use.

VIM   See Vacuum induction melting.

VIM–VAR   A production route for very high-quality steel combining *Vacuum induction melting* and *Vacuum arc remelting*.

Virgin metal   A metal produced from ore and not contaminated by remelted *Scrap*.

Viscoelastic   The phenomenon whereby the reversible relationship between *Stress* and *Strain* is time related.

Viscous flow   A mode of plastic deformation, particularly the early stages of *Creep*, in which the stress/strain relationship is time dependent and in which tensile loading does not initiate *Necking*.

Vitreous   Glass-like, particularly **Vitreous enamels** applied by fusing a glass powder on the surface of a (usually steel) component.

Vitrification   The process of forming a glassy, i.e., amorphous, non-*Crystalline* solid.

Vitriol   Sulphuric acid, H_2SO_4.

Void   A cavity within a solid or liquid.

Voidage   The measure, usually a volume percentage, of voids in a material.

Voltaic cell   An *Electrolytic* cell in which an electric current flows between a pair of electrodes that are immersed in an electrolyte and electrically connected externally.

Volumetric analysis   A form of chemical analysis in which the material in question is dissolved in known concentration and the solution reacted with other solutions of suitable reagents in standardized concentration.

Volumetric strain   The algebraic sum of the strains induced by the three *Principal stresses*.

von Laue technique   See Back reflection X-ray technique.

Vortex (shedding) fatigue   Cracking by *Fatigue* caused by the shedding of vortices arising from the rapid flow of fluid over a surface. The example usually quoted is of slender metal chimney stacks which in strong winds shed vortices on alternate sides causing the stack to bend repeatedly across the airstream, causing fatigue cracking. The effect may be countered by fitting spiral fins around the stack.

Walking beam furnace A furnace in which the *Charge* is carried and moved forward by a system of beams running the length of the furnace. In a simple case two stationary beams are positioned close to the furnace sides. For most of the time these carry the charge, which would typically be long billets. Two moving beams, positoned just within the stationary pair, are carried on cranks which from time to time raise the beams so that they lift the charge off the stationary beams and move it forward an increment.

Wallner lines Groups of parallel lines observed during scanning electron microscopical examination of some *Brittle* fracture surfaces. Groups may intersect producing V patterns. They are probably caused by some interaction between the high-velocity *Cleavage* crack front and associated shock waves. They are sometimes mistaken for fatigue *Striations*.

Warm hardening The term is occasionally used as an alternative to *Warm working* but more usually refers to *Precipitation hardening* at elevated temperature.

Warm work Deformation at a temperature above ambient but below that at which *Recrystallization* occurs. The purpose may be to avoid some brittle range, to benefit from the lower yield strength at higher temperature or to minimize scaling. It induces work hardening similar to *Cold working* and unlike *Hot working*.

Wastage Loss of section by general *Corrosion*, *Erosion* or *Abrasion*.

Water drip test See Drip test.

Water droplet erosion The loss of material by erosion caused by the impingement of water droplets usually entrained in a high-velocity stream of air or steam. Also see Steam erosion as this term is also used in reference to a particular case of water droplet erosion.

Water side The surface in contact with water, such as the bore of a heat exchanger tube with a hot gas heat source on its exterior.

Water side corrosion Severe corrosion on the *Water side* of a tube. The term is commonly used in the context of externally fired steel tubes in steam-raising plant. Steel can develop a protective magnetite oxide on its surface when exposed to water of adequate purity. However, deviations from correct water conditions, for example excessive oxygen, contamination with sea water or other chemicals, can cause severe attack leading to large internal scabs, deep corrosion and a form of *Hydrogen embrittlement*.

Water tube boiler See Boiler tube.

Water wall See Boiler tube.

Waterline corrosion Corrosive attack at the water to air interface of a partially submerged structure. It usually results from the highly oxygenated conditions in the area and hence is a form of *Differential aeration corrosion* but additional factors like wave action or marine organisms may be involved.

Wave soldering See Soft solder.

Wax vent A narrow passage through the *Core* of a casting that allows venting of any gases formed in the core during the casting process. It is formed by incorporating into the core a string thickly coated with wax. When the core is baked the wax evaporates or melts into the core and the string either burns or is pulled out to leave the passage.

Wear The removal of material from a surface by frictional or other mechanical effects as it moves relative to another contacting surface. Two broad categories are recognized. See Abrasive wear and Adhesive wear but also see Fatigue and Pitting.

Weaving (of a weld) The deposition of weld metal with significant side to side deflection of the arc or flame together with any *Filler* wire, relative to the main line of progress (see Figure 49).

Weld A union formed between two components as a result of local mutual *Fusion* or *Diffusion* resulting from heating and/or pressure. Where additional molten material is added to form or fill a joint the term 'weld' is still used provided there is fusion of the main components and provided the melting point of the *Filler* is approximately similar to the melting point of the parent materials. Where the filler melting point is significantly lower the terms *Soldering* and *Brazing* are appropriate. In these cases there is no melting of the parent metals although there may be diffusion between

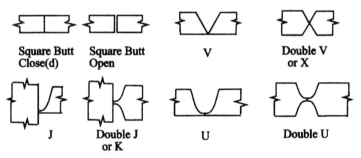

Square Butt Close(d)   Square Butt Open   V   Double V or X

J   Double J or K   U   Double U

The various V, X, U, J, K preparations may be open or close(d), i.e. with or without a root gap. Any fusible insert is intended to fill the gap.

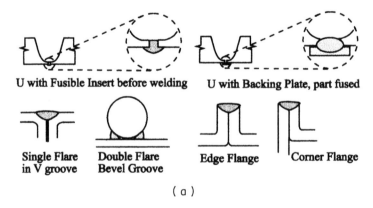

U with Fusible Insert before welding   U with Backing Plate, part fused

Single Flare in V groove   Double Flare Bevel Groove   Edge Flange   Corner Flange

(a)

Figure 47 (a) Weld joint preparations and assemblies in cross-section; (b) more types of joint, weld and preparation

the various components of the joint. Many welding processes are in use. See Friction-, Fusion-, Arc-, Gas-, TIG-, MIG-, welding, etc.

Weld bead The material deposited in a single pass of the welding arc or torch across a joint. It may also be termed a Run. See Figure 50.

Weld decay A form of intergranular corrosion and resultant cracking occurring in some *Austenitic stainless steels* that have been heated in the approximate temperature range 500–900 °C. This causes deposition of chromium carbides in the grain boundaries. The adjacent material is thereby depleted in chromium and consequently prone to corrosion in appropriate environments. Material in this condition is termed *Sensitized*. The attack is a form of Electrolytic corrosion (see Electrochemistry) in which small *Anodic* areas at the grain boundary are surrounded by large *Cathodic* grains. Consequently, although the volume of material corroded may not be large the rates of penetration and

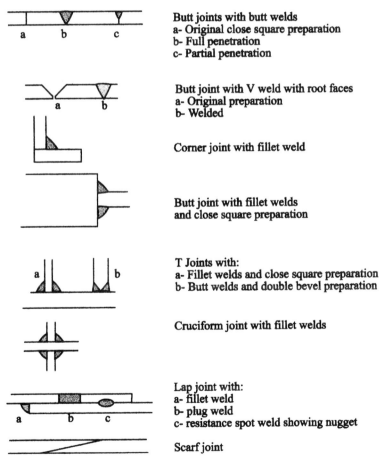

Butt joints with butt welds
a- Original close square preparation
b- Full penetration
c- Partial penetration

Butt joint with V weld with root faces
a- Original preparation
b- Welded

Corner joint with fillet weld

Butt joint with fillet welds
and close square preparation

T Joints with:
a- Fillet welds and close square preparation
b- Butt welds and double bevel preparation

Cruciform joint with fillet welds

Lap joint with:
a- fillet weld
b- plug weld
c- resistance spot weld showing nugget

Scarf joint

(b)

hence cracking can be very rapid. Most welding processes expose a narrow band of the *Heat affected zone* to the damaging temperature range so weld decay can be observed as a line of cracking close to the edge of welds in susceptible steels. However, any heating in the appropriate temperature range can cause the same form of microstructural damage and the consequent attack may, perhaps confusingly, still be described as weld decay even if no welding has been carried out. Attack can be prevented by modification of the composition or other means. See Stabilized steel. No *Stress* is necessary for corrosion of sensitized steel so the mechanism is not a form of *Stress corrosion*.

Weld deposit Fused weld metal comprising mainly *Filler* but including some fused parent metal.

Weld face The exposed surface of the last metal to be deposited. See Figure 50.

Weld filler Material added to a joint during welding. It is usually implicit that the filler is fused to become a load-carrying component.

Weld fusion zone Parent material that has fused. See Figure 50.

Weld heat affected zone See Heat affected zone and Figure 50.

Weld junction The interface between weld metal and heat affected zone. See Figure 50.

Weld metal The metal that has been fused or otherwise manipulated to form a joint.

Weld nugget The welded zone of a resistance weld. It is often termed the zone of fusion but little or none of the metal may have been above its melting point. See Figure 47.

Weld pass A single movement of the welding equipment across the component during which the material is continuously fused. Also the material fused or deposited in a single pass.

Weld preparation or **'Prep'** Generally, any activity associated with preparing an item for welding. More specifically, the geometry of the faces of a component that will receive the weld deposit. For example, the edges of two plates to be welded together may have a Vee preparation in which a single chamfer is formed on each edge. Figures 47(a), 47(b) and 48 illustrate various preparation terms.

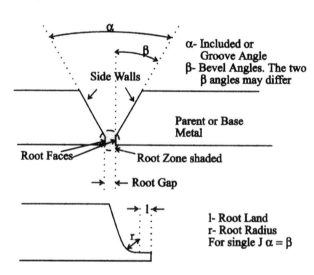

α- Included or
    Groove Angle
β- Bevel Angles. The two
    β angles may differ

Side Walls

Parent or Base
Metal

Root Faces

Root Zone shaded

Root Gap

l- Root Land
r- Root Radius
For single J α = β

Figure 48 Weld preparation terms

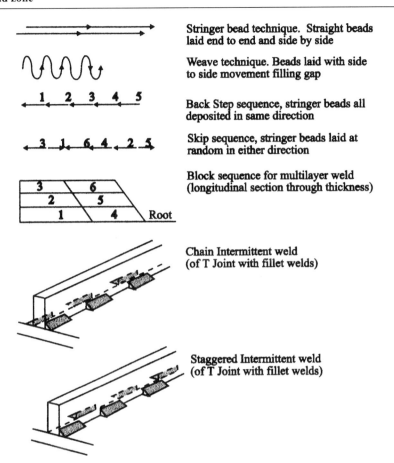

Figure 49   Weld sequences and deposition terms

Weld procedure   The total prescribed system for performing a weld. It will define all significant variables such as materials, joint geometry, equipment set-up, electrical and gas settings, pre-, interpass- and post-heating temperatures.

Weld root   Of a weld to be made, it is the area of the parent materials where the first deposit is to be made. Of a completed weld, it is the face of the first weld run that is remote from the operator. See Figures 48 and 50.

Weld rotation   The angle, relative to the vertical, of a line bisecting the angle between the fusion faces. Compare with Weld slope.

Weld run   Same as Weld pass.

Weld slope   The angle, relative to the horizontal, of a line along the weld root. Compare with Weld rotation.

Weld terminology   See Welding terminology.

Weld zone   The neighbourhood of a weld comprising all material involved in the weld operation including the weld deposit, the tip of any *Electrode* or *Filler* and the adjacent parent materials (see

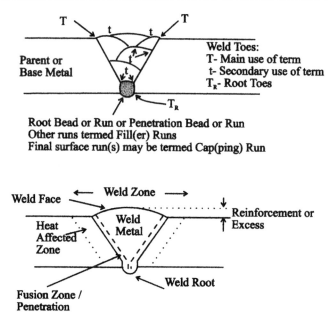

Figure 50   Weld deposit terms

Figure 50). More generally, the area that may be affected by any aspect of welding such as oxidation, tarnishing or splatter but not normally the larger area affected by residual welding stresses.

Weldability This term, in its wildest sense, refers to the ability of a material to form a satisfactory weld. An obvious requirement is freedom from cracking in the weld metal or parent material during or immediately after welding including any associated heat treatment. However, there may be other considerations such as restrictions on hardness (increase or reduction), limits on distortion or other factors critical for the application. The concept of weldability is relevant to all metals being welded but the major role of *Steel* as an engineering and construc-

tional material justifies particular attention. The first opportunity for cracking is immediately after deposition of the molten weld metal when **Solidification cracks** can develop. These are typically located along the centre line of the weld and arise because the high-temperature weld metal is too weak to carry the contraction stresses. The risk of cracking is increased by high levels of joint restraint, poor welding technique and by large amounts of impurities or some alloying elements. In steels, the common deleterious elements are sulphur or phosphorus as impurities and carbon, all of which can be introduced from either the welding electrode or from the parent materials.

The second serious cracking risk in

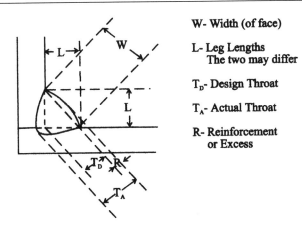

W- Width (of face)

L- Leg Lengths
   The two may differ

T$_D$- Design Throat

T$_A$- Actual Throat

R- Reinforcement
   or Excess

Figure 51   Fillet weld terms

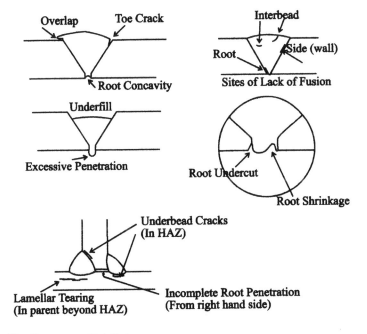

Figure 52   Common weld defects

steel arises from its tendency to harden as a consequence of the rapid cooling of the weld zone by the surrounding mass of metal. The hard martensitic and bainitic structures which develop at a late stage of cooling, and hence at fairly low temperature, are susceptible to **Hydrogen cracking**, sometimes referred to as **Hard zone cracking**. The damage is usually located in the *Heat affected zone* and in some cases may not develop for hours or days after welding, hence the alternative term **Delayed hydrogen cracking**. The term **Heat affected zone cracking** is occasionally used but it may also be used for damage developed in other circumstances. The problem can be serious and the term 'weldability' is often used in a narrow sense referring to the need to determine whether the numerous factors involved in each case collectively introduce a risk of this form of cracking and whether the risk can be eliminated by modifying the welding procedure or by **Preheating**.

Various procedures have been developed for assessing these matters but the basic factors usually taken into account are as follows. (1) The amount of damaging *Hydrogen* introduced into the weld zone, which is primarily a characteristic of the *Electrode* coating. In this context it is vital that no avoidable hydrogen be introduced from moisture, oil or other contaminants on any of the materials involved in the welding operation. (2) The *Hardenability* of the steel which is a function of composition, usually calculated as the *Carbon equivalent*. In one popular system the carbon equivalent and the electrode characteristics are corrrelated to determine a **Weldability index**. (3) The **Joint severity** which is a function of the total cross-section of parent material available to conduct heat from the joint and which therefore determines the cooling rate. This is often calculated as a *Thermal severity number.* (4) The degree of *Restraint* because this limits the ability if the joint to deform to minimize the contraction stress. (5) The rate of heat input, sometimes termed **Arc energy**, arising from the proposed electrode size and burn-off rate. (6) The levels of preheating to be imposed. Preheating reduces the cooling rate and ensures that transformation to martensite will be avoided fully or partially depending upon the temperature level relative to the martensite transformation range. Briefly, both of these benefits

reduce the hardness of the heat affected zone but for further explanation see entries on Steel and Isothermal transformation diagram. The other major benefit of reheating is that it allows more rapid diffusion and escape of any hydrogen.

Some of these factors may be inevitable consequences of the joint material and design but others can be readily varied by changing the welding procedure. For example, some combination of factors could demand a preheat of 200°C but a change from rutile to basic, low-hydrogen electrodes might reduce this to 125°C and if, in addition, the electrode size could be increased from 3 mm to 5 mm then no preheat would be necessary. It is implicit that the preheat temperature will be maintained throughout the welding operation which may be protracted where the weld is substantial and a series of runs has to be deposited. The temperature level during this period is termed the **Interpass temperature**.

Furthermore, the continued requirement to allow hydrogen to escape may require the temperature to be maintained for a specified period after the completion of the weld deposit, hence the terms **Hold temperature** and **Hold time**.

A final factor in assessing weldability is that some welds will require **Post-welding heat treatment**. This treatment, at temperatures ranging from about 500°C for mild steels to about 700°C for alloy steels, is intended to reduce residual stresses, reduce hardness levels and allow high levels of hydrogen to diffuse away. However, it can also induce **Reheat cracking** typically located in the heat affected zone. Its likelihood is reduced by selecting the material composition to be less susceptible, minimizing stresses during welding, avoiding adverse geometry such as the weld zone coinciding with section changes and grinding the weld profile to remove stress-raising notches at the weld toes. Also see entry on Reheat cracking.

Weldability index See previous entry.

Weldability test Any test for establishing whether a satisfactory joint can be produced by a particular *Welding procedure.*

Welder A person making a weld. Colloquially, a machine for welding.

Welding consumables The materials used in producing a welded joint. The term includes *Fillers, Electrodes, Fluxes,* etc., and possibly electricity and gases, etc., but not parent material.

Welding cycle The total series of activities and operations involved in a machine, or less commonly an operative, producing a single weld and returning to the start condition.

Welding flux Material applied to the weld zone primarily to chemically clean the surface, form a slag with impurities and prevent oxidation. Fluxes may contain components with other characteristics. See Electrode coating.

Weld(ing) position The orientation of components and weld deposits during the welding operation. A very wide range of terms and definitions has been used, often with some confusion. The definition used in this book generally accord with those favoured by BS 499, Part I, 1983, Welding terms and symbols. This defines a limited number of positions by reference to the combination of *Weld slope* and *Weld rotation.* See Horizontal–vertical, Flat, Vertical and Overhead, these being the basic welding positions, and also Inclined. Other authorities such as the American Welding Society Publication A.30-80 *Welding Terms and Symbols* offer alternative and more numerous definitions.

Welding primer A paint or other coating applied to protect material until such time as it may be welded. It is implicit that the coating does not have any adverse effect on the weld and hence does not have to be removed prior to welding.

Welding rod Material in the form of wire or rod that provides *Filler* material. It may be bare solid rod, as used for most *Gas welding,* or coated rod or

filled tube as used for *Electric arc welding.* see Electrode.

Weld(ing) sequence The order in which a series of welding operations is performed. In particular, the order and direction in which a multi run weld is deposited. See Figure 49.

Welding technique The way in which the operative holds and manipulates the welding equipment such as the *Electrode, Blowpipe* and *Filler* rod as they form the weld.

Welding terminology Many terms are used in a specialized sense in welding. Those with a metallurgical aspect are included at their alphabetical location in this book and are illustrated in Figures 47(a) to 52 inclusive. American Welding Society Publication A.30-80 *Welding Terms and Symbols* and BS 499, *Welding terms and symbols* detail many more terms of interest to the welding specialist.

Weldment An assembly of components joined by welding.

Wetting The process whereby a liquid makes an intimate contact with and readily spreads over a surface. This occurs when the attraction between the liquid and the surface exceeds the surface tension of the liquid. If the liquid tends to form discrete globules it is not wetting the surface. The angle formed between the two exposed surfaces is a measure of wetting; an acute angle reflects poor wetting and an angle approaching 180° indicates good wetting.

Wheelabrator A proprietary machine for grit or *Shot blasting.* In essence it is a rotating wheel which flings the shot from its periphery to impact the target component.

Whirl The deflection of a rotating shaft such that the axis of deflection precesses around the axis of rotation.

Whisker Fine, hair-like *Single crystal* grown under controlled conditions and usually having only a single axial *Dislocation.*

Whistlers Narrow passages in a mould to allow escape of pockets of entrapped air. The high velocity of the heated escaping air can cause a whistling noise.

White (cast) iron   *Cast iron* with a structure of pearlite and cementite with little or no graphite.

White copper   Usually copper–nickle alloy, in particular *Nickle silver*.

White gold   Various alloys based on *Gold* and having its corrosion resistance but being various shades of silvery white rather than gold colour. They may be gold–platinum, gold–silver or more complex alloys containing copper, zinc, nickle or chromium.

White lead   A white pigment, basic lead carbonate, $2PbCo_2Pb(OH)_2$.

White metal   Alloys based on tin and or lead usually with other elements such as antimony and copper and used for plain bearings, printers type etc. See Babbitt metals. Simple alloys of tin and lead are usually termed *Soft solders*.

White rust   The corrosion products formed on zinc, mainly zinc oxide and carbonate.

Whiteheart (cast) iron   See Cast iron.

Widmanstätten structure   A distinctive, angular microstructure formed when a *Phase* grows, or precipitates, along specific crystallographic planes of the prior solid solution.

Wiped (joint)   a joint formed by applying molten *Filler* metal and, while it is in a pasty state, shaping it by wiping with a cloth or other material. Traditionally, plumbers formed soldered joints to produce a neat, rounded deposit by wiping with a 'moleskin'. See *Solder*.

Wire cable/rope   Cable made from multiple strands of wire. Depending upon the application, the wires may be assembled, referred to as the 'lay', in superimposed helical layers or by helical assemblies of helical strands. See Lang's lay as an example. Some steel cables have a central core of hemp or such material and may be prelubricated.

Wirebar   High-purity copper ingot intended for *Drawing* for electrical conductors.

Wiredrawing (1)   The process of reducing the diameter of wire by drawing it through a *Die*.

Wiredrawing (2)   Linear grooves eroded by steam leaking through narrow joints or across nearly closed valve seats.

Wireframe models   Physical or graphic models of crystals in which atoms are located at the intersections of three-dimensional wire lattices. See Crystal structure and Dislocation for examples.

Witness mark   A surface mark revealing or confirming some action or process.

Wöhler fatigue test   A *Fatigue* test in which the specimen is a rotating cantilever, driven at the supported end and loaded via a bearing at the other end. One rotation produces a single reversed stress cycle.

Wolfram   An old name for tungsten metal but now usually referring to tungsten ores.

Wood's metal   An alloy of 50% bismuth, 27% lead, 13% tin and 10% cadmium, which melts at 66 °C.

Woody fracture   A fracture surface having a coarsely fibrous texture. It usually reflects an elongated grain structure or large quantities of stringers of *Inclusions*.

Wootz process   An ancient steelmaking process which utilized the natural wind, suitable channelled, as an air blast for crucible *Smelting*. The iron was then subjected to a *Cementation process*.

Working   Causing permanent deformation to material. The terms 'work' and 'plastic deformation' are effectively synonymous; the former tends to be used in a manufacturing context and the latter in an academic or mechanical testing context. **Cold working** refers to processes where no *Recrystallization* occurs and the component **Work hardens**, i.e. becomes progressively harder, up to some limiting level, and less *Ductile*. **Hot working** refers to processes, usually at elevated temperature, where deformation is accompanied by recrystallization, i.e. no hardening occurs. **Warm working** is a special case of cold working involving deliberate heating, perhaps to assist deformation (relative to cold working) or minimize oxidation (relative to hot working), but the temperature reached is insufficient to allow recrystallization.

Working stress/load This usually means the *Safe working load* or the associated *Stress*.

Wormhole Elongated forms of gas *Pores* or *Blow holes*, particularly in welds.

Wrapping test A test in which wire or similar section is wound round a former or itself a specified number of times to assess its *Ductility*.

Wrinkle/wrinkling A deviation from flatness, usually multiple undulations.

Wrinkle bend A bend, usually in a tube, produced by forming a series of semi-circumferential corrugations along the side forming the bend inner surface. The process typically involves locally heating a narrow band on the intended inside of the bend. The tube is then bent a small amount. This is repeated a short distance along the tube until the required profile is achieved.

Wrought The same as *Worked*. Any material that has been shaped by a de-formation process, as opposed to a casting.

Wrought iron (1) Originally *Iron* of fairly high purity made by various melting techniques followed by repeated forging. This first usage of the term was therefore in contrast to *Cast iron*. This form of wrought iron was a tough, fairly corrosion-resistant material suitable for major structures as well as decorative purposes. The *Puddling process* was the most common manufacturing process in the UK. The *Aston process* was more common in the USA. Originally, wrought iron was a major structural material but it was superseded by steel made by processes such as the *Open hearth* and the *Bessemer converter* which provided material at lower cost, higher strength and available in much larger sizes. Consequently, these older processes are now commercially obsolete.

Wrought iron (2) Modern 'wrought iron' usually refers to *Mild steel* bar bent in an aesthetically pleasing manner for functional or decorative applications.

Wrung fit See Interference fit.

Wüstite Iron oxide, FeO.

Xenon A gaseous element, one of the noble or inert gases. See Table 15.

Xerography A technique for producing photocopies.

X-ray Electromagnetic radiation of less than 500 Ångström units emitted by decelerated electrons.

X-ray diffraction Techniques utilizing the diffraction of X-rays, particularly to investigate *Crystal structures*. The wavelength of X-rays is of the same order as the lattice spacing of metal crystals so if an X-ray beam is directed at a crystal the regular spacing of the planes of atoms causes diffraction effects which can be recorded on photographic film. The diffraction patterns can provide information on lattice spacings, extent of cold working and material composition etc. See Bragg and von Laue.

X-ray fluorescence spectroscopy An analytical technique in which a beam of X-rays excites the characteristic radiation of elements in the item of interest. The radiation from the individual constituents is then measured in a spectrometer.

Yellow metal/brass   Usually the term refers specifically to brass with 60% copper and 40% zinc but it is also used loosely to refer to any *Brass*.

Yield   To suffer permanent deformation. See Tensile test.

Yield extension   The amount of extension between the upper and lower yield points. See Tensile test.

Yield point   The level of stress and strain at which a material commences to suffer permanent deformation. See Tensile test.

Yield stress   The stress at *Yield*. See Tensile test.

Young's modulus   The elastic modulus. See Tensile test.

Ytterbium   A metallic element, one of the *Rare earth* group. See Table 15.

Yttrium   A metallic element. See Table 15.

Z

Zinc A metallic element having major applications as a corrosion-resisting surface on *Steel*. It can be applied by hot dipping (termed *Galvanizing*), electroplating (termed *Electrogalvanizing*), spraying or diffusion. It is also the basis for some *Die casting* alloys and is an alloying addition in *Brass* and some other alloys. See Tables 13 and 15.

Zinc blend Zinc sulphide, ZnS, usually in reference to the ore.

Zinc chromate An orange-coloured constituent of some primer paints. It is not simply a pigment but provides some sacrificial *Cathodic protection* particularly for aluminium in marine environments.

Zinc coating See Galvanizing.

Zinc (based) die casting alloy Various alloys, principally zinc but with additions of aluminium, typically about 4%, possibly with small amounts of copper, lead, cadmium, etc. These materials have relatively low melting points and provide castings having a good finish with fine detail. See Table 13.

Zinc tinsel An alloy of about 60% zinc, 40% lead which is bright and reasonably tarnish resistant for decorative applications. It has been largely replaced by plastics.

Zinc white Zinc oxide, ZnO.

Zircon A zirconium oxide/silicon oxide compound, $ZrO_2.SiO_2$. It is used in powder form, often termed **Zircon sand**, as a refractory material with good thermal shock resistance up to about 1700 °C.

Zirconia Zirconium oxide, ZrO_2. It is often used as a *Refractory* material usually with the addition of small amounts of calcium oxide or magnesium oxide to improve dimensional stability.

Zirconium A metallic element with good strength and ductility. It reacts rapidly, violently in finely divided form, with oxygen and so is used for flash bulbs and detonators. However, the oxide ZrO_2 is protective and the metal and its alloys are corrosion resistant. It has a low thermal neutron capture cross-section and hence is used as a canning and structural material for nuclear applications. See Table 15.

Zone(d) furnace A furnace in which the *Charge* progresses through a number of zones where the heating rate and/or intended temperature vary. For example, the first zone might be intended to slowly preheat, the second to quickly raise to the peak temperature and the third to allow initial cooling at some controlled rate. The zones are not normally separated by any physical barrier but have individual heating input and control systems.

Zone levelling A technique for *Homogenizing* a material. It is similar to *Zone refining* except that the heated zone sweeps back and forth repeatedly. This has the effect of redistributing impurities evenly along the length except at the extremities, which can be discarded.

Zone melting The process of melting a narrow zone as in *Zone levelling* and *Zone refining*.

Zone refining A technique for refining metals that are already nearly pure. It utilizes a local heat source which slowly and repeatedly sweeps a narrow molten zone along a bar of the metal. In the appropriate cases where the impurity element depresses the melting point of the solvent the molten zone collects some of the impurity on each sweep and deposits it at the bar ends.

Tables

The information presented in Tables 1 to 14 provides a broad overview of typical materials and their properties. However, it is emphasized that the data in all tables are approximate and no responsibility can be accepted for their accuracy. In particular, the data should not be used for specification or design purposes.

List of tables

Table 1   267

Table 1 Typical carbon Steels for engineering applications
See also Table 4 for some physical properties of steels similar to those listed below

Common name	Nominal composition (%)		Condition	UTS (MPa)	YS (MPa)	Elongation (%)	Hardness (HB)	LRS (mm)
	Carbon	Manganese						
	0.15	0.8	Normalized	350	175	22	100	63
	(0.12/0.18)	(0.6/1)	CW	450 / 400	330 / 300	10 / 13	125 / 110	13 / 100
	0.2	0.8	Normalized	400	200	21	115	250
	(0.16/0.24)	(0.5/0.9)	CW	560 / 450	440 / 325	10 / 14	170 / 125	13 / 76
Mild Steel / Low Carbon	0.25	0.8	Normalized	430	215	20	123	250
	(0.22/0.3)	(0.5/0.9)	CW	590 / 490	465 / 355	9 / 13	177 / 146	13 / 76
	0.3 (0.26/0.34)	0.8 (0.6/1)	Normalized	460	230	19	135	250
			CW	620 / 530	480 / 385	9 / 12	185 / 156	13 / 76
			Hardened	550 / 625	340 / 460	18 / 12	152 / 180	63 / 19

continued overleaf

Table 1 (*continued*)

Common name	Nominal composition (%)		Condition	UTS (MPa)	YS (MPa)	Elongation (%)	Hardness (HB)	LRS (mm)
	Carbon	Manganese						
Medium carbon	0.35 (0.32/0.4)	0.8 (0.6/1)	Normalized	490	245	18	145	250
			CW	640 / 550	505 / 400	8 / 11	190 / 164	13 / 76
			Hardened	590 / 660	385 / 465	17 / 16	166 / 200	63 / 19
	0.4 (0.36/0.44)	0.8 (0.6/1)	Normalized	510	245	17	145	250
			CW	660 / 570	530 / 430	7 / 10	196 / 172	13 / 76
			Hardened	625 / 700	385 / 465	16 / 16	180 / 200	63 / 19
High carbon	0.5 (0.45/0.55)	0.8 (0.6/1)	Normalized	570	295	14	165	250
			CW	740 / 650	590 / 510	7 / 10	215 / 195	13 / 76
			Hardened	625 / 850	390 / 600	15 / 9	180 / 250	150 / 13

Table 1   269

			Normalized	CW	Hardened	Softened
600	310	13	170	224 / 197	200 / 250	285 Max
250				13 / 76	100 / 19	
				6 / 9	14 / 9	
760 / 670	610 / 530		700 / 850	415 / 595		

0.55 (0.5/0.6)	0.8 (0.5/0.9)					
'Silver steel'	0.95/1.25	0.25/0.45	0.5 Chromium max.			

Typical physical properties of these steels at 20 °C.
Coefficient of thermal expansion 11.3×10^{-6} per °C   Density 7.85 g/cm^3
Thermal conductivity 0.132 Cal cm/cm^2/°C/s   Electrical resistivity 19.4 microhm cm.

Notes

1 The table indicates the minimum properties that can be expected from commercial steels.

2 The common steel names, mild, medium and high carbon, are imprecise and are offered as a guide only.

3 The nominal carbon compositions are quoted as these are often referred to in terms such as 'point three carbon steel' or more casually in the steel industry as 'thirty carbon steel'. The figures in parentheses are the usual commercial range.

4 Condition: Normalized—material that has been wrought and subsequently normalized. CW—cold worked., i.e. material hardened by cold rolling etc. Hardened—material that has been hardened by quenching and tempering. Two examples of CW and hardened are included to illustrate the influence of ruling section within the normal commercial range.

5 UTS—ultimate tensile strength. YS—yield strength. LRS—limiting ruling section: Also termed ruling section.

Table 2   271

Table 2 Typical carbon manganese steels for engineering applications
See also Table 4 for data on physical properties of steels similar to those listed below

Common name	Nominal composition (%)		Condition	UTS (MPa)	YS (MPa)	Elongation (%)	Hardness (HB)	LRS (mm)
	Carbon	Manganese						
Carbon manganese steels	0.35 (0.32/0.4)	1.2 (1/1.4)	Normalized	590	355.	15	175	150
			CW	710 620	565 480	6 9	210 180	13 76
			Hardened	625 775	415 580	18 10	180 223	100 19
	0.2 (0.15/0.23)	1.5 (1.3/1.7)	Normalized	550	325	18	152	150
			Hardened	550 700	340 520	18 12	152 200	150 29
	0.35 (0.32/0.4)	1.5 (1.3/1.7)	Normalized	620	385	14	180	150
			Hardened	625 850	400 665	18 9	180 248	150 13

Notes

1 The table indicates the minimum properties that can be expected from commercial steels.

2 Virtually all carbon steels contain manganese and are sometimes loosely termed 'carbon manganese' rather than 'plain carbon'. In Tables 1 and 2 the term is used in the narrow sense of steels in which manganese contents deliberately in excess of 1% significantly improve mechanical properties and hardenability. The common names mild, medium and high carbon are subjective, vary from industry to industry and are offered as a guide only.

3 The column on nominal composition includes, in parentheses, the usual commercial range

4 Condition: normalized refers to material that has been wrought and subsequently normalized. CW indicates cold worked and refers to steels that have been hardened by cold rolling, drawing, etc. Hardened indicates steels that have been hardened by quenching followed by tempering. Some grades may also receive subsequent cold working to achieve the indicated properties. Two examples of CW and hardened grades are included to indicate the effect of limiting ruling section

5 UTS—ultimate tensile strength. YS—yield strength. LRS—limiting ruling section. Also termed ruling section.

Table 3 Typical alloy steels for general engineering applications: range of mechanical properties in hardened (quenched and tempered) condition

See also Table 4 for data on physical properties of steels similar to those listed below

Common name	Nominal composition (%)						UTS (MPa)	YS (MPa)	Elongation (%)	Hardness (HB)	LRS (mm)
	C	Mn	Cr	Mo	Ni	Other					
(a) Preferred steels based on BA 970: 1983											
1% Cr	0.4	0.8	1	–	–	–	700 / 850	525 / 680	17 / 13	200 / 248	100 / 29
Manganese–moly.	0.35	1.5	–	0.25	–	–	700 / 1000	495 / 850	15 / 12	200 / 295	250 / 19
1% Cr–Mo	0.4	0.8	1	a) 0.2 b) 0.3	–	–	700 / 1075	495 / 940	15 / 12	200 / 310	250 / a)13, b) 19
3% Cr–Mo	0.25	0.6	3.2	0.5	–	–	850 / 925	650 / 770	13 / 9	248 / 270	250 / 150
1.5% Ni–Cr–Mo	0.4	0.6	1.2	0.27	1.5	–	850 / 1500	650 / 1250	13 / 3	250 / 445	250 / 29
2.5% Ni–Cr–Mo	a) 0.3 b) 0.4	0.6	0.65	0.55	2.5	–	925 / 1500	740 / 1250	12 / 5	270 / 445	250 / a) 63, b) 100
Low alloy	0.38	1.4	0.5	0.2	0.75	–	700 / 1000	495 / 865	15 / 9	200 / 295	250 / 29
1.5 Cr–Mo–Al nitriding	0.4	0.5	1.5	0.2	–	Al 1.0	700 / 850	525 / 680	17 / 13	200 / 250	150 / 63
3 Cr–Mo–V nitriding	0.4	0.5	3.2	1	–	V 0.2	1310	1160	8	375	63

Table 3 273

(b) Additional common steels included in obsolete BS 970: 1955

	C		Cr	Mo	Ni		UTS	YS	El		LRS
Carbon Cr	0.6	0.6	0.6	—	—	—	1000	850	12	290	63
1% nickel	0.37	1.5	—	—	0.8	—	625	450	22	180	150
3% nickel	0.3	0.5	—	—	3	—	775	525	20	223	63
3% Ni–Cr	0.3	0.5	0.7	0.6 (Optional.)	3	—	775	580	20	225	150
							1000	835	16	293	63
3.5% Ni–Cr–Mo	0.33	0.6	1	0.4	3.5	—	925	755	17	270	150
							1225	1095	14	363	63
3% Cr–Mo	0.25	0.6	3	0.5	—	—	700	495	22	200	150
							1540	1230	10	444	63
4.25 Ni–Cr	0.3	0.5	1.2	—	4.2	—	1540	1230	10	444	150
4.25% Ni–Cr–Mo	0.3	0.5	1.2	0.3	4.2	—	1540	1230	15	444	150

Notes

1 The common names are offered as a guide only.

2 The compositions quoted are approximately mid range of steels specified in BS 970: 1983 or 1955 as indicated.

3 The properties quoted are for the steel in the quenched and tempered condition in which it would normally be used for general engineering applications. Where a steel is used in a range of strength levels the usual minimum and maximum are quoted.

5 UTS—ultimate tensile strength. YS—yield strength. LRS—limiting ruling section.

Table 4 Typical steels for high-temperature applications
(a) Steel types: typical composition, physical properties and elastic modulus at temperature

Type and common name	Composition %							
	Carbon	Manganese	Silicon	Molybdenum	Chromium	Vanadium	Nickel	Other
Ferritic and martensitic steels								
0.15% C	0.15 Max	0.4/0.7	0.1–0.35	–	–	–	–	–
0.2% C	0.2 Max	0.4/0.7	0.35 Max	–	–	–	–	–
0.25% C	0.2/0.25	0.4/0.7	0.35 Max	–	–	–	–	–
0.15% C Mn	0.12/0.18	0.9/1.2	0.1/0.35	–	–	–	–	–
0.5% Mo/B	0.16 Max	0.7 Max	0.4 Max	0.4/0.6	–	–	–	0.001/0.005 B
0.5% Mo	0.15 Max	0.4/0.8	0.1/0.35	0.4/0.7	–	–	–	–
1% Cr Mo	0.1/0.15	0.4/0.7	0.1/0.35	0.45/0.65	0.7/1.1	–	–	–
1% Cr Mo Higher C	0.35/0.45	0.4/0.7	0.35 Max	0.5/0.8	1/1.5	–	–	–
$2\frac{1}{4}$% Cr Mo	0.08/0.15	0.4/0.7	0.5 Max	0.9/1.2	2.0/2.5	–	–	–
9% Cr Mo	0.15 Max	0.3/0.6	0.2/1	0.9/1.1	8/10	–	–	–
$\frac{1}{2}$% Cr Mo V	0.15 Max	0.4/0.7	0.1/0.35	0.5/0.7	0.25/0.5	0.22/0.28	–	–
1% Cr Mo V	0.3/0.45	0.4/0.7	0.35 Max	0.5/0.8	1/1.5	0.2/0.3	–	–
1% Cr 1Mo $\frac{3}{4}$V (B & Ti)	0.15/0.25	0.35/0.75	0.1/0.35	0.85/1.1	0.9/1.3	0.6/0.8	–	B 0.001/0.01 Ti 0.05/0.2
12% Cr Mo V	0.08/0.15	0.4/1	0.3 Max	1.5/2	11/13	0.2/0.45	2/3	N 0.02/0.04
12% Cr Mo V Nb	0.1/0.2	0.75/1.2	0.35 Max	0.5/0.7	11/13	0.2/0.4	0.75/1.25	Nb 0.1/0.4 N 0.05/0.1
Austenitic steels								
18/12 Cr Ni Ti stabilized	0.04/0.09	0.5/2	0.2/1	–	17/19	–	10/13	Ti 5x C/0.7 Max
18/12 Cr Ni Mo	0.04/0.09	1/2	0.8 Max	2/2.75	16/18	–	11/14	–
15 CrNiMnMo	0.15 Max	5.5/7	0.2/1	0.8/1.2	14/16	0.15/0.4	9/11	Nb 0.75/1.25

Table 4 275

Type and common name	Elastic modulus GPa (at °C)	Electrical resistivity ($\mu\Omega$ cm at °C)		Thermal conductivity (cal cm/cm²/s at °C)		Mean coefficient of thermal expansion per °C $\times\ 10^{-6}$ for temperature range °C		
		20	500	20	500	20–100	20–500	20–700
0.15% C	210 (20)	15	58	0.14	0.1	12.2	14.3	
0.2% C	181 (400)	16	58	0.13	0.1	12.2	14.3	
0.25% C	165 (600)	19	6	0.12	0.09	12.2	13.9	
0.15% C Mn	210 (20) 175 (500)	21	61	0.11	0.09	10.6	14.2	
0.5% Mo/Boron	211 (20) 165 (600)	22	89	0.11	0.09	12	14.3	14.8
0.5% Mo	211 (20) 165 (600)	22	89	0.11	0.09	12.5	14.4	15
1% Cr Mo	212 (20) 165 (600)	23	90	0.11	0.09	12	13.8	14.4
1% Cr Mo Higher C	211 (20) 164 (600)	24	58	0.1	0.08	11.8	14.6	15
$2\frac{1}{4}$% Cr Mo	213 (20) 168 (600)	28	68	0.09	0.08	11.4	13.6	14
9% Cr Mo	215 (20) 166 (600)	53	89	0.04	0.04	10.3	12.2	12.8
$\frac{1}{2}$% Cr Mo V	212 (20) 166 (600)	21	66	0.11	0.09	11.2	13.8	14.4
1% Cr Mo V	215 (20) 165 (600)	27	66	0.1	0.8	11.2	13.5	14.2
1% Cr 1Mo $\frac{3}{4}$V (B & Ti)	211 (20) 164 (600)	27	68	0.1	0.08	11.1	13.9	14.6
12% Cr Mo V	213 (20) 159 (600)	67	96	0.05	0.06	10.6	11.9	
12% Cr Mo V Nb	213 (20) 168 (600)	61	93	0.06	0.06	9.5	11.8	
Austenitic steels								
18/12 Cr Ni Ti	199 (20) 150 (600)	72	101	0.04	0.05	17	18.5	19.3
18/12 Cr Ni Mo	200 (20) 153 (600)	75	105	0.04	0.06	16	18.4	19.2
15 CrNiMnMo	201 (20) 158 (600)	74	104	0.03	0.05	16.1	18.6	19.3

Table 4 (*continued*)
(b) Strength and creep rupture strength of typical steels

Steel type See Table 4(a) for composition	Cond.	Strength (MPa) and creep strength (MPa at 10 000 and 100 000 hours) at temperature (°C)											
		20 °C		200 °C	350°	400 °C			450 °C			500 °C		
		UTS	PS (Y)	PS	PS	PS	10K	100K	PS	10K	100K	PS	10K	100K
Ferritic and martensitic steels														
0.15% C	N	330	Y 185	145	113	107	210	136	102	118	70		59	32
0.2% C	N	356	Y 210	176	130	117	210	153	114	118	70		59	32
0.25% C	N	420	Y 250	190	144	137	210	153	130	118	70		60	33
0.2% C Mn	N	440	Y 245	220	160	150	225	165	150	130	85		70	37
0.5% Mo B	N	540	Y 430	385	340	324	502	394	310	370	225		210	50
0.5% Mo	N	460	300	290	240	200			195		200	195		100
1% Cr Mo	N	440	235		195	180			170	432	314	162	260	150
1% Cr Mo higher C	H	850	635	600	570	555	570	525	525	450	310	490	230	
$2\frac{1}{4}$% Cr Mo	N	490	260		216	205			195			187	225	170
9% Cr Mo	N	416	Y 210		270	265			255		286	235		158
$\frac{1}{2}$% Cr Mo V	N	465	Y 295	246	215	205			195	385	325	185	260	185
1% Cr Mo V	H	920	690	665	600	555				495	370	510	280	170
1 Cr 1Mo$\frac{3}{4}$V	H	850	700	680	640	620		190	590		400	545	360	290
12% Cr Mo V	H	925	740	650	650	615			550			500	340	215
12 Cr MoVNb	H	925	770	650	630	610			570			505	400	375
Austenitic steels														
18Cr 10Ni Ti	S	510	Y 170	130	110	100			95			86		
18Cr 12Ni Mo	S	510	Y 185	130	110	100			98			95		
15CrNiMn Mo	S	540	210	205	190	185			182			180		

Table 4 277

	550°C			600°C			650°C			700°C			Steel type
	PS	10K	100K	PS	10K	100K	PS	10K	100K	PS	10K	100K	
													0.15% C
													0.2% C
													0.25% C
													0.2% C Mn
		53	28	28									0.5% Mo B
	190		35	190									0.5% Mo
		120	50										1% Cr Mo
													1% Cr Mo Higher C
	180	127	85		65	35							$2\frac{1}{4}$% Cr Mo
			80			40							9% Cr Mo
		145	85		71	45							$\frac{1}{2}$% Cr Mo V
	435												1% Cr Mo V
	515	250	190										1% Cr 1Mo $\frac{3}{4}$V
	400	140	100										12% Cr Mo V
	400	270	290	250	160	70							12% Cr Mo V Nb
	83	235	190	82	170	130	80	110	70	75	52	25	18 Cr 10Ni Ti
	90				80	175	100	110	70	95	63	30	18 Cr 12Ni Mo
	177			175	265	220	168	200	160	160	100	47	15 Cr Mn Mo

Notes
1 UTS—ultimate tensile strength. PS—0.2% proof stress except for the austenitic steels where 1% proof stress is quoted and where Y-indicates yield strength.
2 Cond—condition: N—normalized. H—hardened, i.e. quenched and tempered: S—solution treated.

Table 5 Typical stainless steels for engineering applications
See Table 4 for data on Physical Properties of steels similar to some of those listed below. Properties are minima unless stated

Common name (type)	Composition (%) Maxima unless stated					Condition	UTS (MPa)	YS (MPa)	Elongation (%)	Hardness (HB)	LRS (mm)
	C	Cr	Ni	Mo	Other						
Ferritic stainless steels.											
13 Cr/(403)	0.08	12/14	0.5	–	Mn 1, Si 1	Softened	420	280	20	170 max	150
17 Cr/(430)	0.08	16/18	0.5	–		Softened	430	280	20	170 max	63
Martensitic stainless steels											
12 Cr (Low C) (410)	0.09/0.15	11.5/13.5	1			Softened	–	–	–	207 max	–
						Hardened	550 / 700	370 / 525	20 / 15	152 / 200	150 / 63
12 Cr (Med C) (420)	0.14/0.2	11.5/13.5	1	–	Mn 1, Si 1	Softened	–	–	–	217 max	–
						Hardened	700 / 775	525 / 585	15 / 13	200 / 223	150 / 29
12 Cr (High C) (420)	0.2/0.28	12/14	1	–		Softened	–	–	–	229 max	–
						Hardened	700 / 775	525 / 585	15 / 13	200 / 223	150 / 150
17 Cr (Martensitic) (431)	0.12/0.2	15/18	2/3	–		Softened	–	–	–	277 Max	–
						Hardened	850	680	11	248	150

Table 5 279

Austenitic stainless steels

	C%	Cr	Ni	Mo	Other	Condition	UTS	PS	Elong %	Hardness
18/8 (302)	0.12	17/19	8/10	—	Mn 2 Si 1		510	PS 190	40	183 max
18/10 Low C (304)	0.03	17/19	9/12	—			480	PS 180	40	
18/10 Ti Stab* (321)	0.08	17/19	9/12	—	Mn 2, Si 1 Ti 5x C%	Softened	510	PS 200	35	183 max
18/10 Nb Stab* (347)	0.08	17/19	9/12	—	Mn 2, Si 1 Nb 10x C%		510	PS 205	30	
18/12 Mo (316)	0.07	16.5/18.5	11/14	2.5/3	Mn 2 Si 1	For cold drawn see below	510	PS 205	40	160
25/20 Cr/Ni (310)	0.15	24/26	19/22	—	Mn 2 Si 1.5		510	PS 205	40	207 max
All above austenitic steels						Cold drawn	865	PS 695	12	19
							725	PS 450	20	32
							650	PS 310	28	45

Notes

1 The martensitic steels are capable of air hardening in some sections. *Softened* therefore usually implies some form of annealing. *Hardened* refers to steels cooled sufficiently fast to become fully martensitic and then tempered. The properties quoted are for the alloy in the hardened condition in which it would normally be used for general engineering applications. Where as alloy is used in a range of strength levels the usual minimum and maximum levels are quoted.

2 UTS—ultimate tensile strength. YS—yield strength. LRS—limiting ruling section. PS—0.2% proof stress. Stab—stabilized.

Table 6 Typical cast irons for engineering applications

Type and sub-type	C (total)	Composition (%) Maxima unless stated					UTS (MPa)	YS (MPa)	Elongation (%)	Hardness (HB)
		Si	Mn	P	S	Other				
Grey iron, tested as 30 mm bar										
Low strength	3.4/3.6	2.3/2.5	0.5/0.7	0.2/0.6	0.1/0.8		125/165	–	–	160/180
High strength	2.5/2.9	1.9/2.1	0.7/1	0.05/0.15	0.05/0.1		410/450	–	–	250/290
Malleable iron, tested as 15 mm bar										
Blackheart, ferritic	2.3/2.65	0.9/1.65	0.25/0.55	0.18	0.05/0.18		300/350	190/200	10	150
Blackheart, pearlitic	2/2.65	0.9/1.65	0.25/1.25	0.18	0.05/0.18		450/700	270/530	6/2	160/290
Whiteheart	3.0/3.7	0.4/0.9	0.2/0.4	0.1	0.3		360/480	170/280	3/4	200/220
Spheroidal graphite (SG)/nodular Iron										
Typical plain	3.35	2.6	0.56	0.03	0.01	0.025 Mg	415	400	1	140
Nickel containing	3.54	2.65	0.56	0.04	0.01	0.04 Mg, 1.2 Ni	730	450	3	210
White, abrasion resistant										
Unalloyed	3.3/3.6	0.4/1	0.5/0.7	0.3	0.15					400
Malleable white	2.2/2.5	1/1.6	0.3/0.5	0.15	0.15					320
Martensitic	2.9	0.4/0.7	0.4/0.7	0.4	0.15	1.5/3.5 Cr, 4/4.8 Ni				530
High chromium	2.2/2.8	0.2/1	0.4/1.2	0.4	0.15	24/30 Cr				500

Table 6 281

Corrosion-resistant iron								
High silicon	0.4/1	14/17	0.4/1	0.3	0.15	Mo 3	90/125	480/520
High chromium	1.5/4	0.5/3	0.3/1.5	0.3	0.15	12/35 Cr 5 Ni, 3 Cu, 4 Mo	210/820	250/750
High nickel. Austenitic	1.8/3	1/2.7	0.4/1.5	0.3	0.15	1.5/5 Cr 15/30 Ni 7 Cu	170/300	130/250
Heat-resistant iron								
Silicon	1.6/2.5	4/6	0.4/0.8				170/310	170/250
Chromium	2/3	0.5/2.5	0.4/1.5			15/35 Cr 5 Ni	200/600	250/500
Nickel, chromium, silicon	1.7/2.5	5/6	0.5/1			13/30 Ni 2/5 Cr, 1 Cu	140/310	100/200

Notes
1 Carbon contents are for as-cast material, i.e. prior to any malleabilizing heat treatment.
2 UTS—ultimate tensile strength. YS—yield strength.

Table 7 Aluminium and typical alloys

Aluminium and its alloys are commonly defined in accordance by the alpha-numeric system comprising a four-figure number specifying the composition followed by a temper designation, e.g. 1060 H18.

The compositions are grouped as follows:

1xxx—commercial purity.

2xxx—copper addition, hardenable by heat treatment.

3xxx—manganese addition, not hardenable by heat treatment.

4xxx—silicon addition, usually for brazing filler,

5xxx—magnesium addition, not hardenable by heat treatment.

6xxx—silicon and manganes additions, hardenable by heat treatment.

7xxx—zinc addition, hardenable by heat treatment.

8xxx—other additions

The temper designation comprises an initial letter indicating the basic condition followed by one or more digits providing more detail of condition and properties as follows:

F   As fabricated, i.e. as-cast, extruded, etc.

O   Annealed, fully softened.

H   Strain hardened, i.e. cold worked. Always followed by two digits. The first indicates the thermal and mechanical history as follows: H1x—strain hardened only. H2x—strain hardened and partially annealed. H3x—strain hardened and stabilised by heating to a low temperature.

The second digit indicates the degree of strain hardening for the alloy in question. This digit ranges up to 8, the highest strength, for example H14.

W   As solution treated. Unstable.

T   Stabilized. Followed by a digit indicating the thermal and mechanical history as follows:

T2—annealed (castings). T3—solution heat treated, cold worked and naturally aged. T4—solution heat treated and naturally aged. T5—artificially aged only, no prior solution treatment. T6—solution heat treated and artificially aged. T7—solution heat treated and over-aged. T8—solution heat treated, cold worked and then artificially aged. T9—solution heat treated, artificially aged then cold worked.

Table 7   283

Alloy type	Condition	Typical composition (%)		Mechanical properties					Physical properties	
		Al	Other	UTS (MPa)	0.2% PS (MPa)	Elongation (%)	Hardness (HB)	El mod (GPa)	Elec resist ($\mu\Omega$-cm)	Therm exp (°C $\times$ 10^{-6})
Wrought materials										
Alloy 1060	O	99.6	—	69	28	40	19	69	2.78	⎫ 21.8
	H18			130	125	6	35	69	2.83	⎬
Alloy 5050	O	Rem	0.8 Mg	145	55	24	35	⎱ 69	3.4	21.8
	H38			220	200	6	63	⎰		
Alloy 6063	O	Rem	0.4 Si, 0.7 Mn	90	50	7	25	⎱ 65.6	2.97	⎱ 21.8
	T6	Rem		195	160	7	70	⎰	3.3	⎰
Alloy 6082	T6	Rem	1 Si, 0.7 Mn, 0.9 Mg	295	255	7	95	68.9		21.8
Alloy 7075	T6	Rem	5 Zn, 2.5 Mg, 1.5 Cu, 0.3 Cr	570	500	11	150	71.6	5.7	21.8
Cast materials										
12% Si	F (die)	Rem	12 Si	270	145	3		71	5.6	20.4
Cu Si	F (sand)	Rem	4 Cu, 3 Si	140	90	2	55	71	5.6	22
Si, Mg	T7	Rem	7 Si, 0.3 Mg	230	150	3	70	72	4.2	21.5

Notes

1 Mechanical and physical properties at 20°C.

2 El Mod—elastic modulus. Elec resist—electrical resistivity. Therm exp—thermal expansion.

3 Typical density: Type 1060—2.7 g/cm^3, 12% Sl-2.66 g/cm^3.

Table 8 Copper and typical alloys for engineering applications.
The data presented are derived from a variety of public sources, including Standard Specifications and manufacturers' brochures, with the aim of illustrating the range of properties typically available. However, these sources interpret differently terms such as 'hard' and 'maximum hard' and clearly the mechanical properties that can be achieved in, say, a 30 mm thick rolled plate will differ from those achievable in a fine wire of the same material.

Alloy type, common name	Condition	Typical composition (%)		Mechanical properties					Physical properties		
		Cu	Other	UTS (MPa)	0.2% PS (MPa)	Elong'n (%)	Hardness	El Mod (GPa)	Elec resist ($\mu\Omega$-cm)	Thermal conductivity (W/mC)	Therm exp (°C $\times$ 10^{-6})
Plain copper—wrought											
Electrolytic tough pitch	Annealed	99.95	0.4 O	220	69	45	RF 40	117	1.7	399	16.8
	Hard			345	310	6	RF 90				
Phosphorus deoxidized	Annealed	99.95	0.02 P	220	69	45	RF 40	117	1.7	340 – 395	17.7
	Hard			380	345	8	RF 95				
Copper–zinc alloys (brasses)—wrought											
95/5 Gliding metal	Annealed	Rem	5 Zn	260	69	45	RF 55	117	3.1		18.5
	Hard			420	370	5	RB 70				
	Max Hard			440	400	4	RB 73				
90/10 Gilding metal	Annealed	Rem	10 Zn	260	83	45	RF 57	117	3.9	188	18.4
	Hard			420	372	5	RB 70				
	Max Hard			495	425	3	RB 78				
80/20	Annealed	Rem	20 Zn	290	90	10	RF 57	110	5.4	138	19.1
	Hard			510	405	7	RB 82				
	Max Hard			625	445	3	RB 91				

Table 8   285

Material	Condition		Composition			% Elong.	Hardness				
70/30 Cartridge brass (Admiralty brass similar)	Annealed Hard Max Hard	Rem	30 Zn (Admiralty brass in addition has 1% Sn)	305 525 680	75 435 500	65 8 3	RF 54 RB 82 RB 93	110	6.2	121	19.9
60/40 Muntz metal	Annealed Hard	Rem	40 Zn	370 480	115 345	45 15	RF 80 RB 75	103	6.2	126	20.8
60/40/Sn Naval brass	Annealed Hard	Rem	38 Zn, 0.8 Sn	410 550	185 395	45 45	RB 60 RB 85	103	6.6	117	21.2
High-tensile brass (manganese bronze)	Annealed Hard	Rem	39 Zn, 1.5 Fe, 1 Sn, 0.1 Mn	450 560	200 410	35 25	RB 35 RB 90	103	7.2		21.2
Copper–zinc alloys (brasses)—cast											
Leaded 70/25 brass		Rem	24 Zn, 1 Sn, 3 Pb	260	80	35	HB 45	96	10		20.7
Leaded 60/40 brass		Rem	37 Zn, 1 Pb, 1 Sn	275	95	20	HB 65	96	8		21.6
High-tensile brass/(Manganese bronze)		Rem	26 Zn, 5 Al, 4 Mn, 3 Fe	790	480	15	HB 210	107	8		19.8
Copper–tin alloys (bronzes)—wrought											
5% phosphor bronze	Annealed Hard	Rem	5 Sn	445 565	205 410	35 25	RB 65 RB 90	120	11	75	17.8
10% phosphor bronze	Annealed Hard Full hard	Rem	10 Sn	430 625 800	180 590 780	68 17 4	RB 49 RB 94 RB 102	110	16	62	18.4

continued overleaf

Table 8 (*continued*)

Alloy type, common name	Condition	Typical composition (%)		Mechanical properties					Physical properties		
		Cu	Other	UTS (MPa)	0.2% PS (MPa)	Elong'n (%)	Hardness	El Mod (GPa)	Elec resist (μΩ-cm)	Thermal conductivity (W/mC)	Therm exp (°C × 10⁻⁶)
Copper–tin alloys (bronzes)—cast											
10% tin/phosphor bronze		Rem	10 Sn	370	190	20	HB 90	111	15.5		10.2
88/10/2 leaded gunmetal		Rem	10 Sn, 2 Zn, 1 Pb	241	125	18	HB 60	100			18
Copper aluminium alloys—wrought											
5% Al bronze	Annealed / Hard	Rem	5 Al	410 / 690	170 / 430	60 / 8	RB 48 / RB 93	126	11	85	
10% Al bronze	Annealed / Hard	Rem	10 Al	450 / 600	175 / 470	40 / 10	HB 90 / HB 160	123	13	80	16
Copper–aluminium—cast											
10% Al bronze	Annealed	Rem	10 Al, 1 Fe	480	170	24	HB 120	113	13		17
Copper–silicon—wrought											
Silicon–bronze	Annealed / Hard	Rem	3 Si 1 Mn	380 / 650	145 / 400	63 / 8	RB 40 / RB 93	103	21	50	
Copper–silicon—cast											
Silicon–bronze	As cast	Rem	4 Si, 4 Sn, 2 Zn, 1 Al, 1 Mn	350	80	10	HB 100	105	22		

Therm exp (°C × 10⁻⁶)

Table 8   287

Copper–nickel—wrought

Material	Condition	Cu	Composition								
90/10 Cupronickel	Annealed	Rem	10 Ni, 1.2 Fe	300	110	40	RB 65	150	17	42	17
	Hard			410	390	10	RB 100				
70/30 Cupronickel	Annealed	Rem	30 Ni	280	125	45	RB 36	124	37	21	16
	Hard			570	540	3	RB 85				
65/18 Nickel silver	Annealed	Rem	18 Ni, 17 Zn	400	170	40	RB 40		29	28	16.2
	Hard			580	510	3	RB 87				
65/12 nickel silver or German silver	Annealed	Rem	12 Ni, 23 Zn	380	145	40	RB 37		22	30	16
	Hard			580	515	4	RB 89				
	Full hard			640	544	2	RB 92				

Copper–nickel-cast

Material	Cu	Composition							
12% nickel silver or leaded Ni brass	Rem	12 Ni, 20 Zn, 9 Pb, 2 Sn	245	125	15	HB 55	110	30	
25% nickel silver or leaded Ni bronze	Rem	25 Ni, 2 Zn, 1 Pb, 5 Sn	380	200	20	HB 140		45	

Copper–beryllium alloy—wrought

Material	Condition	Cu	Composition								
Beryllium–copper	Soln. treat	Rem	1.8 Be, 0.25 Co	480	220	45	HB 100	159	9	84	17
	Cold rolled, precipitation hardened			1300	1200	2	360				

Notes

1 Hardness: HB—Brinell. RB, RC, RF—Rockwell Scales B, C or F.
2 Mechanical and physical properties at 20 °C.
3 UTS—ultimate tensile strength. PS—proof strength. El mod-elastic modulus. Elec resist—electrical resistivity. Therm exp—coefficient of thermal expansion. Therm conductivity. Soln. treat—solution treated.
4 The mechanical and, to a large extent, the physical properties of Admiralty brass are similar to plain 70/30 brass.

Table 9

Table 9 Magnesium and typical alloys for engineering applications

The alloy designations follow the ASTM system which is generally accepted internationally. In this system the first two letters indicate the two principal alloying elements as detailed below and the next digits indicate their percentage. The following letter differentiates between alloys with similar principal alloying elements, the letter being allocated alphabetically in order of registration as an alloy. The final letter and digit indicate the condition and properties as described in Table 7, Aluminium alloys.

Alloy designation	Condition	Typical composition (%)		Mechanical properties					Physical properties	
		Mg	Other	UTS (MPa)	0.2% PS (MPa)	Elongation (%)	Hardness (Brinell)	El Mod (GPa)	Elec resist ($\mu\Omega$-cm)	Density (g/cm^3)
AZ81A	Sand cast	Rem	10 Al, 0.15 Mn, 0.7 Zn	275	83	15	56	45	15	1.8
ZK51A	Sand cast	Rem	5 Zn, 0.7 Zr	275	165	8	65	45	6.2	1.8
AZ91A	Die cast	Rem	9 Al, 0.13 Mn, 0.7 Zn	230	160	3	60	45	17	1.81
ZK60A-T5	Extrusion	Rem	5.5 Zn, 0.5 Zr	360	300	11	85	45	6	1.83
AZ31B-H24	Plate/sheet	Rem	3 Al, 1 Zn	290	220	15	73	45	9.2	1.77

Notes

1 Mechanical and physical properties at 20 °C.

2 UTS—ultimate tensile strength. PS—proof strength. El mod—elastic modulus. Elec resist—electrical resistivity.

3 A—aluminium. B—bismuth. C—copper. D—cadmium. E—rare earth. F—iron. H—thorium. K—zirconium. L—Beryllium. M—manganese. N—nickel. P—lead. Q—silver. R—chromium. S—silicon. T—tin. Z—zinc.

Table 10   289

Table 10 Nickel and typical alloys for engineering applications

Alloy type or trade name	Condition	Typical composition (%)		Mechanical properties				Physical properties		
		Ni	Others	UTS (MPa)	0.2% PS (MPa)	Elongation (%)	Hardness (HB)	El mod (GPa)	Elec resist ($\mu\Omega$-cm)	Therm exp (°C $\times$ 10^{-6})
Commercial purity, As strip	Annealed / Hard rolled	99.5	Total nickel includes cobalt	340/550 620/900	70/200 480/800	35/50 2/15	90/120 RB 96	207	9.5	13
Monel® As sheet	Annealed / Hard rolled	70	30 copper	480/580 685/830	170/310 620/770	30/50 2/15	RB 61/73 RB 94	179	48	14
K Monel® As sheet	Annealed / Hard roll, aged	67	30 copper 3 aluminium	620/830 1170/1380	275/415 860/1000	30/45 2/10	RB 140/180 RC 25/32	179	58	14
Inconel® As sheet	Annealed / Hard rolled	76	16 chromium 8 iron	550/685 1000/1175	200/310 830/1105	35/50 2/10	RB 65/85 RC 27/30	214	98	11.5
Hastelloy® B As sheet	Solution treated	62	28 molybdenum 5 iron	900	400	30	200	182	135	10
Nichrome® As wire	Annealed / Hard drawn	80	20 Chromium	650 1140		55 1	RB 85 RB 100	213	108	17.3
Nimonic® 80 As bar	Soln. treated & precipitation hardened	Rem	20 chromium 2.25 titanium 1.25 aluminium	1200	800	20	370		117	11.1

Notes
1 Alloy names indicated ® are registered trade marks.
2 Mechanical and physical Properties at 20 °C.
3 UTS—ultimate tensile strength. El mod—elastic modulus. Elec resist—electrical resistivity. Therm exp—thermal expansion.
4 Hardness is Brinell unless otherwise stated. RB, RC—Rockwell scales B or C.

Table 11 Tin and typical alloys for engineering and other applications

Alloy designation or common name	Condition	Typical composition (%) Tin	Other	Mechanical properties UTS (MPa)	0.2% PS (MPa)	Elongation (%)	Hardness (Brinell)	El mod (GPa)	Physical properties Elec resist ($\mu\Omega$-cm)	Density (g/cm^3)
'Pure'	Cast	Rem	–	20		60	5	50	12	5.76
Hard tin	Rolled strip	Rem	0.4 copper	22						5.7
Solder (plumber's)	Cast	Rem	30 lead	45		35	12		14	8.32
Solder (electrical)	Cast	Rem	37 lead	50		10	15		14	8.42
Babbit	Cast on steel	Rem	7 antimony, 3.5 copper	76	65		HV$_5$ 27	43		7.34
Pewter (one of a variety)	Rolled, annealed	Rem	7 antimony, 2 copper	60		40	9	53		7.3
	Cast						24			

Notes

1 Mechanical and Physical Properties at 20 °C

2 UTS—ultimate tensile strength. PS—proof strength. El mod—elastic modulus. Elec resist—electrical resistivity.

3 Hardness Brinell unless stated. HV—Vickers Hardness.

Table 12 Titanium and typical alloys for engineering applications

Alloy designation	Condition	Typical composition (%)		Mechanical properties					Physical properties	
		Ti	Other	UTS (MPa)	0.2% PS (MPa)	Elongation (%)	Hardness (Brinell)	El mod (GPa)	Elec resist (μΩ-cm)	Density (g/cm³)
Comm. pure	Sheet annealed	99	–	600	275	28		105	50	4.5
317	Sheet annealed	Rem	5 Al, 2.5 Sn	860	825	18				4.46
318	Sheet annealed	Rem	6 Al, 4 V	1120	1060	12		106	162	
	Rod aged	Rem	6 Al, 4 V	1180	1050	10				
680	Rod quenched and aged	Rem 11	Sn, 4 Mo, 2.2 Al, 0.2 Si	1350	1200	12		106		

Notes
1 IMI designations
2 Mechanical and physical Properties at 20 °C.
3 UTS—ultimate tensile strength. PS—proof strength. El mod—elastic modulus. Elec resist—electrical resistance.

Table 13 Zinc and typical alloys for engineering applications

Alloy designation	Condition	Typical composition (%)		Mechanical properties					Physical properties	
		Zn	Other	UTS (MPa)	0.2% PS (MPa)	Elongation (%)	Hardness (Brinell)	El mod (GPa)	Elec resist (μΩm-cm)	Density (g/cm³)
Commercially pure zinc	Cold rolled	Rem	0.06 Pb, 0.06 Cd	175		50	43	105	6	7.14
B.S.3436, Zn 4	Hard rolled	Rem	1.35 Pb, limits on others	241		2.5				
Cu, Mg	Cold rolled	Rem	1 Cu, 0.01 Mg	250		20	61		6.3	7.18
Al, Mg 'MAZAC 3' ®	Die cast	Rem	4 Al, 0.04 Mg	280		10	82		6.3	6.6

Notes
1 Mechanical and physical Properties at 20 °C.
2 UTS—ultimate tensile strength. PS—proof strength. El mod—elastic modulus. Elec resist—electrical resistance.
3 ® —trade name.

Table 14 Typical properties of non-metallic materials

Material Alternatives and abbreviations	Density (g/cm³)	Melting point or service range (°C)	Therm exp (per°C × 10^{-6})	Therm con (W/m°C)	Resisistivity ($\mu\Omega$-cm)	Elast mod (GPa)	TS (MPa)
Asbestos (hard packed)	0.58			0.19			
Brick	1.8		3–9	0.6			
Carbon as diamond	3.51	>3600	1.3	165	10^{12}		
Carbon as graphite	2.25	3500	2	160	10^{-5}		
Concrete—aged	1.5–2.4		10	1.7		20–40	
Cork	0.18			0.11			
Ebonite	1.15		85	0.17	10^{8}		
Glass (window)	2.5		8	0.93	10^{12}	50–100	
Marble	2.7		12	2–4	10^{9}	40–50	
Mica	2.8		3	0.4–0.6	10^{14}	160–200	
Paper (dry)	0.7–1.2			0.18	10^{10}		
Plaster of Paris	0.97		80	1.3	10^{19}		
Porcelain	2.4	1900	2–5	1	10^{12}	70–80	
Quartz	2.2	1700	0.4	0.22	10^{20}	70	
Wood, general, along grain	0.4–0.8		5	0.3	10^{8}–10^{14}	10–20	
Wood, general, across grain			40	0.1–0.2			
Softwoods							
Douglas fir 12% moisture (&41%)	0.5 (0.6)					12.7 (10.4)	93 (54)
Scots Pine 12% moisture (& 89%)	0.38					10 (7.3)	89 (46)
Hardwoods							
European oak 12% moisture (& 89%)	0.69 (1.07)					10.1 (8.3)	97 (59)
African mahogany 12% moist (& 64%)	0.53 (0.7)					9 (7.4)	78 (54)
European birch 12% moisture (& 76%)	0.66					13 (9.9)	123 (63)
Malay kerung 12% moisture (& 52%)	0.7					17 (16)	144 (83)
Burma teak 10.6% moisture (& 48%)	0.6 (0.8)					10 (8.8)	106 (84)
Balsa	0.04/0.3					3.3	20

Table 14   293

Material	Density	Service range (SR) / Melt point	Resistivity	Therm con (0.15)	Therm exp	Elast mod (10^{-3})	TS
Rubbers							
Natural	1.2–1.3	SR −50/+80	10^{13}	0.15			20
Butyl rubber, (isobutylene isoprene)		SR −50/+100					20
Butadiene styrene rubber (SBR)(GR-S)		SR −50/+80					24
Nitrile rubber, (butadiene acrylonitrile)		SR −50/+125					28
Neoprene rubber, (polychloroprene)		SR −50/+100					25
Polyurethane rubber		SR −55/+125					36
Silicone rubber		SR −105/+300					10
Acrylonite–butadiene–styrene (ABS)	1.04	SR < 110		0.13–0.3	80–120	1.5–3.5	30–60
Acrylic (PMMA)	1.18			0.13–0.2	60–70	2.7–3.2	50–80
Phenol formaldehyde (bakelite(PF))	1.3	SR < 120	10^5	0.19	30	5.2–7	35–55
Ditto wood filled	1.32–1.45	SR < 150				5.5–8	40–50
Ditto asbestos filled	1.6–1.8	SR < 180				5–10	30–55
Epoxy resin (EP)	1.1–1.4	SR < 200 / 265		0.2–0.9	60–110	2.2–40	30–70
Polyamide (e.g. Nylon 66) (PA)	1.14	265		0.17–0.24	80–100	2–2.8	65–85
Polycarbonate (PC)	1.2	SR < 150		0.14–0.2	40–70	2.1–2.4	55–65
Polyethylene (polythene) (PE)	0.93	115–135		0.25–0.6	130–250	0.2–1.4	8–35
Polyethylene trephthalate (polyester) (PET)	1.3	265		0.15–0.25	70–120	2.1–4.4	50–70
Ditto +50–80% (weight) long fibre glass	2					20–50	400–1200
Ditto +58% (vol) carbon fibre	1.5					189	1000
Polyformaldehyde (polyacetal)	1.4	180		0.16–0.24	80–150	2.1–4.4	50–70
Polypropylene (PP)	0.9	175		0.12–0.17	60–100	1–1.6	25–40
Polystyrene (PS)	1.05	SR < 100		0.13–0.17	70–80	3	45–65
Polystyrene, toughened	1.05					2–2.8	26–38
Polytetrafluoroethylene. (PTFE) (Teflon)	2.2	327 SR −250/+250	10^{20}	0.2–0.25	100–140	0.3–0.6	20–35
Polyvinylchloride, unplasticized (UPVC)	1.2	SR < 80		0.12–0.22	50–100	2.4–4	52–58

Notes
1 Properties at 20 °C.
2 Melt point—melting point. Service range—useful service temperature range. Therm exp—Linear coefficient of thermal expansion. Therm con—thermal conductivity. Elast mod—elastic modulus.
TS—tensile strength/fracture stress.
3 For material other than wood the properties quoted are typical figures. For woods the strengths are maximum modulus of rupture measured in three-point bending. For permissible safe working stresses etc see BS 5268. Structural use of timber.

Table 15   Physical properties of the elements

Element	Symbol	Atomic number	Atomic weight	Density (g/cm^3)	Melting point (°C)	Boiling point (°C)	Specific heat $(cal/gm/°C)$	Coefficient of linear thermal expansion $(°C \times 10^{-6})$	Thermal conductivity $(cal/cm^2/s/°C)$	Electrical resistivity $(\mu\Omega\text{-cm})$	Elastic modulus (kN/mm^2)
Actinium	Ac	89	227	10.06	1230	3200					
Aluminium	Al	13	26.98	2.7	660	2350	0.22	23.6	0.53	2.65	67
Americium	Am	95	243	13.7	990	2600					
Antimony	Sb	51	121.76	6.69	630.5	1600	38.3	9	0.05	35	78
Argon	A	18	39.9	1.784×10^{-3}	−189.4	−185.8	0.13		0.4×10^{-4}		
Arsenic	As	33	74.92	5.77	817*	613 S	0.08	4.7		33.3	
Astatine	At	85	211	1.85	300	350					
Barium	Ba	56	137.36	3.6	714	1640	0.07	18		50	
Berkelium	Bk	97	247	14.79	986						
Beryllium	Be	4	9.01	1.85	1277	2470	0.45	11.6	0.35	4	290
Bismuth	Bi	83	209	9.8	271.3	1564	0.03	13.3	0.02	106.8	32
Boron	B	5	10.82	2.34	2030	3700	0.31	8.3		1.8×10^{12}	
Bromine	Br	35	79.9	3.12	−7.2	59	0.07				
Cadmium	Cd	48	112.4	8.65	320.9	765	0.06	29.8	0.22	6.83	60
Calcium	Ca	20	40.08	1.55	838	1440	0.15	22.3	0.3	3.91	23
Caesium	Cs	55	132.91	1.9	28.7	690	0.05	97		20	
Californium	Cf	98	251								
Carbon, graphite	C	6	12.01	2.26	3700 S	5000 S	0.17	0.6–4	0.06	1375	4.8
Cerium	Ce	58	140.13	6.77	804	3470	0.05	8	0.03	75	41
Chlorine	Cl	17	35.46	3.214×10^{-3}	−101	−34.7	0.12		0.17×10^{-4}		

Table 15 295

Chromium	Cr	24	52.01	7.19	1875	2665	0.11	6.2	0.16	12.9	248
Cobalt	Co	27	58.94	8.85	1495	2900	0.1	13.8	0.17	6.24	206
Columbium*	Cb	41	92.91	8.57	2450	4927	0.06	7.31	0.13	12.5	110
Copper	Cu	29	63.54	8.94	1083	2590	0.09	16.5	0.94	1.69	
Curium	Cm	96	247	13.3	1340						
Dysprosium	Dy	66	162.51	8.55	1407	2330	0.04	9	0.02	57	83
Einsteinium	E	99	254								
Erbium	Er	68	167.27	9.15	1497	2630	0.04	9	0.02	107	110
Europium	Eu	63	152	5.25	826	1490	0.04	26	0.02	90	
Fermium	Fm	100	253								
Flourine	F	9	19	1.70×10^{-3}	−219.6	−188.2	0.18				
Francium	Fr	87	223		27	650					
Gadolinium	Gd	64	157.26	7.86	1312	2730	0.07	4	0.02	140.5	55/96
Gallium	Ga	31	69.72	5.91	29.7	2237	0.08	18	0.08 at M Pt	17.4	
Germanium	Ge	32	72.6	5.32	937	2830	0.07	5.7	0.14		
Gold	Au	79	197	19.32	1063	2970	0.03	14.2	0.71	2.35	80
Hafnium	Hf	72	178.58	13.09	2224	5400	0.04	5.9	0.22	35.1	
Helium	He	2	4	0.179×10^{-3}	−269.7	−268.9	1.25		3.32×10^{-4}		76
Holmium	Ho	67	164.94	6.79	1461	2330	0.04			87	
Hydrogen	H	1	1	0.089×10^{-3}	−259.19	−252.7	3.45		4.06×10^{-4}		
Indium	In	49	114.82	7.31	156.2	2070	0.06	33	0.06	8.37	11.7
Iodine	I	53	126.91	4.94	113.7	183	0.05	93	10.4×10^{-4}	1.3×10^{15}	
Iridium	Ir	77	192.2	22.5	2454	5300	0.03	6.8	0.14	5.3	520
Iron	Fe	26	55.85	7.87	1536.5	2860	0.11	11.76	0.18	9.71	196.5
Krypton	Kr	36	83.8	3.7×10^{-3}	−157.3	−152	0.11		0.21×10^{-4}		

continued overleaf

Table 15 *(continued)*

Element	Symbol	Atomic number	Atomic weight	Density (g/cm³)	Melting point (°C)	Boiling point (°C)	Specific heat (cal/gm/°C)	Coefficient of linear thermal expansion (°C $\times 10^{-6}$)	Thermal conductivity (cal/cm²/s/°C)	Electrical resistivity ($\mu\Omega$-cm)	Elastic modulus (kN/mm²)
Lanthanum	La	57	138.92	6.19	920	3470	0.05	5	0.03	57	72
Lead	Pb	82	207.21	11.36	327.4	1750	0.03	29.3	0.08	20.64	16
Lithium	Li	3	6.94	0.53	180.5	130	0.79	56	0.17	8.55	
Lutetium	Lu	71	174.99	9.85	1652	1930	0.04			79	
Magnesium	Mg	12	24.32	1.74	650	1102	0.25	27.1	0.37	4.45	44
Manganese	Mn	25	54.94	7.43	1245	2150	0.12	22		185	159
Mendelevium	Mv	101	256								
Mercury	Hg	80	200.61	13.55	−38.86	357	0.03		0.02	98.4	
Molybdenum	Mo	42	95.95	10.22	2610	4800	0.07	4.9	0.34	5.2	324
Neodymium	Nd	60	144.27	7	1019	3180	0.05	6	0.03	64	
Neon	Ne	10	20.18	0.900×10^{-3}	−248.6	−246			1.1×10^{-4}		
Neptunium	Np	93	237	20.45	640						
Nickel	Ni	28	58.71	8.9	1453	2730	0.11	13.3	0.22	6.84	207
Niobium*	Nb	41	92.91	8.57	2468	4927	0.07	7.31	0.13	12.5	
Nitrogen	N	7	14.01	1.25×10^{-3}	−201	−195.8	0.25		6×10^{-5}		
Nobelium	No	102	247								
Osmium	Os	76	190.2	22.57	3000	5500	0.03	4.6		9.5	558
Oxygen	O	8	16	143×10^{-3}	−218.83	−183	0.22		5.9×10^{-5}		
Palladium	Pd	46	106.7	12.02	1552	3000	0.06	11.76	0.17	10.8	112

Table 15   297

Element											
Phosphorus, white	P	15	30.97	1.83	44.25	280	0.18	125	0.18	1×10^{17}	152
Platinum	Pt	78	195.09	21.54	1769	4530	0.03	8.9	0.17	10.6	
Plutonium	Pu	94	242	19.5	640	3235	0.03	55	0.02	141	
Polonium	Po	84	210	9.4	245	960					
Potassium	K	19	39.1	0.86	63.7	770	0.18	83	0.24	6.15	
Praseodymium	Pr	59	140.92	6.79	919	3020	0.05	4	0.03	68	48/97
Promethium	Pm	61	145		1030	2700					
Proactinium	Pa	91	231.1	15.4	1230	4000					
Radium	Ra	88	226.05	5	700	1500					
Radon	Rn	86	222	9.96×10^{-3}	−71	−61.8					
Rhenium	Re	75	186.22	21.04	3180	5900	0.03	6.7	0.17	19.3	460
Rhodium	Rh	45	102.91	12.44	1966	3700	0.06	8.3	0.21	4.51	290
Rubidium	Rb	37	85.48	1.53	38.9	688	0.08	90		12.5	
Ruthenium	Ru	44	101.1	12.2	2310	4100	0.06	9.1		7.6	413
Samarium	Sm	62	150.35	7.49	1072	1630	0.04			88	
Scandium	Sc	21	44.96	2.99	1539	2730	0.13			61	55
Selenium	Se	34	78.96	4.79	217	685	0.08	37	$7{-}18 \times 10^{-4}$	12	58
Silicon	Si	14	28.09	2.33	1410	2680	0.16	2.8–7.3	0.2	10	113
Silver	Ag	47	107.88	10.49	961	2160	0.06	19.68	1	1.59	76
Sodium	Na	11	22.99	0.97	97.82	892	0.3	71	0.32	4.2	
Strontium	Sr	38	87.63	2.6	768	1380	0.18	100		23 (0 °C)	
Sulphur, yellow	S	16	32.07	2.07	119	445	0.18	64	6.3×10^{-4}		
Tantalum	Ta	73	180.95	16.6	3000	5425	0.03	6.5	0.13	12.45	186

continued overleaf

Table 15 *(continued)*

Element	Symbol	Atomic number	Atomic weight	Density (g/cm³)	Melting point (°C)	Boiling point (°C)	Specific heat (cal/gm/°C)	Coefficient of linear thermal expansion (°C × 10⁻⁶)	Thermal conductivity (cal/cm²/s/°C)	Electrical resistivity (μΩ-cm)	Elastic modulus (kN/mm²)
Technetium	Tc	43	98		2200	4600					
Tellurium	Te	52	127.61	6.24	449.5	989.8	0.05	16.75	0.01	4.36×10^5	41
Terbium	Tb	65	158.93	8.25	1356	2530	0.04	7			
Thallium	Tl	81	204.39	11.85	303	1457	0.03	28	0.09	18	
Thorium	Th	90	232.05	11.66*	1750	4790	0.03	12.5	0.9	13	73
Thulium	Tm	69	168.94	9.31	1545	1720	0.04			79	
Tin	Sn	50	118.7	7.3*	231.9	2700	0.05	23	0.15	11	42.5
Titanium	Ti	22	47.9	4.5	1670	3260	0.02	8.4	6.6	42	116
Tungsten	W	74	183.86	19.3	3410	5500	0.03	4.6	0.4	5.65	345
Uranium	U	92	238.07	19.07	1132.2	4000	0.03	6.8–14	0.07	30	165
Vanadium	V	23	50.95	6.1	1900	3400	0.12	8.3	0.07	25	130
Xenon	Xe	54	131.3	5.89×10^{-3}	−111.9	−108			1.24×10^{-4}		
Ytterbium	Yb	70	173.04	6.96	824	1530	0.35	25		29	
Yttrium	Y	39	88.92	4.47	1509	3030	0.07		0.04	57	390
Zinc	Zn	30	65.38	7.13	419.5	787.11	0.09	39.7	0.27	5.92	97
Zirconium	Zr	40	91.22	6.49	1852	4400	0.07	5.85	0.21	40	93

Notes
1 Density, specific heat, coefficient of expansion and thermal conductivity measured at 20 °C unless otherwise stated.
2 Columbium (the preferred American term) and niobium (The preferred European term) are the same element. Both are in the table.
3 Thorium density—arc melted iodide.
4 Tin density—metallic (beta) tin.
5 Arsenic melting point at 28 atmospheres.
6 S—sublimes.

Table 16   299

Table 16 Periodic Table of the elements
Groups, periods, descriptions (where applicable), atomic numbers and atomic weights

Metals — To heavy line including lanthanides and actinides

Transition metals — Including lanthanides and actinides

Reactive metals

Semi metals (six shaded) — Non metals*

Halogens

Noble or Rare gases

	Group 1A	Group 2A	Group 3B	Group 4B	Group 5B	Group 6B	Group 7B	Group 8			Group 1B	Group 2B	Group 3A	Group 4A	Group 5A	Group 6A	Group 7A	Group 0
Period 1	1 H 1.00797																	2 He 4.0026
Period 2	3 Li 6.939	4 Be 9.012											5 B 10.81	6 C 12.01	7 N 14.007	8 O 15.999	9 F 18.998	10 Ne 20.183
Period 3	11 Na 22.90	12 Mg 24.312											13 Al 26.98	14 Si 28.086	15 P 30.97	16 S 32.064	17 Cl 35.453	18 Ar 39.95
Period 4	19 K 39.102	20 Ca 40.08	21 Sc 44.96	22 Ti 47.90	23 V 50.94	24 Cr 52.00	25 Mn 54.94	26 Fe 55.85	27 Co 58.93	28 Ni 58.71	29 Cu 63.54	30 Zn 65.37	31 Ga 69.72	32 Ge 72.59	33 As 74.92	34 Se 78.96	35 Br 79.91	36 Kr 83.80
Period 5	37 Rb 85.47	38 Sr 87.62	39 Y 88.91	40 Zr 91.22	41 Nb 92.91	42 Mo 95.94	43 Tc 99.0	44 Ru 101.07	45 Rh 102.91	46 Pd 106.4	47 Ag 107.87	48 Cd 112.4	49 In 114.82	50 Sn 118.69	51 Sb 121.75	52 Te 127.6	53 I 126.9	54 Xe 131.3
Period 6	55 Cs 132.91	56 Ba 137.34	57 La 138.91 *La	72 Hf 178.49	73 Ta 180.95	74 W 183.85	75 Re 186.2	76 Os 190.2	77 Ir 192.2	78 Pt 195.1	79 Au 196.97	80 Hg 200.59	81 Tl 204.37	82 Pb 207.19	83 Bi 208.98	84 Po 210	85 At 210	86 Rn 222
Period 7	87 Fr 223	88 Ra 226	89 Ac 227 °Ac	104	105	106	107	108	109									

Lanthanide series or rare earths *La

58 Ce 140.12	59 Pr 140.91	60 Nd 144.24	61 Pm 145	62 Sm 150.35	63 Eu 151.96	64 Gd 157.25	65 Tb 158.92	66 Dy 162.50	67 Ho 164.93	68 Er 167.26	69 Tm 168.93	70 Yb 173.04	71 Lu 174.97

Actinide series °Ac

90 Th 232.04	91 Pa 231	92 U 238.03	93 Np 237	94 Pu 242	95 Am 243	96 Cm 247	97 Bk 249	98 Cf 251	99 Es 254	100 Fm 253	101 Md 256	102 No 253	103 Lw 257

Note

1 In this context metals are defined as elements which readily release valency electrons; non-metals accept or share electrons. The elements of Groups 4A to 6A, particularly those having higher atomic numbers, have some other characteristics of metals.

Table 17   Strength-to-hardness coversions
The correlations between strength and hardness and between the various hardness systems can be erratic. They are probably most reliable for carbon and low-alloy steels in the heat-treated condition. The conversions are less reliable for work hardened material and for non-ferrous materials

MPa	Tonsf/in^2	PSI × 1000	Brinell	Rockwell	Vickers
380	24.6	55.11	105		111
400	25.9	58.02	110		116
420	27.19	60.92	116		122
440	28.49	63.82	123	B 74	130
460	29.78	66.72	130	B 76	137
480	31.08	69.62	138	B 79	145
500	32.37	72.52	146	B 82	153
520	33.67	75.42	154	B 84.5	163
540	34.96	78.32	163	B 87	172
560	36.26	81.22	171	B 89	180
580	37.55	84.12	176	B 90	186
600	38.85	87.02	179	B 91	189
620	40.14	89.92	184	B 92	193
640	41.44	92.82	189	B 93.5	200
660	42.73	95.72	195	B 94	206
680	44.03	98.63	202	B 96	213
700	45.32	101.53	208	B 97	219
720	46.62	104.43	214	B 98	225
740	47.91	107.33	218	B 99	229
760	49.21	110.23	224	C 20	236
780	50.5	113.13	231	C 21	243
800	51.8	116.03	237	C 22	250
820	53.09	118.93	241	C 23	253
840	54.39	121.83	245	C 24	257
860	55.68	124.73	251	C 25	265
880	56.98	127.63	257	C 26	271
900	58.27	130.53	264	C 26.5	278
920	59.57	133.43	271	C 28	285
940	60.86	136.34	276	C 29	291
960	62.16	139.24	282	C 29.5	297
980	63.45	142.14	289	C 30	305
1000	64.75	145.04	296	C 31.5	312
1020	66.04	147.94	302	C 32	319
1040	67.34	150.84	307	C 32.5	325
1060	68.63	153.74	313	C 33.5	331
1080	69.93	156.64	319	C 34	337
1100	71.22	159.54	325	C 35	345
1120	72.52	162.44	331	C 35.5	350
1140	73.81	165.34	236	C 36	355
1160	75.11	168.24	341	C 36.5	360
1180	76.4	171.14	349	C 37.5	368
1200	77.7	174.05	355	C 38	375
1225	79.32	177.67	360	C 39	380

Table 17   (*continued*)

MPa	Tonsf/in^2	PSI × 1000	Brinell	Rockwell	Vickers
1250	80.94	181.3	369	C 40	389
1275	82.55	184.92	377	C 41	398
1300	84.17	188.55	384	C 41.5	405
1325	85.79	192.17	392	C 42.5	413
1350	87.41	195.8	399	C 43	421
1375	89.03	199.43	406	C 43.5	430
1400	90.65	203.05	411	C 44	435
1425	92.27	206.68	416	C 44.5	441
1450	93.89	210.3	425	C 45	450
1475	95.5	213.93	431	C 46	457
1500	97.12	217.56	436	C 46.5	466
1525	98.74	221.19	443	C 47	473
			461	C 48.5	491
At these levels the strength to hardness			477	C 49.5	508
correlation is poor.			495	C 51	528
The Brinell figures above 450 BH refer to tests			514	C 52	547
utilizing a tungsten carbide ball, the so-called			534	C 53.5	569
'modified Brinell test'.			555	C 54.5	591
			578	C 56	615
			601	C 57	640

Table 18   Useful conversion factors
The International System of Units (SI), prefixe

Name	Symbol	Multipliers	
tera	T	10^{12}	1 000 000 000 000
giga	G	10^9	1 000 000 000
mega	M	10^6	1 000 000
kilo	k	10^3	1 000
hecto	h	10^2	100
deca	da	10	10
deci	d	10^{-1}	0.1
centi	c	10^{-2}	0.01
milli	m	10^{-3}	0.001
micro	μ	10^{-6}	0.000 001
nano	n	10^{-9}	0.000 000 001
pico	p	10^{-12}	0.000 000 000 001
femto	f	10^{-15}	0.000 000 000 000 001
atto	a	10^{-18}	0.000 000 000 000 000 001

continued overleaf

Table 18 (*continued*)

Length

1 mile	=	1.609 km
1 yard	=	0.914 m
1 foot	=	0.305 m
1 inch	=	25.4 m
1 'thou', 'mil'*, 'milli-inch'	=	25.4 μm ('micron')
1 nautical mile	=	1.853 km
1 angstrom	=	10^{-10} m

*'mil' in this context refers to a thousandth of an inch but in other contexts it may refer to a thousandth of a metre. It is a term to be viewed with caution.

Area

1 mile^2	=	2.59 km^2 or 259.0 ha
1 acre (4840 yd^2)	=	4046.86 m^2 or 0.4047 ha
1 yd^2	=	0.836 m^2
1 ft^2	=	0.0929 m^2 or 929 cm^2
1 in^2	=	6.452 cm^2 or 645.2 mm^2
1 ha (hectare)	=	10^4 m^2

Volume

1 yd^3	=	0.765 m^3
1 ft^3	=	0.028 m^3 or 28.31 dm or 1 (litre)
1 in^3	=	16.39 cm^3
1 gallon (US)	=	3.785 l
1 gallon (British)	=	4.546 l
1 quart	=	1.137 l
1 pint	=	0.568 l

Velocity and acceleration

1 mph	=	1.609 km/h or 0.4470 m/s
1 ft/s	=	0.3048 m/s
1 knot	=	1.853 km/h
1 g (gravity) 32 ft/s^2	=	9.81 m/s^2

Mass

1 ton	=	1016 kg
1 cwt (112 lb)	=	50.80 kg
1 lb (pound)	=	0.454 kg
1 oz	=	28.35 g
1 oz tr or apoth	=	31.1 g
1 tonne	=	1000 kg

Volume/distance ratios, i.e. fuel consumption

1 gal/mile (US)	=	2.352 l/km
1 gal/mile (British)	=	2.824 l/km
1 mile/gal (US)	=	0.425 km/l
1 mile/gal (British)	=	0.35 km/l

Density

1 lb/ft^3	=	16.019 kg/m^3
1 lb/in^3	=	27.68 g/cm^3
1 lb/gal	=	0.09978 kg/l

Table 18   303

Table 18   (*continued*)

Force		
1 tonf	=	9.964 kN
1 lbf	=	4.448 N
1 ozf	=	0.278 N
1 pdl (poundal)	=	0.138 N
1 kgf or kp (kilopond)	=	9.806 N

Torque (moment of force)

1 tonf ft	=	3.04 kNm
1 lbf ft	=	1.356 Nm
1 lbf in	=	0.1130 Nm

Pressure (stress)

1 tonf$/$ft^2	=	107.2 kPa
1 tonf$/$in^2	=	15.444 MPa or N$/$mm^2
1 lbf$/$in^2 (psi)	=	6.895 kPa or 68.95 mbar
1 kgf$/$mm^2	=	9.81 MPa
1 atm (standard atmosphere)	=	101.3 kPa
1 bar	=	10^5 Pa
1 inH$_2$O	=	249.089 Pa or 2.490 89 mbar
1 inHg	=	3.386 kPA or 33.86 mbar
1 mmHg (torr)	=	133.3 Pa
1 N$/$m^2	=	1 Pa

Energy and power etc.

1 therm	=	105.5 MJ
1 calorie (IC)	=	4.19 J
1 erg	=	10^7 J
1 kWh	=	3.6 MJ
1 Btu	=	1.055 kJ
1 hp	=	745.7 W
1 Btu$/$lb	=	2326 J$/$kg
1 Btu$/$ft^3	=	37.26 J$/$l
1 ft lbf	=	1.356 J

Thermal conductivity

1 Btu ft$/$ft^2 h °F	=	1.730 73 W$/$m°C or Jm$/$m^2 s °C
1 Btu in$/$ft^2 s °F	=	519.22 W$/$m °C or Jm$/$m^2 s °C
1 kcal m$/$m^2 h °C	=	1.163 W$/$m°C (J$/$m^{2}s°C)
1 cal cm$/$cm^2 s °C	=	418.68 W$/$m°C (J$/$m^{2}s°C)

Thermal capacity per unit mass = specific heat

1 Btu$/$lb °F	=	4186.8 J$/$g °C
1 cal$/$g °C	=	4.1868 J$/$g °C

Heat flow rate

1 cal$/$s	=	4.1868 W (J$/$s)
1 Btu$/$hr	=	0.293 071 W (J$/$s)

Table 19 Stress conversions

Basic unit is MPa $=$ MN/m^2 $=$ N/mm^2. MPa $\times$ 0.064 749 $=$ Tf/in^2. MPa $\times$ 145.038 $=$ psi. MPa $\times$ 0.101 971 6 $=$ kgf/mm^2.

MPa	Tonsf/in^2	PSI $\times$ 1000	kgf/mm^2
1	0.06	0.15	0.1
2	0.13	0.29	0.2
3	0.19	0.44	0.31
4	0.26	0.58	0.41
5	0.32	0.73	0.51
7	0.45	1.02	0.71
10	0.65	1.45	1.02
15	0.97	2.18	1.53
20	1.29	2.9	2.04
25	1.62	3.63	2.55
30	1.94	4.35	3.06
35	2.27	5.08	3.57
40	2.59	5.8	4.08
45	2.91	6.53	4.59
50	3.24	7.25	5.1
60	3.88	8.7	6.12
70	4.53	10.15	7.14
80	5.18	11.6	8.16
90	5.83	13.05	9.18
100	6.47	14.5	10.2
110	7.12	15.95	11.22
120	7.77	17.4	12.24
130	8.42	18.85	13.26
140	9.06	20.31	14.28
150	9.71	21.76	15.3
160	10.36	23.21	16.31
170	11.01	24.66	17.33
180	11.65	26.11	18.35
190	12.3	27.56	19.37
200	12.95	29.01	20.39
210	13.6	30.46	21.41
220	14.24	31.91	22.43
230	14.89	33.36	23.45
240	15.54	34.81	24.47
250	16.19	36.26	25.49
260	16.83	37.71	26.51
270	17.48	39.16	27.53
280	18.13	40.61	28.55
290	18.78	42.06	29.57
300	19.42	43.51	30.59
310	20.07	44.96	31.61
320	20.72	46.41	32.63
330	21.37	47.86	33.65
340	22.01	49.31	34.67

Table 19 305

Table 19 (*continued*)

MPa	Tonsf/in^2	PSI $\times$ 1000	kgf/mm^2
350	22.66	50.76	35.69
360	23.31	52.21	36.71
380	24.6	55.11	38.75
400	25.9	58.02	40.79
420	27.19	60.92	42.83
440	28.49	63.82	44.87
460	29.78	66.72	46.9
480	31.08	69.62	48.94
500	32.37	72.52	50.98
520	33.67	75.42	53.02
540	34.96	78.32	55.06
560	36.26	81.22	57.1
580	37.55	84.12	59.14
600	38.85	87.02	61.18
620	40.14	89.92	63.22
640	41.44	92.82	65.26
660	42.73	95.72	67.3
680	44.03	98.63	69.34
700	45.32	101.53	71.38
720	46.62	104.43	73.42
740	47.91	107.33	75.46
760	49.21	110.23	77.5
780	50.5	113.13	79.53
800	51.8	116.03	81.57
820	53.09	118.93	83.61
840	54.39	121.83	85.65
860	55.68	124.73	87.69
880	56.98	127.63	89.73
900	58.27	130.53	91.77
920	59.57	133.43	93.81
940	60.86	136.34	95.85
960	62.16	139.24	97.89
980	63.45	142.14	99.93
1000	64.75	145.04	101.97
1020	66.04	147.94	104.01
1040	67.34	150.84	106.05
1060	68.63	153.74	108.09
1080	69.93	156.64	110.12
1100	71.22	159.54	112.16
1120	72.52	162.44	114.2
1140	73.81	165.34	116.24
1160	75.11	168.24	118.28
1180	76.4	171.14	120.32
1200	77.7	174.05	122.36
1220	78.99	176.95	124.4
1240	80.29	179.85	126.44
1260	81.58	182.75	128.48
1280	82.88	185.65	130.52

continued overleaf

Table 19 (*continued*)

MPa	Tonsf/in²	PSI × 1000	kgf/mm²
1300	84.17	188.55	132.56
1320	85.47	191.45	134.6
1340	86.76	194.35	136.64
1360	88.06	197.25	138.68
1380	89.35	200.15	140.71
1400	90.65	203.05	142.75
1420	91.94	205.95	144.79
1440	93.24	208.85	146.83
1460	94.53	211.76	148.87
1480	95.83	214.66	150.91
1500	97.12	217.56	152.95
1520	98.42	220.46	154.99
1540	99.71	223.36	157.03
1560	101.01	226.26	159.07
1580	102.3	229.16	161.11
1600	103.6	232.06	163.15
1620	104.89	234.96	165.19
1640	106.19	237.86	167.23
1680	108.78	243.66	171.3
1700	110.07	246.56	173.34
1720	111.37	249.46	175.38
1740	112.66	252.37	177.42
1760	113.96	255.27	179.46
1780	115.25	258.17	181.5
1800	116.55	261.07	183.54
1820	117.84	263.97	185.58
1840	119.14	266.87	187.62
1860	120.43	269.77	189.66
1880	121.73	272.67	191.7
1900	123.02	275.57	193.74
1920	124.32	278.47	195.78
1940	125.61	281.37	197.82
1960	126.91	284.27	199.86
1980	128.2	287.17	201.89
2000	129.5	290.08	203.93
2020	130.79	292.98	205.97
2040	132.09	295.88	208.01
2060	133.38	298.78	210.05
2080	134.68	301.68	212.09
2100	135.97	304.58	214.13
2120	137.27	307.48	216.17
2140	138.56	310.38	218.21
2160	139.86	313.28	220.25
2180	141.15	316.18	222.29
2200	142.45	319.08	224.33
2220	143.74	321.98	226.37

Table 19 307

Table 19 (*continued*)

MPa	Tonsf/in²	PSI × 1000	kgf/mm²
2240	145.04	324.88	228.41
2260	146.33	327.79	230.45
2280	147.63	330.69	232.49
2300	148.92	333.59	234.52
2320	150.22	336.49	236.56
2340	151.51	339.39	238.6
2360	152.81	342.29	240.64
2380	154.1	345.19	242.68
2400	155.4	348.09	244.72
2420	156.69	350.99	246.76
2440	157.99	353.89	248.8
2460	159.28	356.79	250.84
2480	160.58	359.69	252.88
2500	161.87	362.59	254.92
2520	163.17	365.5	256.96
2540	164.46	368.4	259
2560	165.76	371.3	261.04
2580	167.05	374.2	263.08
2600	168.35	377.1	265.11
2620	169.64	380	267.15
2640	170.94	382.9	269.19
2660	172.23	385.8	271.23
2680	173.53	388.7	273.27
2700	174.82	391.6	275.31
2720	176.12	394.5	277.35
2740	177.41	397.4	279.39
2760	178.71	400.3	281.43
2780	180	403.2	283.47
2800	181.3	406.11	285.51
2820	182.59	409.01	287.55
2840	183.89	411.91	289.59
2860	185.18	414.81	291.63
2880	186.48	417.71	293.67
2900	187.77	420.61	295.7
2920	189.07	423.51	297.74
2940	190.36	426.41	299.78
2960	191.66	429.31	301.82

Table 20   Greek mathematical symbols

A	α	Alpha	N	ν	Nu
B	β	Beta	Ξ	ξ	Xi
Γ	γ	Gamma	O	o	Omicron
Δ	δ	Delta	Π	π	Pi
E	ε	Epsilon	P	ρ	Rho
Z	ζ	Zeta	Σ	σ	Sigma
H	η	Eta	T	τ	Tau
Θ	θ	Theta	Y	υ	Upsilon
I	ι	Iota	Φ	ϕ	Phi
K	κ	Kappa	X	χ	Chi
Λ	λ	Lambda	Ψ	ψ	Psi
M	μ	Mu	Ω	ω	Omega